Lecture Notes in Computer Science 13144

More information about this subseries at https://link.springer.com/bookseries/7411

Vladimir M. Vishnevskiy ·
Konstantin E. Samouylov ·
Dmitry V. Kozyrev (Eds.)

Distributed Computer and Communication Networks

Control, Computation, Communications

24th International Conference, DCCN 2021
Moscow, Russia, September 20–24, 2021
Revised Selected Papers

 Springer

Editors
Vladimir M. Vishnevskiy (iD)
V.A. Trapeznikov Institute of Control
Sciences of Russian Academy of Sciences
Moscow, Russia

Dmitry V. Kozyrev (iD)
V.A. Trapeznikov Institute of Control
Sciences of Russian Academy of Sciences
Moscow, Russia

RUDN University
Moscow, Russia

Konstantin E. Samouylov (iD)
Applied Probability and Informatics
Department
RUDN University
Moscow, Russia

ISSN 0302-9743   ISSN 1611-3349 (electronic)
Lecture Notes in Computer Science
ISBN 978-3-030-92506-2   ISBN 978-3-030-92507-9 (eBook)
https://doi.org/10.1007/978-3-030-92507-9

LNCS Sublibrary: SL5 – Computer Communication Networks and Telecommunications

This Springer imprint is published by the registered company Springer Nature Switzerland AG
The registered company address is: Gewerbestrasse 11, 6330 Cham, Switzerland

Preface

This volume contains a collection of revised selected full-text papers presented at the 24th International Conference on Distributed Computer and Communication Networks (DCCN 2021), held in Moscow, Russia, during September 20–24, 2021.

The conference is a continuation of the traditional international conferences of the DCCN series, which have taken place in Sofia, Bulgaria (1995, 2005, 2006, 2008, 2009, 2014); Tel Aviv, Israel (1996, 1997, 1999, 2001); and Moscow, Russia (1998, 2000, 2003, 2007, 2010, 2011, 2013, 2015, 2016, 2017, 2018, 2019, 2020) in the last 24 years. The main idea of the conference is to provide a platform and forum for researchers and developers from academia and industry from various countries working in the area of theory and applications of distributed computer and communication networks, mathematical modeling, and methods of control and optimization of distributed systems, by offering them a unique opportunity to share their views, as well as discuss the prospective developments, and pursue collaboration in this area. The content of this volume is related to the following subjects:

1. Communication networks, algorithms, and protocols
2. Wireless and mobile networks
3. Computer and telecommunication networks control and management
4. Performance analysis, QoS/QoE evaluation, and network efficiency
5. Analytical modeling and simulation of communication systems
6. Evolution of wireless networks toward 5G
7. Internet of Things and fog computing
8. Machine learning, big data, and artificial intelligence
9. Probabilistic and statistical models in information systems
10. Queuing theory and reliability theory applications
11. High-altitude telecommunications platforms

The DCCN 2021 conference gathered 151 submissions from authors from 26 different countries. From these, 105 high-quality papers in English were accepted and presented during the conference. The current volume contains 29 extended papers which were recommended by session chairs and selected by the Program Committee for the Springer post-proceedings. Thus, the acceptance rate is 27.6%.

All the papers selected for the post-proceedings volume are given in the form presented by the authors. These papers are of interest to everyone working in the field of computer and communication networks.

We thank all the authors for their interest in DCCN, the members of the Program Committee for their contributions, and the reviewers for their peer-reviewing efforts.

September 2021

Vladimir M. Vishnevskiy
Konstantin E. Samouylov
Dmitry V. Kozyrev

Organization

DCCN 2021 was jointly organized by the Russian Academy of Sciences (RAS), the V.A. Trapeznikov Institute of Control Sciences of RAS (ICS RAS), the Peoples' Friendship University of Russia (RUDN University), the National Research Tomsk State University, and the Institute of Information and Communication Technologies of Bulgarian Academy of Sciences (IICT BAS).

Program Committee Chairs

V. M. Vishnevskiy (Chair)	ICS RAS, Russia
K. E. Samouylov (Co-chair)	RUDN University, Russia

Publication and Publicity Chair

D. V. Kozyrev	ICS RAS and RUDN University, Russia

International Program Committee

S. M. Abramov	Program Systems Institute of RAS, Russia
S. D. Andreev	Tampere University of Technology, Finland
A. M. Andronov	Transport and Telecommunication Institute, Latvia
N. Balakrishnan	McMaster University, Canada
S. E. Bankov	Kotelnikov Institute of Radio Engineering and Electronics of RAS, Russia
A. S. Bugaev	Moscow Institute of Physics and Technology, Russia
S. R. Chakravarthy	Kettering University, USA
T. Czachorski	Institute of Theoretical and Applied Informatics of Polish Academy of Sciences, Poland
D. Deng	National Changhua University of Education, Taiwan, China
S. Dharmaraja	Indian Institute of Technology, Delhi, India
A. N. Dudin	Belarusian State University, Belarus
A. V. Dvorkovich	Moscow Institute of Physics and Technology, Russia
Yu. V. Gaidamaka	RUDN University, Russia
P. Gaj	Silesian University of Technology, Poland
D. Grace	University of York, UK
Yu. V. Gulyaev	Kotelnikov Institute of Radio-engineering and Electronics of RAS, Russia
J. Hosek	Brno University of Technology, Czech Republic
V. C. Joshua	CMS College Kottayam, India

H. Karatza	Aristotle University of Thessaloniki, Greece
I. A. Kochetkova	RUDN University, Russia
N. Kolev	University of São Paulo, Brazil
J. Kolodziej	NASK, Poland
G. Kotsis	Johannes Kepler University Linz, Austria
A. E. Koucheryavy	Bonch-Bruevich Saint-Petersburg State University of Telecommunications, Russia
Ye. A. Koucheryavy	Tampere University of Technology, Finland
T. Kozlova Madsen	Aalborg University, Denmark
U. Krieger	University of Bamberg, Germany
A. Krishnamoorthy	Cochin University of Science and Technology, India
N. A. Kuznetsov	Moscow Institute of Physics and Technology, Russia
L. Lakatos	Budapest University, Hungary
E. Levner	Holon Institute of Technology, Israel
S. D. Margenov	Institute of Information and Communication Technologies of Bulgarian Academy of Sciences, Bulgaria
N. Markovich	ICS RAS, Russia
A. Melikov	Institute of Cybernetics of the Azerbaijan National Academy of Sciences, Azerbaijan
E. V. Morozov	Institute of Applied Mathematical Research of the Karelian Research Centre RAS, Russia
V. A. Naumov	Service Innovation Research Institute (PIKE), Finland
A. A. Nazarov	Tomsk State University, Russia
I. V. Nikiforov	Université de Technologie de Troyes, France
P. Nikitin	University of Washington, USA
S. A. Nikitov	Kotelnikov Institute of Radio Engineering and Electronics of RAS, Russia
D. A. Novikov	ICS RAS, Russia
M. Pagano	University of Pisa, Italy
E. Petersons	Riga Technical University, Latvia
V. V. Rykov	Gubkin Russian State University of Oil and Gas, Russia
K. E. Samouylov	RUDN University, Russia
L. A. Sevastianov	RUDN University, Russia
M. A. Sneps-Sneppe	Ventspils University College, Latvia
A. N. Sobolevski	Institute for Information Transmission Problems of RAS, Russia
P. Stanchev	Kettering University, USA
S. N. Stepanov	Moscow Technical University of Communication and Informatics, Russia
S. P. Suschenko	Tomsk State University, Russia
J. Sztrik	University of Debrecen, Hungary
H. Tijms	Vrije Universiteit Amsterdam, The Netherlands
S. N. Vasiliev	ICS RAS, Russia

V. M. Vishnevskiy	ICS RAS, Russia
M. Xie	City University of Hong Kong, Hong Kong, China
A. Zaslavsky	Deakin University, Australia
Yu. P. Zaychenko	Kyiv Polytechnic Institute, Ukraine

Organizing Committee

V. M. Vishnevskiy (Chair)	ICS RAS, Russia
K. E. Samouylov (Vice Chair)	RUDN University, Russia
D. V. Kozyrev	ICS RAS and RUDN University, Russia
A. A. Larionov	ICS RAS, Russia
S. N. Kupriyakhina	ICS RAS, Russia
S. P. Moiseeva	Tomsk State University, Russia
T. Atanasova	IIICT BAS, Bulgaria
I. A. Kochetkova	RUDN University, Russia

Organizers and Partners

Organizers

Russian Academy of Sciences (RAS), Russia
V.A. Trapeznikov Institute of Control Sciences of RAS, Russia
RUDN University, Russia
National Research Tomsk State University, Russia
Institute of Information and Communication Technologies of Bulgarian Academy of Sciences, Bulgaria
Research and Development Company "Information and Networking Technologies", Russia

Support

Information support was provided by the Russian Academy of Sciences. The conference was organized with the support of the RUDN University Strategic Academic Leadership Program.

Contents

Analytical Modeling of Distributed Systems

Computer and Communication Networks

Analysis of Cognitive Radio Networks with Balking and Reneging

Hamza Nemouchi⬤, Mohamed Hedi Zaghouani$^{(\boxtimes)}$⬤, and János Sztrik⬤

Doctoral School of Informatics, University of Debrecen, Debrecen, Hungary
sztrik.janos@inf.unideb.hu

Abstract. In this paper, we investigate the concept of balking and reneging on Cognitive Radio Networks. The concept of balking and reneging is quite common in the networking world nowadays. The more occupied the system, the more discouraged are the new coming customers, on the other hand, impatient users will leave the system after a maximum waiting time.

Keywords: Finite source queuing systems · Simulation · Cognitive Radio Networks · Performance measures · Balking · Reneging

1 Introduction

The primary goal of our model "Cognitive Radio Network" is to make use of the free spaces of the primary frequency band for the benefit of the secondary one. [1–6] and [7] provide more details.

Two elements are considered in our queuing system, the first one is designed for Primary Users (PU) with a limited number of sources that generate primary calls after an exponentially distributed time. All the generated calls will join a FIFO queue to get served. The service time is exponentially distributed. The second subsystem is dedicated to the jobs of the Secondary Users (SU) generated by a finite number of sources and headed to the Secondary Channel Service (SCS) to be serviced. The arrival time of these calls is exponentially distributed, however, their service time is generally distributed using hypo-exponential, hyperexponential and gamma distributions.

The generated licensed calls will check the availability of the Primary Channel Service (PCS), if it is free, the service might start straight away, if it is occupied with a primary call the later call will join the FIFO queue. However, if this PCS is taken by a secondary customer, its service stops immediately and will be directed back to the SCS. Depending on the current situation of the secondary

The research work of János Sztrik is supported by the EFOP–3.6.1–16–2016–00022 project. The project is co-financed by the European Union and the European Social Fund. Mohamed Hedi Zaghouani is supported by the Stipendium Hungaricum Scholarship.

V. M. Vishnevskiy et al. (Eds.): DCCN 2021, LNCS 13144, pp. 3–13, 2021.
https://doi.org/10.1007/978-3-030-92507-9_1

service unit, the aborted call will restart from the beginning its service or will be added to the retrial queue (orbit).

On the other hand, SCS receives unlicensed requests. Supposing the intended server is idle, SU is allowed to start the service, if it is occupied, they might try opportunistically to start their service in the PCS. If the last service channel is not occupied, the low priority might have the opportunity to start the service, otherwise, if it is occupied, such call will be joining the orbit automatically, from which they retry to get served, after an exponentially distributed time.

Several studies have studied the CRN according to various scenarios. Authors of [3] for example investigated the effect of server unreliability on the CRN. In [6] the same system was used involving abandonment, SUs were forced to leave the system once their total waiting time exceeds a random maximum waiting time. Balking and/or reneging were studied in several queuing systems such as [7–17] and [18]. However, after a thorough search of several related topics and reports, we were unable to locate any articles that addressed this model in the case of balking and reneging, which is the novelty of our investigation.

2 System Model

Figure 1 shows a queueing cognitive radio system based on the following assumptions. Consider two linked subsystems, in which primary requests are created by a finite number of sources N_1 and sent to the first server based on an exponentially distributed time with an average value of $1/\lambda_1$. If the unit is free, the service might start, if it is busy, the call resides at the preemptive priority queue. The service time of the primary users is a exponentially distributed random variable with parameter μ_1.

The number of sources of the secondary subsystem is denoted by N_2. According to an exponentially distributed time with parameter λ_2/N_2, each source produces low priority jobs. With a rate μ_2, the service time of SUs is generally distributed using hypo-exponential, hyper-exponential and gamma distributions, having the same mean and different variances. The secondary customer's retrial time is assumed to be an exponentially distributed random variable with a parameter of ν.

New arriving secondary customers might balk (refuse to join the server) with probability n/N_2 where n is the number of customers in the system and N_2 is the number of sources. After joining, they might also renege (leave the orbit after entering) if service does not begin by a certain random time, which is exponentially distributed with parameter τ.

3 Simulation Results

In this section, the effect of the service times distributions and the impact of the cognitive technology on the key performance measures of our system are investigated. Assuming that all random variables included in the system are exponentially distributed except the services, we created a stochastic simulation

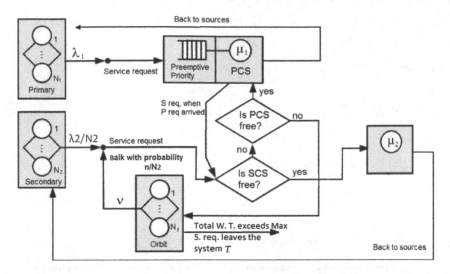

Fig. 1. Finite-source retrial queuing system: Modeling the Cognitive Radio Network with balking and reneging.

program written in C coding language with SimPack [19]. All the numerical results were collected by the validation of the simulation outputs. Table 1 shows the numerical values of the simulation main class input parameters while Table 2 defines the numerical values of the statistical class of the simulation program.

Table 1. Simulation input parameters

N_1	N_2	λ_1	λ_2/N_2	μ_1	μ_2	ν	τ
20	50	0.1	x-axis	1	1	0.1	0.1

Table 2. Parameters of the general distributions

Distribution	Gamma, $c_x^2 < 1$	Hyper	Hypo	Gamma, $c_x^2 > 1$
Parameters	$\alpha = 1{,}7857$ $\beta = 1{,}7857$	$p = 0{,}3309$ $\lambda_1 = 0{,}66198$ $\lambda_2 = 1{,}33803$	$\lambda_1 = 1{,}4854$ $\lambda_2 = 3{,}06$	$\alpha = 0{,}3906$ $\beta = 0{,}3906$
Mean	1	1	1	1
Variance	0.56	2.56	0.56	2.56
c_x^2	0.56	2.56	0.56	2.56

Service Times Are Generally Distributed. Figure 2 illustrates the influence of primary and secondary service times distribution on the mean residence time of SUs versus secondary request time generation. A high distributions sensitivity

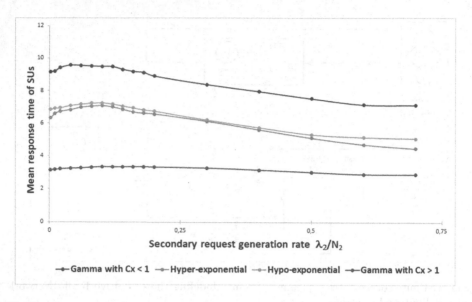

Fig. 2. The impact of primary and secondary service times distribution on the mean residence time of SUs vs secondary request time generation

can be observed when service times are gamma distributed with a squared coefficient of variation greater than one. The same clear sensitivity is seen in Fig. 3, where the effect of primary and secondary service times distribution on the mean reneging time of SUs vs secondary request time generation was displayed, while gamma with a c_x^2 greater than one. This confirms the same behaviour noticed in the previous figure. Furthermore, as anticipated, by increasing the arrival intensity of SUs, it involves greater reneging rate, we note an important number of customers that leave the system, especially in the hypo-exponential case. Table 2 states the parameters of the used general distributions.

The impact of the service times distribution of the primary and secondary subsystems on the mean balking rate versus λ_2 can be observed in Fig. 4. Increasing the secondary arrival rate involves a higher discouragement for new arriving secondary customers, this can be seen clearly in the case of Gamma distribution. It is well known according Gamma distribution function that when $c_x^2 > 1$ the generated random service time is great which leads to an overloading of the system, hence, the figure shows (Table 3).

Figure 5 illustrates The impact of primary and secondary service times distribution on the mean response time of SUs vs primary request time generation. An obvious impact can be seen in this figure especially in the case of having a squared coefficient of variation greater than one. The phenomenon of dealing with primary arrival intensity and investigating the effect of the service time distribution can be also seen in cite.......

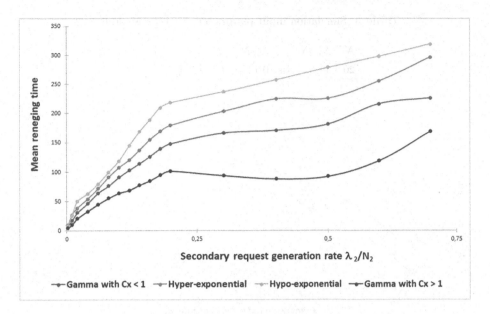

Fig. 3. The impact of primary and secondary service times distribution on the mean reneging time of SUs vs secondary request time generation

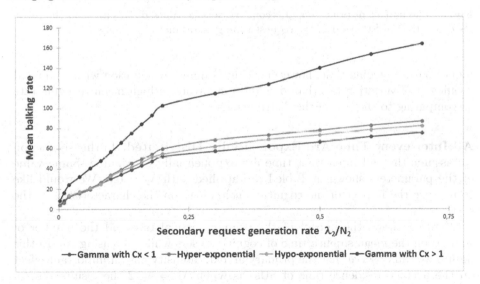

Fig. 4. The impact of primary and secondary service times distribution on the mean balking rate of SUs vs secondary request time generation

Figure 6 was generated to investigate the impact of primary and secondary service times distribution on the mean reneging time of SUs vs primary request time generation. Increasing the primary request generation rate involves a

Table 3. Simulation input parameters for Figs. 5 and 6

N_1	N_2	λ_1	λ_2/N_2	μ_1	μ_2	ν	τ
20	50	x-axis	0.14	1	1	0.1	0.1

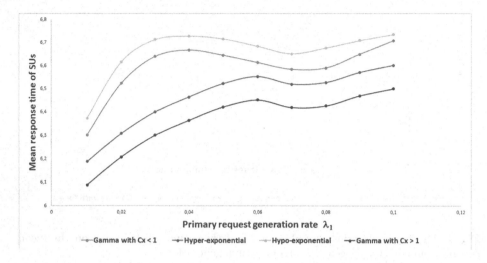

Fig. 5. The impact of primary and secondary service times distribution on the mean residence time of SUs vs primary request time generation

higher mean reneging time, however, using Gamma distribution with a squared coefficient of variation less than one, did not involve a high mean reneging rate as comparing to the rest of the distributions.

All Inter-event Time Are Exponentially Distributed. In this subsection we assume that all inter-event time are exponentially distributed. Same value of the parameters shown in Table 1 are applied with $\lambda_2 = 0.5$. We would like to analyze the impact of the cognitive technology on the characteristics of the system.

Figure 7 shows the impact of the primary arrival rate and the number of sources on the mean sojourn time of cognitive users while increasing N_2. In this figure, we can observe that the primary arrival intensity has an important effect on the average residence time of SUs, as when $\lambda_1 = \lambda_2/2$ the results show a smaller value of the mean than when $\lambda_1 = \lambda_2 \cdot 2$. Unlike the primary number of sources which does not have any effect, as when N_2 is high the traffic intensity in the primary sub-subsystem is bigger. However, in Fig. 8 different impact can be seen, when the primary number of sources is bigger. This is due to more secondary customers are reneging from the system.

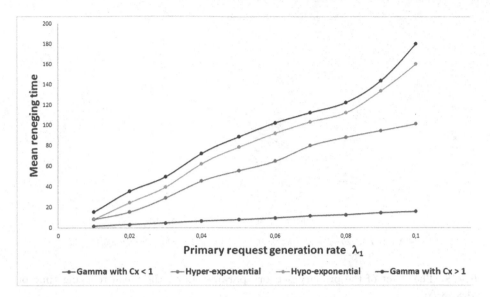

Fig. 6. The impact of primary and secondary service times distribution on the mean reneging time of SUs vs primary request time generation

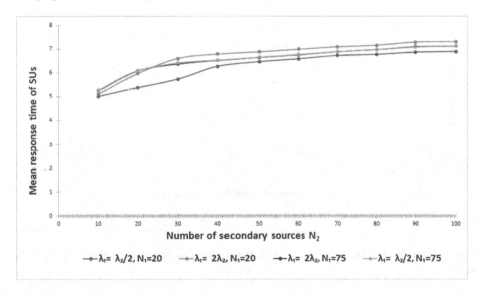

Fig. 7. The effect of the primary network parameters on the mean response time of SUs vs N_2

Figure 9 demonstrates how the primary network parameters can effect the mean response time of SUs vs N_2, the only impact than can be observed when the primary arrival rate is half the secondary's with the few primary sources. This is due to the opportunistic utilization of PCS by SUs, therefore, less customers balk.

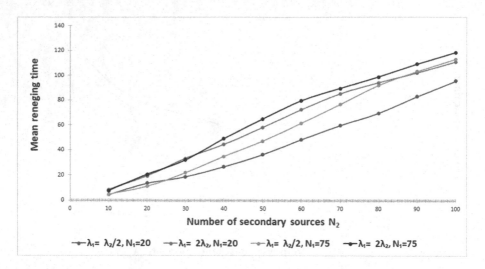

Fig. 8. The effect of the primary network parameters on the mean reneging time of SUs vs N_2

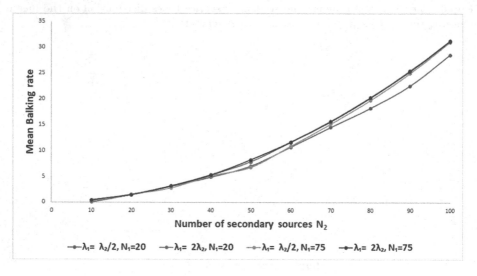

Fig. 9. The effect of the primary network parameters on the mean balking rate of SUs vs N_2

Figure 10 shows the effect of the primary subsystem rates on the mean residence time of cognitive customers vs secondary number of sources. Increasing N_2 provides a higher utilization of the secondary system, however, at same point it reaches the maximum and the server becomes totally full. A clear difference could be found when the primary arrival rate is at its maximum or minimum, $\lambda_1 = \lambda_2 * 2$ and $\lambda_1 = \lambda_2/2$, respectively.

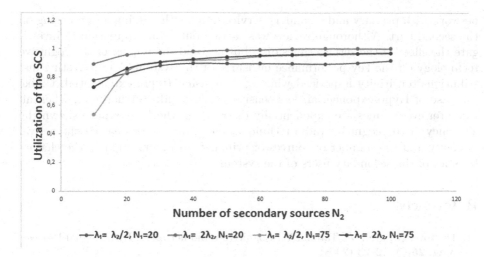

Fig. 10. The effect of the primary network parameters on the utilization of SCS vs N_2

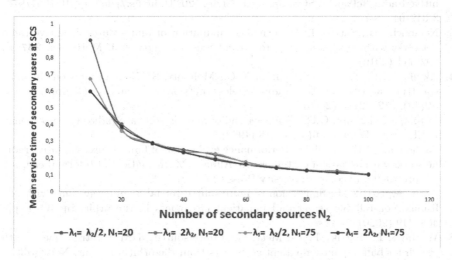

Fig. 11. The effect of the primary network parameters on the mean service time of SUs vs N_2

In Fig. 11 an impact can be seen when we are dealing with a small number of sources ($N_1 = 20$ and $N_2 = 10$). This figure shows how the primary network parameters can effect the mean response time of SUs vs N2.

4 Conclusion

A finite-source retrial queuing system that contains two non-independent parts was introduced in this paper. Our system was built to model a cognitive radio network with primary and secondary service units with balking and reneging on the second part. A thorough review was carried out using simulation to investigate the effect of the service times distributions and the impact of the cognitive technology on the key performance measures of the system. An interesting distributions sensitivity is noticed while $c_x^2 > 1$ using Hyper-exponential. Unlike the case of Hypo-exponential, less sensitivity is found. Furthermore, when all the inter-event times are exponentially distributed, the figures have shown the efficiency of the cognitive radio technique, increasing or decreasing the arrival intensity and the number of sources of primary customers impacts the characteristics of the secondary users of the system.

References

1. Devroye, N., Mai, V., Tarokh, V.: Cognitive radio networks. IEEE Signal Process. Mag. **25**(6), 12–23 (2008)
2. Gunawardena, S., Zhuang, W.: Modeling and Analysis of Voice and Data in Cognitive Radio Networks. SCS, Springer, Cham (2014). https://doi.org/10.1007/978-3-319-04645-7
3. Nemouchi, H., Sztrik, J.: Performance simulation of finite-source cognitive radio networks with servers subjects to breakdowns and repairs. J. Math. Sci. **237**(5), 702–711 (2019)
4. Akyildiz, I.F., Lee, W.-Y., Vuran, M.C., Mohanty, S.: Next generation/dynamic spectrum access/cognitive radio wireless networks: a survey. Comput. Netw. **50**(13), 2127–2159 (2006)
5. Mitola, J., Maguire, G.Q.: Cognitive radio: making software radios more personal. IEEE Pers. Commun. **6**(4), 13–18 (1999)
6. Zaghouani, M.H., Sztrik, J.: Performance evaluation of finite-source cognitive radio networks with impatient customers. In: Annales Mathematicae et Informaticae, vol. 51, pp. 89–99. Lyceum University Press (2020)
7. Tóth, Á., Sztrik, J.: Simulation of finite-source retrial queuing systems with collisions, non-reliable server and impatient customers in the orbit. In: ICAI, pp. 408–419 (2020)
8. Wuchner, P., Sztrik, J., De Meer, H.: Finite-source M/M/S retrial queue with search for balking and impatient customers from the orbit. Comput. Netw. **53**(8), 1264–1273 (2009)
9. Kumar, R., Som, B.K.: An M/M/1/N queuing system with reverse balking and reverse reneging. Adv. Model. Optim. **1**, 339–353 (2014)
10. Nemouchi, H., Zaghouani, M.H., Sztrik, J.: Simulation analysis in cognitive radio networks with unreliability and abandonment. In: Dudin, A., Nazarov, A., Moiseev, A. (eds.) ITMM 2020. CCIS, vol. 1391, pp. 31–45. Springer, Cham (2021). https://doi.org/10.1007/978-3-030-72247-0_3
11. Danilyuk, E.Yu., Moiseeva, S.P., Sztrik, J.: Asymptotic analysis of retrial queueing system M/M/1 with impatient customers, collisions and unreliable server. 218–230 (2020)
12. Satin, Y., Zeifman, A., Sipin, A., Ammar, S.I., Sztrik, J.: On probability characteristics for a class of queueing models with impatient customers. Mathematics **8**(4), 594 (2020)

13. Kuki, A., Bérczes, T., Tóth, Á., Sztrik, J.: Numerical analysis of finite source Markov retrial system with non-reliable server, collision, and impatient customers. In: Annales Mathematicae et Informaticae, vol. 51, pp. 53–63. Lyceum University Press (2020)
14. Inoue, Y., Boxma, O., Perry, D., Zacks, S.: Analysis of $m^x/g/1$ queues with impatient customers. 89(3–4), 303–350 (2017)
15. Hasenbein, J., Perry, D.: Introduction: queueing systems special issue on queueing systems with abandonments. Queueing Syst. 75(2), 111–113 (2013). https://doi.org/10.1007/s11134-013-9376-4
16. Ke, J.-C., Chang, F.-M., Liu, T.-H.: M/M/C balking retrial queue with vacation. Qual. Technol. Quant. Manag. 16(1), 54–66 (2019)
17. Zirem, D., Boualem, M., Adel-Aissanou, K., Aïssani, D.: Analysis of a single server batch arrival unreliable queue with balking and general retrial time. Qual. Technol. Quant. Manag. 16(6), 672–695 (2019)
18. Kotb, K.A.M., El-Ashkar, H.A.: Quality control for feedback M/M/1/N queue with balking and retention of reneged customers. Filomat 34(1), 167–174 (2020)
19. Fishwick, P.A.: Simpack: getting started with simulation programming in C and C++. In: Proceedings of the 24th Conference on Winter Simulation, pp. 154–162 (1992)

Recent Advances in Scheduling Theory and Applications in Robotics and Communications

Eugene Levner[1] and Vladimir Vishnevsky[2(⊠)]

[1] Holon Institute of Technology, Holon, Israel
`levner@hit.ac.il`
[2] V. A. Trapeznikov Institute of Control Sciences of Russian Academy of Sciences,
Moscow, Russia
`vishn@inbox.ru`

Abstract. Scheduling theory is a major field in operations research and discrete applied mathematics. This paper focuses on several recent developments in scheduling theory and a broad range of new applications – from multiagent scheduling to robots in communication networks. The survey presents a personal view on current trends, critical issues, strengths and limitations of this advantageous field.

1 Introduction and Brief History

Scheduling theory is a field in operations research and discrete applied mathematics which is concerned with the *optimal allocation* of scarce resources (for instance, machines, processors, robots, operators, etc.) to activities *over time*, with the objective of optimizing one or several performance measures. The systematic study and efficient computer-aided solution of scheduling problems started about 70 years ago, being initiated by seminal papers by Johnson (1954) [1], Bellman (1956) [2], and Smith (1956) [3].

Since then a great diversity of scheduling models and solution techniques have been developed. Vast applications are found in industry, communications, transport, planning of military operations, healthcare, etc. Today, scheduling theory is a rapidly evolving area fruitfully contributing to computer science, artificial intelligence, industrial engineering and data science. The interested reader can find many nice examples of scheduling problems and elegant algorithms in surveys, textbooks, and monographs by Tanaev et al. [4], Lee et al. [5], Chen et al. [6], Pinedo [7], Blazewicz et al. [8], Levner [9], and Lenstra and Shmoys [10].

This paper describes the current state and prospects for the development of theoretical and practical research in the field of scheduling theory. The article is organized as follows. The second section discusses recent theoretical and algorithmic advances, including models and algorithms for multiagent scheduling, issues of integrating scheduling theory and queueing theory, etc. Section 3 is devoted to the description of scheduling models in artificial intelligence and

V. M. Vishnevskiy et al. (Eds.): DCCN 2021, LNCS 13144, pp. 14–23, 2021.
https://doi.org/10.1007/978-3-030-92507-9_2

scheduling of flying unmanned aerial vehicles and broadband wireless communication networks based on them. The final Sect. 4 briefly describes the directions for further research.

2 Recent Theoretical and Algorithmic Advances

2.1 Multiagent Schedulings

Research on multi-agent scheduling started about 20 years ago with papers by Baker and Smith [11] and Agnetis et al. [12] in which two-agent scheduling was introduced. In multiagent scheduling problems, activities (jobs) are allocated to resources (machines) similarly as in conventional single-agent single-objective scheduling problem but the major difference is that each job is controlled by an (usually human) agent who has his/her own criteria which may be competing and which is to be optimized by each agent.

Extensive surveys of recent multiagent scheduling models and algorithms are given in Agnetis et al. [13,14], and Cheng et al. [15].

The multiagent scheduling, and, in particular, corresponding coordination mechanisms, are studied in a novel area in game theory known as *job scheduling games*, in which multiple users (agents) wish to utilize multiple machines, the incentive of each user being to optimize his/her own objective function [16].

2.2 Integrating Scheduling Theory and Queueing Theory

For many years queueing theory and scheduling theory have been developed independently and, as a consequence, have different performance measures, problem formulations and techniques. However, in recent years, the scheduling community has noticed that the long-term stochastic reasoning as studied in queueing theory can be usefully combined, both theoretically and practically, with short-term algorithmic reasoning studied in scheduling theory, and this fact can be used in applications when solving practical artificial intelligence and big data processing problems.

Interesting that as early as in 1956, Richard Bellman has noticed in [2]: *"If we fix the [job] order and fasten our attention upon the distribution of idle times, the distribution of waiting times, and similar questions, we enter the domain of queuing theory (see Kendall, 1951 [49]). In this connection, we would like to point out that the explicit formulas of Johnson may be of some utility in determining limiting distributions".*

The *queueing system* consists of a single or several lines of waiting jobs (customers) and the available number of servers. Factors that are to be studied include: the line length(s), number of lines and the queue servicing discipline. Queue servicing discipline is the priority rule, similar to that in the scheduling theory, for determining the order of service to customers in waiting lines. Similar to what happens in the scheduling theory, the most common used are the following greedy priority rules: "first come, first served" (FCFS, FIFO); highest-profit or "best" customer first; largest orders first, longest wait-time first, and many others (see, for instance [8,17–19].

An important aspect of the queueing system is the line structure. Analogous structures are investigated in the scheduling theory. However, different jobs in the queueing systems have different probabilistic characteristics and requirements for processing on different resources, and these characteristics often do not become known until the jobs arrive.

Some basic questions studied in queueing theory are beyond the scope of scheduling theory, like, e.g., "What is the average waiting time for customers in a line and in the system?"

In [20], Terekhov et al. took initial steps toward integrating queueing and scheduling for dynamic scheduling problems. The dynamic scheduling problems are characterized by a varying stream of jobs arriving stochastically over time. Each job requires a combination of resources, sequentially and/or in parallel, using different processing times. The existence of any particular job and its corresponding characteristics are not known until its arrival. However, the authors in [20] assumed, - as in queueing theory but not as in scheduling settings, that stochastic characteristics of the distribution of job arrivals are known. Jobs may require a complex routing through the available resources, which may be heterogeneous but with known and deterministic capacities. To solve a dynamic scheduling problem, the jobs must be assigned to appropriate resources and start times, respecting the resource and temporal constraints. As jobs arrive, there must be an online process to make decisions: it is not possible to solve the entire problem of-line.

Solving dynamic scheduling problems is challenging both due to the combinatorics of the interaction of jobs, resources, and time, and due to the stochastics: to make a decision, one can use only the information that is known with certainty at a decision point and the stochastic properties of scenarios that may occur in the future.

The authors of [20] showed that the well-known queueing theory notion of stability can be used to analyze periodic scheduling algorithms, and that, for each of the problems studied, periodic scheduling algorithms can be maximally stable, that is, no queueing policy or scheduling algorithm can allow the system to operate at a higher load or achieve a higher throughput. For one dynamic scheduling problem, the long-term stochastic reasoning of queueing can be combined with short-term combinatorial reasoning and produces a hybrid scheduling algorithm that achieves better performance than either queueing or scheduling approaches can provide if used alone.

These authors empirically demonstrated that for a comparatively simple dynamic flowshop, the use of combinatorial scheduling has a small impact on schedule quality. In contrast, for more complicated flexible queueing networks, a novel algorithm that combines long-term guidance from queueing theory with short-term decision making outperforms all other tested approaches.

Integrating Scheduling Theory and Queueing Theory has many yet unsolved challenges.

2.3   An Improved Near-Optimal Algorithm for the Traveling Salesman Problem[1]

One of the fundamental problems in scheduling theory and discrete optimization is the traveling salesman problem (TSP), included in Karp's initial list of 21 NP-complete problems [21]. In TSP we are given a set of n nodes (cities) V along with their pairwise symmetric distances, $V \times V \to R \geq 0$. The goal is to find a Hamiltonian cycle of minimum cost. In the *metric* TSP problem, which we shall study in this section, the distances satisfy the triangle inequality.

First, we recall the classical Christofides–Serdyukov (CS) algorithm, also known in literature as the Christofides algorithm. Nicos Christofides and Anatoly Serdyukov have discovered it independently more than four decades ago, in [22] and [23], respectively (see also [24]). This algorithm finds approximate solutions to the metric TSP and guarantees that its solution is always within a factor of 3/2 of the optimal solution value. Such type algorithms are called *near-optimal* or *with performance guarantees*. Let $G = (V, w)$ be an instance of the travelling salesman problem. That is, G is a complete graph on the set V of vertices, and the function w assigns a nonnegative real weight to every edge of G. The CS algorithm works as follows: Given an instance of the metric TSP, choose a minimum spanning tree and then add the minimum cost matching on the odd degree vertices of the tree. In spite of its simplicity, this algorithm is the best known to date polynomial time approximation algorithm with the performance guarantee for the metric traveling salesman problem. It is worth mentioning that there known the 'complementary' *inapproximability* results, for instance, it is NP-hard to approximate TSP within a factor of 123/122 [25].

For many years, nobody could improve the Christofides-Serdukov algorithmic result. And only quite recently, Karlin, Klein, and Gharan [26] introduced a novel approximation algorithm with a slightly better approximation ratio $r = (3/2 - 10^{36})$.

Theorem 1. *For some absolute constant* $>10^{36}$, *there is a randomized algorithm that outputs a tour with expected cost at most* $3/2 - \varepsilon$ *times the cost of the optimum solution.*

The method [26] closely follows the Christofides-Serdyukov's algorithm, but uses a special randomly chosen tree rather than the minimum spanning tree. The authors note that their algorithm makes use of the Held-Karp relaxation [27]. They also remark that although their approximation factor is only slightly better than Christofides-Serdyukov's, in their numerical experiments they did not discover any example where the approximation ratio of the algorithm exceeded 4/3 in expectation. This recent approach is an exciting and promising direction for further study and improvement of the classical Christofides-Serdyukov algorithm.

---

[1] This Section Is Dedicated to the Memory of Dr. Anatoly Ivanovich Serdyukov (1951–2001).

2.4   Almost-Optimal (Fully Polynomial Time Approximation) Scheduling Algorithms

An algorithm **A** for solving an optimization problem P (in particular, a scheduling problem) is called an *almost optimal* algorithm (or a fully polynomial-time approximation scheme FPTAS) if given an input I for P and an $\varepsilon > 0$, the algorithm **A** finds in time polynomial in the size of I and in $1/\varepsilon$, a solution s for I that satisfies: $|OPT(I) - f(s)| \leq \varepsilon OPT(I)$, where $OPT(I)$ is the optimal value of a solution for I.

A polynomial time approximation scheme PTAS is an algorithm which takes an instance of an optimization problem and a parameter $\varepsilon > 0$ and, in polynomial time in the problem size, produces a solution that is within a factor $1 + \varepsilon$ of being optimal for a minimization problem (or $1 - \varepsilon$ for maximization problems). There exists dozens of different types of PTAS and FPTAS for many classes of scheduling problems (see Sahni [28], Babat [29], Gens and Levner [30], Lawler [31], Kovalyov et al. [32], Kovalyov and Kubiak [33], Woeginger [34], Hoesel.and Wagelmans [35], and references therein).

In this section, we give a brief description of several not-trivial FPTAS that have appeared in recent years. Liu and Wu [36] considered a scheduling problem in a flexible supply chain where jobs can be either processed in house, or outsourced to a third-party supplier with the goal of minimizing the sum of holding and delivery costs subject to an upper bound on the outsourcing cost. The problem with identical job processing times has been proved to be binary $\mathcal{NP}$-hard; a fully polynomial time approximation scheme that runs in $O(n^8(1/\varepsilon^2))$ time has been known. This paper derives a faster FPTAS. Kacem and Levner [37] revisited he problem of scheduling a set of proportional deteriorating non-resumable jobs on a single machine subject to maintenance. The maintenance has to be started prior to a given deadline. The jobs as well as the maintenance are to be scheduled so that to minimize the total completion time. For this problem a new dynamic programming algorithm and a faster fully polynomial time approximation scheme are proposed improving a recent result by Luo and Chen [JIMO (2012), 8:2, 271–283]. Yin et al. [38] considered the problem of scheduling n independent and simultaneously available jobs without preemption on a single machine, where the machine has a fixed maintenance activity. The objective is to find the optimal job sequence to minimize the total amount of late work, where the late work of a job is the amount of processing of the job that is performed after its due date. The authors first discussed the approximability of the problem, then developed two pseudo-polynomial dynamic programming algorithms and constructed a fully polynomial-time approximation scheme for the problem. Finally, the authors performed extensive numerical studies to evaluate the performance of the proposed algorithms in practice. Zhao and Hsu [39] considered a single-machine scheduling problem in which the processing time of a job is a linear increasing function of its starting time. The objective is to minimize the weighted number of tardy jobs. A pseudo-polynomial dynamic programming algorithm and a new fully polynomial-time approximation scheme were proposed.

Halman et al. [40] presented a new framework for obtaining fully polynomial time approximation schemes for stochastic univariate dynamic programs with either convex or monotone single-period cost functions. This framework is workable for the stochastic scheduling problems and is developed through the establishment of two sets of computational rules, namely, the calculus of K-approximation functions and the calculus of K-approximation sets. Using this general framework, the novel FPTASs for several NP-hard scheduling problems were obtained.

More details on the properties of the FPTAS algorithms can be found in the tutorial [41].

3   Novel Models and Applications

3.1   Scheduling and Artificial Intelligence. Robots are Everywhere

In recent decades, there is a growing interest on scheduling problems for autonomous robots and robotic systems. There are numerous applications of artificial intelligence (AI) and smart robots in different industries, on the earth and in space (see, e.g., Pinedo [7], Blazewicz et al. [8], Levner et al. [9], Agnetis et al. [14]) and in communications and transport (Vishnevsky et al. [42, 43]). Modern flexible manufacturing systems are integrated with computer-controlled hoists, robots and other automatic devices. Robots have expanded production capabilities in manufacturing world making the technological processes faster, more efficient and precise than ever before. As larger and more complex robotic systems are implemented, more sophisticated scheduling models, methods and algorithms are required for performing and optimizing these processes.

More details on AI and robotic scheduling problems and algorithms are presented in [7–9, 42, 43].

3.2   Scheduling of Flying Unmanned Stations in Airborne Communication Networks

The unmanned aerial vehicle communication networks (UAV-CN) contain a set of unmanned aerial vehicles (UAVs) that constitute together a network which may be used for very many applications. The UAVs autonomously fly in 3-dimensional space in ad-hoc mode and carry out the communications and collaboration missions. The specificity of such communication networks is a high-speed of UAVs constituting the network nodes, which has a crucial impact on efficient routing. Another factor affecting the quality of communication service is the power issue that is limited and should be optimized during the design of the UAV-CNs.

In the last decade, unmanned aerial communication systems have emerged in different areas, such as military and police operations, search and evacuation emergency missions, detection of ecological disasters, border surveillance, traffic monitoring, etc. In these systems, unmanned aerial vehicles communicate with each other based on a flying network to provide services to customers such as live streaming, high-speed access point, etc. (see Vishnevsky et al. [42,43]). The performance of UAV-CNs can be advantageous in emergency, for example, rescuing and searching people in areas where the conventional communication network is unavailable. In [44], the authors propose a method for detecting the coordinates of subscribers with the Wi-Fi signals generated from victims' phones using a flying network for emergencies based on UAV swarms.

According to the IEEE classification (see [45,46]), the wireless communication networks can be divided into two large domains: Infrastructure-based networks (IBN) and infrastructure-less networks (ILN), also called ad hoc networks.

Infrastructure-based network (IBN) has two subgroups, stationary and mobile. The stationary subgroup consists of stationary (static) base stations (BS) while the mobile subgroup consists of mobile nodes (MN) and master stations (MS) which are known as access points (AP).

The communication among MN is done with the help of access points which use different frequencies to establish the communication.

The infrastructure-less (ad hoc) networks are grouped, in turn, into wireless sensor network (WSN), wireless mesh networks (WMN), and mobile ad hoc networks (MANET). Next, the mobile ad hoc networks are classified into two sub groups: vehicular ad-hoc networks VANET and unmanned aerial vehicle communication networks UAV-CN.

As a typical fragment in novel applications, consider a perspective of exploiting classic scheduling models and methods for efficient scheduling of a flying UAV-CNs with the minimum number of UAVs.

We suggest to use the extension of the Kats-Levner model [47,48] for defining the minimum number of vehicles to meet a fixed schedule. In a new setting, we have the following problem. We study a cyclic process in which a set of several sensors perform m operations of data collection. A number of flying unmanned stations (drones) are used to transfer information of sensors from one sensor to another. The durations of data collecting and data transfers are known. The problem that has the key performance measure, the number of drones to be used. The aim is to find the minimum number of drones needed to meet a given cyclic schedule, for a fixed cycle length.

If the transfer times in the considered UAV (drone) scheduling problem satisfy the triangle inequality then the minimal number of UAVs needed to meet a given cyclic schedule of a fixed period length, is equal to the optimum solution of the $m \times m$ assignment problem. The complexity of the algorithm is $O(m^3)$, independently of the range within which the cycle length value may vary. The problem has several practical modifications (see, for instance, [47,48]).

4   Concluding Remark: A Look to the Future

There exist a number of related attractive fields which are not covered in this survey, for instance, knowledge-based scheduling, real time scheduling, scheduling in cloud computing, and scheduling with communication delays. They will be overviewed in our future work.

References

1. Johnson, S.M.: Optimal two- and three-stage production schedules with setup times included. Naval Res. Logist. Q. **1**, 61–68 (1954)
2. Bellman, R.: Mathematical aspects of scheduling theory. J. Soc. Ind. Appl. Math. **4**, 168–205 (1956)
3. Smith, W.E.: Various optimizers for single-stage production. Naval Res. Logist. Q. **3**(1–2), 59–66 (1956)
4. Tanaev, V.S., Gordon, V.S., Shafransky, Y.M.: Scheduling Theory. Single-Stage Systems. Kluwer, Dordrecht (1994)
5. Lee, C.Y., Lei, L., Pinedo, M.: Current trends in deterministic scheduling. Ann. Oper. Res. **70**, 1–41 (1997). https://doi.org/10.1023/A:1018909801944
6. Chen, B., Potts, C.N., Woeginger, G.J.: A review of machine scheduling: complexity, algorithms and approximability. In: Du, D.-Z., Pardalos, P.M. (eds.) Handbook of Combinatorial Optimization, pp. 21–169. Kluwer Academic Publishers, Dordrecht (1998)
7. Pinedo, M.: Scheduling: Theory. Algorithms and Systems. Prentice Hall, Englewood Cliffs (2016)
8. Blazewicz, J., Ecker, K.H., Pesch, E., Schmidt, G., Weglarz, J.: Handbook on Scheduling. Springer, Berlin (2007). https://doi.org/10.1007/978-3-540-32220-7
9. Levner, E. (ed.): Multiprocessor Scheduling Theory and Applications. I-TECH Education and Publishing, Vienna (2007)
10. Lenstra, J.K., Shmoys, D.B. (eds.): Elements of Scheduling. Centrum Wiskunde & Informatica, Amsterdam (2020)
11. Baker, K., Smith, J.C.: A multiple criterion model for machine scheduling. J. Sched. **6**, 7–16 (2003). https://doi.org/10.1023/A:1022231419049
12. Agnetis, A., Mirchandani, P., Pacciarelli, D., Pacifici, A.: Scheduling problems with two competing agents. Oper. Res. **52**, 229–242 (2004)
13. Agnetis, A., Pacciarelli, D., Pacifici, A.: Combinatorial models for multi-agent scheduling problems. Ch. 2 in [9], pp. 21–47 (2007)
14. Agnetis, A., Billaut, J.-C., Gawiejnowicz, S., Pacciarelli, D., Soukhal, A.: Multi-agent Scheduling. Models and Algorithms. Springer, Heidelberg (2014). https://doi.org/10.1007/978-3-642-41880-8
15. Cheng, C.T., Ng, J.J.Y.: Multi-agent scheduling on a single machine with max-form criteria. Eur. J. Oper. Res. **188**, 603–609 (2008)
16. Cohen, J., Dürr, C., Kim, T.N.: Non-clairvoyant scheduling games. Theory Comput. Syst. **49**, 3–23 (2011). https://doi.org/10.1007/s00224-011-9316-9
17. Vishnevsky, V., Semenova, O.: Polling systems and their application to telecommunication networks. Mathematics **9**(2), 117 (2021). https://doi.org/10.3390/math9020117
18. Dudin, A.N., Klimenok, V.I., Vishnevsky, V.M.: The Theory of Queuing Systems with Correlated Flows. Springer, Cham (2020). https://doi.org/10.1007/978-3-030-32072-0
19. Klimenok, V., Dudin, A., Vishnevsky, V.: Priority multi-server queueing system with heterogeneous customers. Mathematics **8**(9), 1501 (2020). https://doi.org/10.3390/math8091501
20. Terekhov, D., Tran, T.T., Down, D.G., Beck, J.C.: Integrating Queueing theory and scheduling for dynamic scheduling problems. J. Artif. Intell. Res. **50**(2014), 535–572 (2014)

21. Karp, R.M.: Reducibility among combinatorial problems. In: Miller, R.E., Thatcher, J.W., Bohlinger, J.D. (eds.) Complexity of Computer Computations. The IBM Research Symposia Series, pp. 85–103. Springer, Boston (1972). https://doi.org/10.1007/978-1-4684-2001-2_9

22. Christofides, N.: Worst-case analysis of a new heuristic for the travelling salesman problem (PDF), Report 388. Graduate School of Industrial Administration, CMU (1976)

23. Serdyukov, A.I.: On some extremal walks in graphs. Upravlyaemye Sistemy **17**, 76–79 (1978). (in Russian)

24. van Bevern, R., Slugina, V.A.: A historical note on the 3/2-approximation algorithm for the metric traveling salesman problem (2020). arXiv:2004.02437v2 [cs.DS]

25. Karpinski, M., Lampis, M., Schmied, R.: New inapproximability bounds for TSP. J. Comput. Syst. Sci. **81**(8), 1665–1677 (2015)

26. Karlin, A.R., Klein, N., Gharan, S.O.: A (slightly) improved approximation algorithm for metric TSP (2020). arXiv:2007.01409

27. Held, M., Karp, R.M.: The traveling salesman problem and minimum spanning trees. Oper. Res. **18**, 1138–1162 (1970). https://doi.org/10.1007/BF01584070

28. Sahni, S.: Approximate algorithms for the 0/1 knapsack problem. J. ACM **22**, 115–124 (1975)

29. Babat, L.G.: Linear functions on the N-dimensional unit cube. Dokl. Akad. Nauk SSSR **222**, 761–762 (1975). (in Russian)

30. Gens, G.V., Levner, E.V.: Fast approximation algorithms for job sequencing with deadlines. Discrete Appl. Math. **3**, 313–318 (1981)

31. Lawler, E.L.: A fully polynomial approximation scheme for the total tardiness problem. Oper. Res. Lett. **1**, 207–208 (1982)

32. Kovalyov, M.Y., Potts, C.N., Van Wassenhove, L.N.: A fully polynomial approximation scheme for scheduling a single machine to minimize total weighted late work. Math. Oper. Res. **19**, 86–93 (1994)

33. Kovalyov, M.Y., Kubiak, W.: A fully polynomial time approximation scheme for minimizing Makespan of deteriorating jobs. J. Heuristics **3**, 287–297 (1998). https://doi.org/10.1023/A:1009626427432

34. Woeginger, G.J.: When does a dynamic programming formulation guarantee the existence of an FPTAS? INFORMS J. Comput. **12**, 57–75 (2000)

35. van Hoesel, C.P.M., Wagelmans, A.P.M.: Fully polynomial approximation schemes for single-item capacitated economic lot-sizing problems. Math. Oper. Res. **26**, 339–357 (2001)

36. Liu, S.C., Wu, C.C.: A faster FPTAS for a supply chain scheduling problem to minimize holding costs with outsourcing. Asia-Pac. J. Oper. Res. **33**, 05 (2016)

37. Kacem, I., Levner, E.: An improved approximation scheme for scheduling a maintenance and proportional deteriorating jobs. J. Ind. Manage. Optim. **12**(3), 811–817 (2016)

38. Yin, Y.Q., Xu, J.Y., Cheng, T.C.E., Wu, C.C., Wang, D.-J.: Approximation schemes for single-machine scheduling with a fixed maintenance activity to minimize the total amount of late work. Naval Res. Logist. **63**(2), 172–183 (2016)

39. Zhao, C.L., Hsu, C.-J.: Fully polynomial-time approximation scheme for single machine scheduling with proportional-linear deteriorating jobs. Eng. Optim. **51**(11), 1938–1943 (2019)

40. Halman, N., Klabjan, D., Li, C.-L., Orlin, J., Simchi-Levi, D.: Fully polynomial time approximation schemes for stochastic dynamic programs. SIAM J. Discrete Math. **28**(4), 1725–1796 (2014)

41. Schuurman, P., Woeginger, G.J.: Approximation schemes, a tutorial. Unpublished book (2002). http://www.math.nsc.ru/LBRT/k5/DEP/P.Schuurman,%20G. Woeginger.pdf

42. Vishnevsky, V.M., Mikhailov, E.A., Tumchenok, D.A., Shirvanyan, A.M.: Mathematical model of the operation of a tethered unmanned platform under wind loading. Math. Models Comput. Simul. **12**(4), 492–502 (2020). https://doi.org/10. 1134/S2070048220040201

43. Vishnevsky, V., Meshcheryakov, R.: Experience of developing a multifunctional tethered high-altitude unmanned platform of long-term operation. In: Ronzhin, A., Rigoll, G., Meshcheryakov, R. (eds.) ICR 2019. LNCS (LNAI), vol. 11659, pp. 236–244. Springer, Cham (2019). https://doi.org/10.1007/978-3-030-26118-4_23

44. Dinh, T.D., Vishnevsky, V., Larionov, A., Vybornova, A., Kirichek, R.: Structures and deployments of a flying network using tethered multicopters for emergencies. In: Vishnevskiy, V.M., Samouylov, K.E., Kozyrev, D.V. (eds.) DCCN 2020. LNCS, vol. 12563, pp. 28–38. Springer, Cham (2020). https://doi.org/10.1007/978-3-030-66471-8_3

45. Nawaz, H., Ali, H.M., Laghari, A.A.: UAV communication networks issues: a review. Arch. Comput. Methods Eng. **28**(3), 1349–1369 (2020). https://doi.org/ 10.1007/s11831-020-09418-0

46. Khan, M.A., Qureshi, I.M., Safi, A., Khan, I.U.: Flying ad-hoc networks (FANETs): a review of communication architectures, and routing protocols. In: 1st International Conference on Latest Trends in Electrical Engineering and Computing Technologies (2017). https://www.researchgate.net/publication/311707613

47. Kats, V., Levner, E.: Minimizing the number of robots to meet a given cyclic schedule. Ann. Oper. Res. **69**, 209–226 (1997). https://doi.org/10.1023/A: 1018980928352

48. Kats, V., Levner, E.: Minimizing the number of vehicles in periodic scheduling: the non-Euclidean case. Eur. J. Oper. Res. **107**, 371–377 (1998)

49. Kendall, D.G.: Some problems in the theory of queues. J. Roy. Statist. Soc. Ser. B **13**, 151–185 (1951)

The PageRank Vector of a Scale-Free Web Network Growing by Preferential Attachment

Natalia M. Markovich[1]([✉])[ID] and Udo R. Krieger[2][ID]

[1] V.A. Trapeznikov Institute of Control Sciences, Russian Academy of Sciences, Profsoyuznaya Street 65, 117997 Moscow, Russia
`markovic@ipu.rssi.ru`
[2] Fakultät WIAI, Otto-Friedrich-Universität, An der Weberei 5, 96047 Bamberg, Germany
`udo.krieger@ieee.org`

Abstract. We consider a scale-free model of the Web network that is evolving by preferential attachment schemes and derive an explicit formula of its PageRank vector. Its ith element indicates the probability that a surfer resides at a related Web page i in a stationary regime of an associated random walk. Considering the growth of a directed Web graph, we apply linear preferential attachment schemes proposed by Samorodnitsky et al. (2016). To express the probability of a connection between two nodes of this Web graph, our derivation allows us to avoid the consideration of complicated paths with random lengths and to cover both self-loops and multiple edges between nodes. An algorithm of the PageRank vector calculation for graphs without loops is provided. The approach can be extended in a similar way to graphs with loops. In this way, our approach enhances existing analysis schemes. It provides a better insight on the PageRank of growing scale-free Web networks and supports the adaptation of the model to gathered network statistics.

Keywords: PageRank vector · Scale-free network · Linear preferential attachment

1 Introduction

Let $G = (V, E)$ be the directed graph of a scale-free Web network with $n = |V|$ vertices $v \in V$, i.e., Web pages, and a growing number $|E|$ of edges $e \in E$, i.e., hyperlinks among these pages. The PageRank is accepted as a popular basic measure to rank the Web pages that are provided by a search engine after a request to the Web. Google's PageRank vector $R = (R_1, ..., R_n)^T \in (0, \infty)^n$ [1] is the unique solution of the following system of linear equations

$$R_i = c \sum_{j:(j,i)\in E} \frac{R_j}{D_j} + (1-c)q_i, \qquad i \in \{1, \dots, n\}.$$

© Springer Nature Switzerland AG 2021
V. M. Vishnevskiy et al. (Eds.): DCCN 2021, LNCS 13144, pp. 24–31, 2021.
https://doi.org/10.1007/978-3-030-92507-9_3

The sum is taken over a number of pages j with incoming links to page i (in-degree), D_j is the number of outgoing links of page j (out-degree), and $c \in (0, 1)$ is a damping factor which was originally set equal to 0.85 by Google. The probability vector $q = (q_1, q_2, ..., q_n)^T$ is a personalization vector of a user's preferences such that $q_i \geq 0$ and $\sum_{i=1}^{n} q_i = 1$ hold.

The in-degree is the simplest guidance to calculate the PageRank since the distributions of both entities are similar, [2]. Given the weak degree correlations in the Web graph, the approximation of the PageRank vector by the in-degrees of nodes can be relatively accurate [2]. A data analysis of Web graphs has revealed that the tails of both the PageRank and in-degree distributions follow power laws $1 - F(x) = c \cdot x^{-\alpha}$ with the same exponent α which is about 1.1, [3]. As a search engine can easily monitor the in-degree arising from the evolution of the Web and the in- and out-degree are available gathered statistics, the latter approximation is useful in practice.

Among other approaches to calculate the PageRank R_i of a randomly chosen Web page are the iteration procedure [4] and the mean field approach [5].

We focus on the PageRank formula proposed in [6, Lemma 3.1] and recalled here as Lemma 1. This formula is only valid for trees and directed acyclic Web graphs and the assumption that the out-degree m for each node is fixed. It is our objective to obtain an explicit formula of the PageRank vector without the constraints of [6]. Our derivation allows us to avoid the consideration of complicated paths with random lengths and to cover both self-loops and multiple edges between nodes. To this end, we use the probabilities of the edge creation by the $\alpha-$, $\beta-$ and $\gamma-$schemes of the linear preferential attachment (PA) proposed in [7] and [8] to determine the probability distribution of accessing the Web pages. The latter schemes are considered as basic growth model of the network. We propose a new formula for the PageRank of a node with a random out-degree m which is obtained after appending new nodes to the scale-free network.

The paper is organized as follows. In Sect. 2 the dynamics of an evolving Web graph is described and in Sect. 3 the calculation of its PageRank vector. The new computational formula for the PageRank vector of evolving Web networks is presented in Sect. 4. The exposition is finalized by some conclusions.

2   The Linear Preferential Attachment Schemes

Let $G(n) = (V(n), E(n))$ denote a directed graph with sets of nodes $V(n)$ and edges $E(n)$, n be the number of edges and $N(n)$ denote the number of nodes in $G(n)$. Let us recall the $\alpha-$, $\beta-$ and $\gamma-$PA-schemes given in [7,8].

A finite directed graph $G(n_0)$ is used as a seed network. It consists of at least one node v_0 and n_0 edges. A new node v is appended to the existing graph $G(n - 1), n > n_0$, by adding a single edge to $G(n - 1)$. The edge creation is provided by flipping a three-sided coin with probabilities α, β and γ. To this end, an i.i.d. sequence of trinomial r.v.s with values 1, 2 and 3 and the corresponding probabilities α, β and γ are generated to select schemes. $I_n(v)$ and $O_n(v)$ denote the in- and out-degree of v. A new node v is selected among nodes of the network and appended to $G(n - 1)$.

The edge $v \to w \equiv (v, w) \in E(n)$ directed from the new node $v \in V(n)$ to an existing node $w \in V(n-1)$ is created with probability α. The existing node $w \in V(n-1)$ is chosen with probability

$$(P_\alpha)_{v,w} = P\{v \to w\} = \frac{I_{n-1}(w) + \delta_{in}}{n - 1 + \delta_{in}N(n-1)} \tag{1}$$

by the α-scheme.

An edge (v, w) is added to $E(n-1)$ with probability β and the existing nodes $v \in V(n-1) = V(n), w \in V(n-1)$ are chosen independently from $G(n-1)$ with probability

$$(P_\beta)_{v,w} = P\{v \to w\} = \left(\frac{I_{n-1}(w) + \delta_{in}}{n - 1 + \delta_{in}N(n-1)} \right) \left(\frac{O_{n-1}(v) + \delta_{out}}{n - 1 + \delta_{out}N(n-1)} \right) \tag{2}$$

by the β-scheme.

An edge (w, v) from the existing node $w \in V(n-1)$ to v is created and the node w is chosen with probability

$$(P_\gamma)_{w,v} = P\{w \to v\} = \frac{O_{n-1}(w) + \delta_{out}}{n - 1 + \delta_{out}N(n-1)} \tag{3}$$

by the γ-scheme.

The parameters of the PA method δ_{in} and δ_{out} can be estimated by the semi-parametric extreme value method (EV) based on the maximum-likelihood method, [8]. It holds $\alpha + \beta + \gamma = 1$.

3   Calculating the PageRank Vector of a Web Network

There are numerous approaches to calculate the PageRank R_i of a randomly chosen page $v = i \in V$ in a Web graph $G = (V, E)$. One of them is determined by the following iteration

$$\widehat{R}_i^{(n,0)} = 1, \quad \widehat{R}_i^{(n,k)} = \sum_{j \to i} \frac{c}{D_j} \widehat{R}_j^{(n,k-1)} + (1 - c), \quad k \in \mathbb{N}, \tag{4}$$

proposed in [4] for a given uniform personalization vector $q_i = 1/n, 1 \le i \le n = |V|$. Then the scale-free PageRank of a node $v = i$ is denoted by $R_i^{(n)} = nR_i$. This iteration (4) is proceeding until the difference between two consecutive iterations $|\widehat{R}_i^{(n,k)} - \widehat{R}_i^{(n,k-1)}|$ will be small enough to reach approximately its limit $R_i^{(n)} = \lim_{k \to \infty} \widehat{R}_i^{(n,k)}$ which is sufficient for a moderate number of iterations k. Here, $j \to i$ implies that node j is linked to node i, i.e. $(j, i) \in E$.

Another important method is the mean field approach [5]. Its idea is to average the PageRanks of nodes that are aggregated within in- and out-degree classes (k_{in}, k_{out}). Such class contains nodes with the same in-degree k_{in} and the same out-degree k_{out}. We focus on the PageRank formula (5) proposed in [6, Lemma 3.1] and recalled further in Lemma 1.

In [6] the following notations are adopted for a Web graph $G = (V(n), E(n))$: $V(n) = \{0, \ldots, n\}$, $E(n) \subseteq V(n) \times V(n)$, $n \in \mathbb{N}$.

Let $\pi_v(n)$ be the PageRank of a node v after the nth step of the network's evolution, $P_v(n)$ be the set of all paths from nodes $v + 1, \ldots, n$ to v and $\ell(p)$ be the length of such a path p.

Lemma 1. *[6] The PageRank of node $v \in V(n)$, $v > 0$ within the realization of a growing network at time step $n > 0$ is given by*

$$\pi_v(n) = \frac{1-c}{n+1} \left(1 + \sum_{p \in P_v(n)} \left(\frac{c}{m} \right)^{\ell(p)} \right) \tag{5}$$

and the PageRank of the initial node $v = 0$ is given by

$$\pi_0(n) = \frac{1}{n+1} \left(1 + \sum_{p \in P_0(n)} \left(\frac{c}{m} \right)^{\ell(p)} \right).$$

Note that $m = 1$ implies a chain as special case of a tree. The proof of the Lemma is based on a formula of the PageRank vector in [9]

$$\pi(n) = \frac{1-c}{n+1} \mathbf{1}^T [I - cP]^{-1} \tag{6}$$

with a scaling term $n+1$ due to a uniform personalization vector $q = 1/(n+1)\mathbf{1} \in (0,1]^{n+1}$. Here, $\mathbf{1}$ is the column vector of all ones.

$P \in [0,1]^{(n+1) \times (n+1)}$ is the square hypermatrix of a random walk defined by the transition probabilities of a Web surfer in the following equation

$$\widetilde{P} = c \cdot P + \frac{1-c}{n+1} \cdot E, \tag{7}$$

$E = \mathbf{1} \cdot \mathbf{1}^T$ is a matrix whose entries are all equal to one. The Web pages can be considered as states of a Markov chain. The $(i, j)^{th}$ element p_{ij} of P is the probability of a surfer moving from Web page i to j in one time step [9]. If there is no outgoing link from page i to j, then $p_{ij} = 0$. If a surfer follows a random walk corresponding to $\widetilde{P}$, then the ith coordinate π_i of the PageRank vector π coincides with the probability in a stationary regime that the surfer stays at page i [6].

This formula (5) is only valid for trees and directed acyclic Web graphs. Other constraints of (5) are as follows. Firstly, the PageRank expression is addressed to a growing directed graph with a fixed value m of the out-degree for each node which is not plausible for real-world networks. The fixed m allows to start from any node (Web page) equally likely with a probability c/m and to follow any of the outgoing links to arrive to another node. Secondly, the PA is considered as a basic growth model of the network's evolution. A new node is appended to the network at each time step and adds m incoming links to existing nodes.

Indeed, not all pairs of nodes may be connected by a direct edge and the path length between them is random. In [6] the probability of a connection between two nodes is taken equal to $(c/m)^{\ell(p)}$. To this end, it is assumed that all per-hop links on the paths from node i to node j (i.e. $i \Rightarrow j$) are independent. This feature may be unrealistic since according to a linear PA each existing node j is chosen randomly from the current network state with a probability proportional to its degree k_j. This means that new nodes prefer to become attached to an existing node with a large node degree. The existence of superstar nodes to which a large proportion of nodes is attached may create a dependence between the links of paths between nodes. The length of the path $i \Rightarrow j$ is also determined by the direction of these link attachments.

4   The PageRank Vector of an Evolving Web Graph

We use (1)–(3) to define the elements of the hypermatrix P in (6). Let us start the PA from a single node and consider elementary graphs like trees and chains to derive the PageRank vector of an arbitrary graph obtained after the nth step of the network's evolution. Nodes are numerated in the order of appending them to the evolving graph. The probability to append a new edge leading from the newly appearing node i to the existing node j is

$$p_{ij} = P\{i \to j\} = \begin{cases} \alpha(P_\alpha)_{ij}, \, i > j, \\ \gamma(P_\gamma)_{ij}, \, i < j, \\ \beta(P_\beta)_{ij}, \, i \in V(n-1) = V(n). \end{cases} \tag{8}$$

For simplicity, graphs without loops are considered. The approach can be extended the same way for graphs with loops and multiple edges.

To obtain a PageRank vector we use (6) and the expansion of the inverse matrix

$$[I - cP]^{-1} = I + cP + c^2 P^2 + \dots \tag{9}$$

by a power series. Let us consider examples to explain our approach.

Example 1. Let the node 2 be appended to the initial node 1 by adding a single edge leading from 1 to 2, Fig. 1(a). The self-loops in the node 2 are impossible on Step 2 due to the application of the PA schemes (1)–(3). If $p_{12} = P\{1 \to 2\} \neq 0$ holds, then the hyperlink matrix at Step 1 is given by $P = \begin{pmatrix} 0 & p_{12} \\ 0 & 0 \end{pmatrix}$, and we get $P^2 = P^3 = \dots = 0$. By (6) and (9) the PageRank vector at Step 1 is

$$\pi(1) = \frac{1-c}{2}((1,1) + (0, cp_{12})).$$

If $p_{21} = P\{2 \to 1\} \neq 0$ holds (see Fig. 1(b)), then

$$\pi(1) = \frac{1-c}{2}((1,1) + (cp_{21}, 0)).$$

If $p_{12} \neq 0$ and $p_{21} = \beta(P_\beta)_{21}$ hold (see Fig. 1(c)), that means a loop, then we get $P = \begin{pmatrix} 0 & p_{12} & 0 \\ p_{21} & 0 & 0 \\ 0 & 0 & 0 \end{pmatrix}$, and by (6) and (9) it holds

$$\pi(2) = \frac{1-c}{3}(1 + \sum_{k=1}^{\infty} c^{2k-1} p_{12}^{k-1} p_{21}^{k}(cp_{12} + 1), 1 + \sum_{k=1}^{\infty} c^{2k-1} p_{12}^{k} p_{21}^{k-1}(cp_{21} + 1), 0)$$

at Step 2. Here, zero relates to the node 3 that is absent.

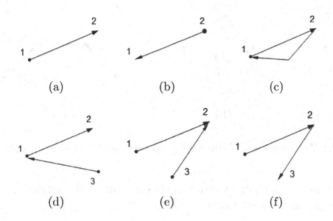

Fig. 1. Simplest graphs on Steps 1 and 2 of the graph evolution.

In the same way we have

$$\pi(2) = \frac{1-c}{3}((1,1,1) + c \cdot (p_{31}, p_{12} + cp_{31}p_{12}, 0)), \tag{10}$$

$$\pi(2) = \frac{1-c}{3}((1,1,1) + c \cdot (0, p_{12} + p_{32}, 0)), \tag{11}$$

$$\pi(2) = \frac{1-c}{3}((1,1,1) + c \cdot (0, p_{12}, p_{23} + cp_{12}p_{23})) \tag{12}$$

corresponding to graphs in Fig. 1(d)–(f). Note that root nodes which have outgoing edges only (see Fig. 1(a)–(f) except Fig. 1(c)) correspond to zeros in the brackets in formulae of $\pi(n)$. If the node i has only incoming edges, e.g. the node 2 in Fig. 1(e), then the probabilities $\{p_{ji}\}$ corresponding to these edges are summarized. The chains incoming to a node correspond to sums like in (10) and (12) reflecting the order of the edges in incoming chains.

Then the subsequent Lemma follows.

Lemma 2. *Let the graph be evolving by a linear PA model (1)–(3) and p_{ij} be defined by (8). Then the PageRank vector at time step n is given by*

$$\pi(n) = \frac{1-c}{n+1}(\mathbf{1}^T + c \cdot (p_{1j_1}(1 + cp_{i1}), p_{1j_2}(1 + cp_{i2}), ...,$$
$$p_{1j_{i-1}}(1 + cp_{i(i-1)}), 0, p_{1j_{i+1}}(1 + cp_{i(i+1)}), ..., p_{N_ij_{N_i}}(1 + cp_{iN_i}), ...))$$

for the Galton-Watson tree (Fig. 2(a)) with the root node i, and

$$\pi(n) = \frac{1-c}{n+1} \left(\mathbf{1}^T + c \cdot (0, p_{12}, p_{23} + cp_{12}p_{23}, ..., p_{n(n+1)} + \sum_{t=1}^{n-1} c^t \prod_{j=n-t}^{n} p_{j(j+1)}) \right)$$

for chains $(1 \rightarrow 2 \rightarrow ... \rightarrow n)$ beginning at node 1.

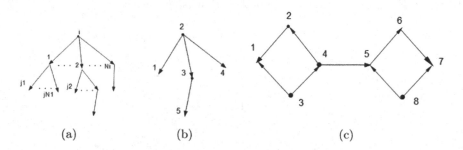

(a)   (b)   (c)

Fig. 2. Examples of Galton-Watson trees and a directed graph.

By analyzing incoming chains to each node one can easily obtain the PageRank vector of any graph without loops.

Algorithm 1. Calculation of the PageRank vector of graphs without loops

- Let a node 1 initiate a seed graph and n be the number of steps.
- Define a PageRank vector as $\pi(n) = \frac{1-c}{n+1}((1, ..., 1) + c \cdot (a_1, a_2, ..., a_n, a_{n+1}))$, where a_i corresponds to the node i.
- Find all incoming chains of the graph to a node i, $i \in \{1, 2, ..., n, n+1\}$, and determine $a_i = \sum_{j \rightarrow i} p_{ji} + \sum_{t=1}^{\ell_i - 1} c^t \sum_{j_t \rightarrow ... \rightarrow i} \underbrace{p_{j_t j_{t-1}} \cdot ... \cdot p_{ji}}_{t+1 \text{ multipliers}}$, where ℓ_i is a maximum length of the incoming chains to the node i.
- If the node i is a root, i.e. it has only outgoing edges, then determine $a_i = 0$.

Example 2. Let us consider the tree (Fig. 2(b)) and the directed graph (Fig. 2(c)). Their corresponding PageRank vectors are the following

$$\pi(4) = \frac{1-c}{5}((1, 1, 1, 1, 1) + c(p_{21}, 0, p_{23}, p_{24}, p_{35}(1 + cp_{23}))),$$

$$\pi(8) = \frac{1-c}{9}(\mathbf{1}^T + c(p_{21} + p_{31}, cp_{21}p_{42}, p_{42} + cp_{42}p_{34}, 0, p_{34}, p_{45} + cp_{34}p_{45}, p_{56}$$

$$+ cp_{45}p_{56} + c^2 p_{34}p_{45}p_{56}, p_{87} + p_{67} + cp_{56}p_{67} + c^2 p_{45}p_{56}p_{67} + c^3 p_{34}p_{45}p_{56}p_{67}, 0)).$$

The PageRank vector of a graph with loops like in Fig. 1(c) may be obtained in a similar way, but this item is out of scope of the paper.

5   Conclusions

Regarding a free-scale directed Web network without loops evolving by the $\alpha-$, $\beta-$ and $\gamma-$linear PA-schemes given in [7,8], the associated PageRank vector at time step n is computed. The PageRank vector is derived avoiding constraints of [6, Lemma 3.1], particularly, a fixed out-degree of all nodes. Our result allows us to avoid the consideration of random length paths to express the probability of a connection between two nodes. The algorithm to calculate the PageRank vector of directed graphs without loops is provided.

Since the $\alpha-$, $\beta-$ and $\gamma-$linear PA-schemes allow us to generate graphs with loops and multiple edges, PageRank vectors of graphs with loops and multiple edges are a subject of our further research.

Acknowledgments. The first author was partly supported by Russian Foundation for Basic Research (grant 19-01-00090).

References

1. Brin, S., Page, L.: The anatomy of a large-scale hypertextual web search engine. Comput. Netw. ISDN Syst. **30**(1–7), 107–117 (1998)
2. Fortunato, S., Boguñá, M., Flammini, A., Menczer, F.: Approximating PageRank from in-degree. In: Aiello, W., Broder, A., Janssen, J., Milios, E. (eds.) WAW 2006. LNCS, vol. 4936, pp. 59–71. Springer, Heidelberg (2008). https://doi.org/10.1007/978-3-540-78808-9_6
3. Litvak, N., Scheinhardt, W.R.W., Volkovich, Y.: In-degree and PageRank: why do they follow similar power laws? Internet Math. **4**(2–3), 175–198 (2007)
4. Chen, N., Litvak, N., Olvera-Cravioto, M.: PageRank in scale-free random graphs. In: Bonato, A., Graham, F.C., Prałat, P. (eds.) WAW 2014. LNCS, vol. 8882, pp. 120–131. Springer, Cham (2014). https://doi.org/10.1007/978-3-319-13123-8_10
5. Fortunato, S., Boguñá, M., Flammini, A., Menczer, F.: On local estimations of PageRank: a mean field approach. Internet Math. **4**(2–3), 245–266 (2007)
6. Avrachenkov, K., Lebedev, D.: PageRank of scale-free growing networks. Internet Math. **3**(2), 207–231 (2006)
7. Samorodnitsky, G., Resnick, S., Towsley, D., Davis, R., Willis, A., Wan, P.: Non-standard regular variation of in-degree and out-degree in the preferential attachment model. J. Appl. Prob. **53**(1), 146–161 (2016)
8. Wan, P., Wang, T., Davis, R.A., et al.: Are extreme value estimation methods useful for network data? Extremes **23**, 171–195 (2020). https://doi.org/10.1007/s10687-019-00359-x
9. Langville, A.N., Meyer, C.D.: Deeper inside PageRank. Internet Math. **1**(3), 335–380 (2005)

Traffic Management Algorithm
for V2X-Based Flying Fog System

Malik AlSweity[1], Ammar Muthanna[1,2(✉)], Ibrahim A. Elgendy[3,4],
and Andrey Koucheryavy[1]

[1] The Bonch-Bruevich Saint-Petersburg State University of Telecommunications,
Pr. Bolshevikov, 22, St. Petersburg 193232, Russia
[2] Peoples' Friendship University of Russia (RUDN University),
6 Miklukho-Maklaya Street, Moscow 117198, Russia
[3] School of Computer Science and Technology, Harbin Institute of Technology,
Harbin, China
ibrahim.elgendy@hit.edu.cn
[4] Department of Computer Science, Faculty of Computers and Information,
Menoufia University, Shibin el Kom 32511, Egypt

Abstract. V2X system support a large number of new services and protocols, therefore, moving to integrated interaction with the surrounding information space, defined as V2X interoperability, should ensure the interoperability of the various networks and systems involved in providing services to users [5]. The data must be transmitted with low latency and high speed between vehicles. This requires the introduction of a new high-speed network, advanced road transport, and telecommunications infrastructure. To ensure the effective functioning of V2X networks, in this paper we propose to use flying fog computing for a reliable and efficient system for simulation model, we used a UAVs as a base station (BS) equipped by a controller (CU) that provide execution of the requests arriving from the users equipment located in the communication range of the BS. Users that are connected with the BS requests flow that arrive at CU, we assume that one cloud server (CS) can serve a number of UAVs. CU is described as a service system which provides execution of some requests simultaneously. We assume that the execution of each request flow from users needs a fixed amount of energy. We've noticed that in the case of low traffic intensity the delay value increases with increasing of the part of traffic forwarded to the cloud, the energy consumption lower if the part of redirected traffic is bigger.

Keywords: Dynamic algorithm · Edge computing · 5G · Core network

1 Introduction

5G typically promises a very high bandwidth, ultra low latency and very high reliability. With these promises, it becomes evident that technologies specifi-

This paper has been supported by the RUDN University Strategic Academic Leadership Program.

cally with 5G that can be leveraged in vehicles. Getting from location A to B has never been easier and more enjoyable than in today's vehicles which are connected to the internet to provide navigation, entertainment and information diagnostics. Yet, this is just the beginning of a revolution of our cars, and in the near future, we'll go from connected to automated and eventually to fully autonomous. They will always be connected, energy efficient and increasingly intelligent, this provides fewer accidents, less congestion and reduces emissions. To realise this vision, it requires V2X technologies in which vehicles are connected with other vehicles, networks, pedestrians, infrastructures, everything. V2X lets vehicles share data in real time to avoid accidents, coordinate traffic and become more aware of the surroundings. V2X leverages sensors and computer vision for non line-of-sight awareness, effectively allowing cars to see what human drivers cannot see [1]. The next generation 5G cellular V2X will not only provide low latency communications but also enhance safety, improve awareness and enable reliable and seamless autonomous driving. Cellular V2X defines two new transmissions modes that work together, the first utilizes existing 5G networks with ubiquitous coverage so that cars can be alerted of accidents kilometers ahead, the second transmission mode is builds on LTE direct, which functions to exchange real-time information directly between vehicles traveling on high speed and high density traffic [2]. The benefits of 5G V2X networks go beyond safety given the ability to share rich real time data combined with powerful cloud computing technologies, the network system will better coordinate the traffic flow and reduce congestion's to a lower level, the result is a journey not only safer but also shorter in time and more efficient [3]. Drones are the main focus for many people because of their unlimited advantages in surveillance, videography, communication, search and rescue, military and numerous public services [4]. Fog computing is a group of edge nodes connected to the cloud and connected to devices and considers mediator among them, sometimes it contains a mini data center and it is a new distributed architecture, on that's band the continuum between the cloud and everything else, this allows for the distribution of critical coral functions [5] e.g., compute, communications, control, storage and decision making, closer to where data is originated, that's what makes Fog not just a common sense architecture, it's a necessary vital architecture for scenarios where latency, privacy and other data intensive issues are creating cons for concern. Fog computing technology can provide many advantages e.g., better low latency support [6], scalability, mobility, location awareness, and efficient integration with other systems such as cloud computing, These advantages can help V2X applications to perform better.

The rest of the paper is organized as follows. In Sect. 2 we present the related work and talk about problem statements. Section 3 presents System model and simulation results . In Sect. 4 we conclude the paper.

2 Problem Statement and Related Works

The task of transmitting data from UAV to autonomous cars and other users is extremely important. In recent years There has been a lot of research about UAV,

UAVs have the ability to perform complex activities with maneuvering flexibility and low-cost flight, UAVs connected with a station that is located in the cloud and allows a high-level processing [7]. An algorithm has been researched to help in transmitting and delivering data from the drones to the users in a short time and at a high speed. In [8] authors discuss Different IoT applications scenarios of using UAVFog and propose UAVFog that provides Fog computing capabilities onboard one or more UAVs that can dynamically support IoT applications at various locations. In [9] authors introduce fog computing into swarm of drones, and construct a task allocation optimization problem which jointly considering the latency, reliability and energy consumption, in order to minimize the energy consumption of the swarm of drones when the latency and reliability requirements are met

Figure 1 shows the general architecture, including elements of the physical world (cars, road, UAVs, people, etc.). As already been discussed, transmitting data from the users to the nearest CU and then to UAVs needs a fixed amount of energy, some requests flow may be lost if the CU is already providing data to some users, requests can be redirected to the CS if the CU is busy or it can be lost if both CU and CS are busy. In this work we use two core models, each one provides execution of the request in one millisecond. The difference between each node is in the number of nodes and the average service time. According to traffic intensity, we have collected data for different cases, high, normal or low traffic rates.

3   System Model and Simulation Results

Simulation model of the considered system was implemented in the event-driven simulation software package AnyLogic.

We have made some assumptions for the simulation model. We suppose that the UAV which is used as a base station (BS) equipped with a controller (CU) that provides execution of the requests arriving from the users equipment located in the communication range of the BS. The traffic originated by users presents requests flow which arrives at the CU. Each request may be served by CU, may be redirected to the cloud server (CS) or may be lost if both CU and CS are too busy. The part of the redirected requests may be changed by correspondent parameter.

We assume that one CS can serve a number of UAVs. We describe the CU as a service system which provides execution of some requests simultaneously. For example, it may be a multi-core processor. We use two core models which provide execution of the request in one millisecond (average service time).

The model of CS differs from CU by a number of cores (8 cores model) and the average service time which is equal to 0.5 ms.

We assume that execution of each request needs a fixed amount of energy E_0. The simulation model structure shown in the Fig. 2.

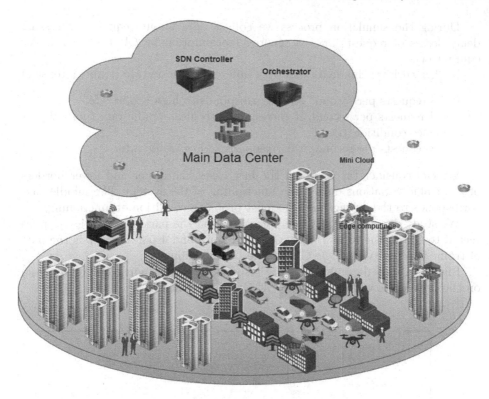

Fig. 1. The common architecture for our task

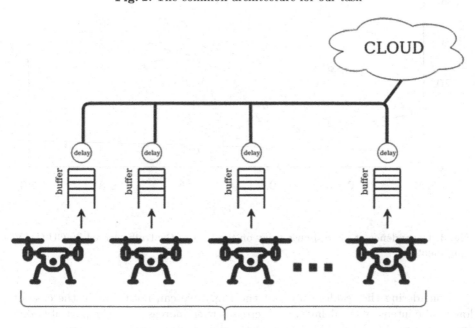

Fig. 2. The structure of the simulation model

During the simulation process we collect data about requests processing delay, losses of requests and the energy consumption by CU (only for UAVs equipment).

In this work we choose 3 cases with different requests rate (traffic intensity)

- a = 8 requests per second, which corresponds to high traffic rate.
- a = 4 requests per second, it corresponds to mean traffic rate (normal performance conditions). and
- a = 2 requests per second, it corresponds to low traffic rate.

We can consider the first and the last cases like upper and lower borders of the traffic variations during the functioning of the system. The middle case corresponds to the average and the most probable condition of functioning.

We study dependence of those parameters on the part of traffic $1 - p$, redirected to CS. In the results, shown in the Figs. 3 and 4 parameter p is the part of traffic which served at the CU (at the UAV equipment).

Figure 3 shows dependence of the response delay (response on the request) on the part of the traffic served by CU (UAV equipment).

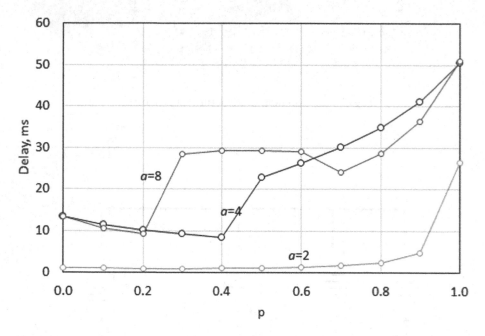

Fig. 3. Dependence of the response delay on the part of the traffic served by CU (UAV equipment)

Considering the results shown at the Fig. 3 we can note that in the case of low traffic intensity the delay value increases with increasing of the part of traffic forwarded to the cloud. This is due to the final throughput of the cloud.

In the cases of $a = 8$ and $a = 4$ we see minimal values of the delay at the points $p = 0.2$ and $p = 0.4$ correspondingly. Therefore, we can note that in the range of average and high traffic rate there are optimal values of the distribution of traffic p. The optimal value depends on the traffic rate. Increasing the traffic rate leads to decreasing the optimal value of p. This result means that in the real traffic values range we can choose optimal distribution of the traffic between the internal UAVs resources and the cloud resources.

Study of the power consumption shown in the Fig. 4.

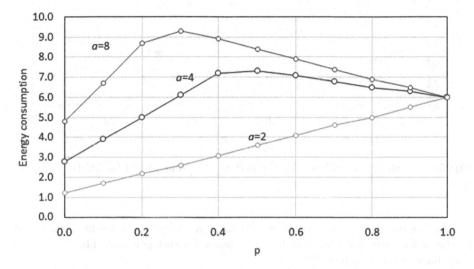

Fig. 4. Dependence of the power consumption on the part of the traffic served by CU (UAV equipment).

Looking at the results shown in the figure we can note that at the low traffic rate $(a = 2)$ energy consumption increases with increasing the part of traffic forwarded to the internal UAVs resources and this dependence is close to line function. We also see the mean and high traffic intensity maximal values of energy consumption. These results mean that the energy consumption in the real traffic range is not linear and has a maximal value. The maximal consumption point (value of p) depends on the traffic intensity. Increasing the traffic decreases the p value for the maximal consumption value.

It is clear that if all the traffic redirected to CS ($p = 0$) the energy spent on the service in the CU is equal to zero. Increasing the part of traffic more than 0.5 (in the case $a = 4$) does not affect energy consumption because the maximum load of CU is reached.

Study results of the loss dependence is shown in the Fig. 5.

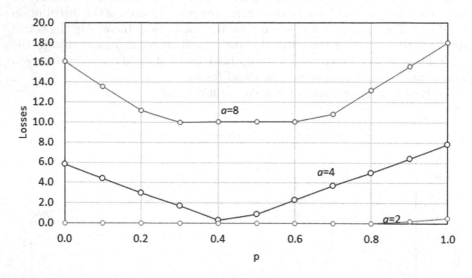

Fig. 5. Dependence of the losses on the part of the traffic served by CU (UAV equipment).

We can see that in the case of small traffic intensity losses are near zero. But in the other cases for $a = 4$ and $a = 8$ losses dependencies are not linear and they have minimal values.

For example in case $a = 4$ we can see that the delay shown at the Fig. 3 this dependence has the minimal value at the point of $p = 0.4$.

Presented results show that response delay and probability of losses depends on the part of the redirected traffic. We see that it may be reached the optimal value of delay and loss probability by choosing the correct part of redirected traffic.

The energy consumption either depends on the part of the redirected traffic but this dependence is smooth. The energy consumption is lower if the part of redirected traffic is bigger.

Figures 6, 7 and 8 show empirical probability distributions of delay, energy and losses.

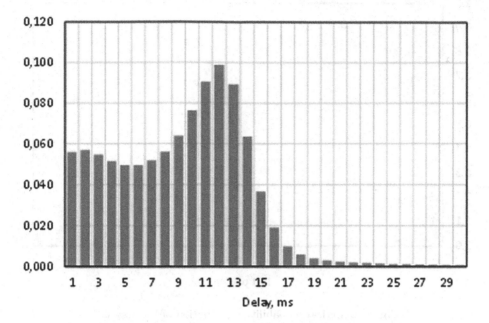

Fig. 6. Empirical probability density of the delay of response

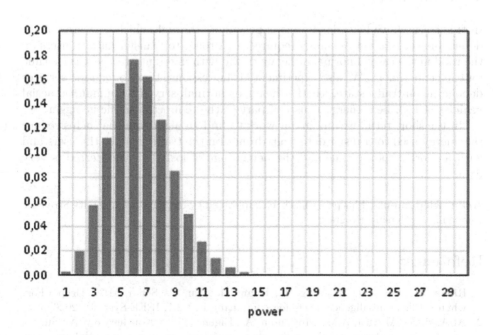

Fig. 7. Empirical probability density of the energy consumption

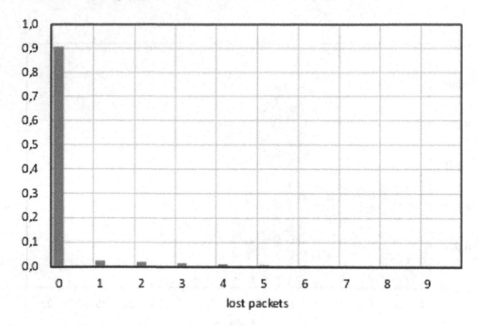

Fig. 8. Empirical probability distribution of the losses.

4   Conclusion

In this research we have done some experiments through which we were able to determine the amount of energy required for drones to transfer data to users on the road in the case of normal or heavy traffic with different average service time for each experiment. We also experimented the response time to requests services depending on traffic state, whether heavy or normal, served by unmanned aerial vehicles (CUs). Less energy is used if part of the redirected traffic is bigger.

In the short term, we aim in our work to find an algorithm that helps reduce the energy used to transfer data from drones on the traffic road to the users and with a short response time and the percentage of lost data transferred is very little.

Acknowledgment. This paper has been supported by the RUDN University Strategic Academic Leadership Program.

References

1. Rihan, M., Elwekeil, M., Yang, Y., Huang, L., Xu, C., Selim, M.M: Deep-VFog: when artificial intelligence meets fog computing in V2X. IEEE Syst. J. (2020)
2. Al-Ansi, A., Al-Ansi, A.M., Muthanna, A., Elgendy, I.A., Koucheryavy, A.: Survey on intelligence edge computing in 6G: characteristics, challenges, potential use cases, and market drivers. Future Internet **13**(5), 118 (2021)

3. Artem, V., Al-Sveiti, M., Elgendy, I.A., Kovtunenko, A.S., Muthanna, A.: Detection and recognition of moving biological objects for autonomous vehicles using intelligent edge computing/LoRaWAN mesh system. In: NEW2AN/ruSMART -2020, Part II. LNCS, vol. 12526, pp. 3–15. Springer, Cham (2020). https://doi.org/10.1007/978-3-030-65729-1_1

4. Ding, G., Wu, Q., Zhang, L., Lin, Y., Tsiftsis, T.A., Yao, Y.-D.: An amateur drone surveillance system based on the cognitive internet of things. IEEE Commun. Mag. **56**(1), 29–35 (2018)

5. Vladyko, A., Khakimov, A., Muthanna, A., Ateya, A.A., Koucheryavy, A.: Distributed edge computing to assist ultra-low-latency VANET applications. Future Internet **11**(6), 128 (2019)

6. Elgendy, I.A., Muthanna, A., Hammoudeh, M., Shaiba, H., Unal, D., Khayyat, M.: Advanced deep learning for resource allocation and security aware data offloading in industrial mobile edge computing. Big Data (2021)

7. Pinto, M.F., Marcato, A.L., Melo, A.G., Honório, L.M., Urdiales, C.: A framework for analyzing fog-cloud computing cooperation applied to information processing of UAVs. Wireless Communications and Mobile Computing (2019)

8. Mohamed, N., Al-Jaroodi, J., Jawhar, I., Noura, H., Mahmoud, S.: UAVFog: a UAV-based fog computing for internet of things. In: 2017 IEEE SmartWorld, Ubiquitous Intelligence and Computing, Advanced and Trusted Computed, Scalable Computing and Communications, Cloud and Big Data Computing, Internet of People and Smart City Innovation (SmartWorld/SCALCOM/UIC/ATC/CBDCom/IOP/SCI), pp. 1–8. IEEE (2017)

9. Hou, X., Ren, Z., Wang, J., Zheng, S., Cheng, W., Zhang, H.: Distributed fog computing for latency and reliability guaranteed swarm of drones. IEEE Access **8**, 7117–7130 (2020)

Investigation of Wireless Hybrid Communication System Reliability Under External Influences

Konstantin Vytovtov$^{(\boxtimes)}$ ⓘ, Elizaveta Barabanova ⓘ,
and Vladimir Vishnevsky ⓘ

V.A. Trapeznikov Institute of Control Sciences RAS, Profsoyuznaya 65 Street,
Moscow, Russia
vytovtov_konstan@mail.ru

Abstract. The transient mode of the hybrid communication system that includes two wireless information channels operating in optical and millimeter domains is considered for the first time. The accurate analytical method for analyzing the transient behavior of the hybrid system is developed. It allows us to analyze the transient behavior of the system in an operating mode, in the case of a failure one of the channels, and periodical man-made radio influences. The expressions for reliability indicators and performance metrics of the hybrid system are presented and numerical calculations demonstrate the proposed method.

Keywords: Hybrid system · Transient mode · Reliability

1 Introduction

Development of high-speed millimeter [1,2] and optical [3,4] wireless communication systems is one of the most important direction in the implementation of 5G/6G networks. This is due to the fact that data transfer rate of these wireless channels can reach 10 Gb/s. However, optical and millimeter-wave systems have low stability in relation to the influence of atmospheric phenomena. To solve this problem, hybrid systems consisting of various combinations of optical, millimeter and radio frequency communication channels have recently developed [5–7]. In general, a hybrid system contains two or more channels [8–10], which can be used simultaneously or separately. For example if the highest quality or highest priority channel is used, another can be used as reserve. Besides this one channel can be used as direct transmitting, and another one can be used for receiving. At the same time, the calculation of system reliability indicators, as well as the probability of operation of each of the channels is the most important theoretical and practical problem. Reliability indices of hybrid system including an optical channel and a radio channel, have been investigated in [5] for a stationary mode.

The reported study was funded by RFBR, project number 19-29-06043.

V. M. Vishnevskiy et al. (Eds.): DCCN 2021, LNCS 13144, pp. 42–54, 2021.
https://doi.org/10.1007/978-3-030-92507-9_5

A similar system has been considered in [9]. In that paper the authors found the steady-state probabilities of the system using the matrix-analytical method, determined conditions for the existence of a stationary mode and calculated system performance metrics. At the same time it should be noted that the transient mode was not considered by the authors of these works. However, transient mode occurs quite often during operation of hybrid systems. For example, a high-speed non-reliable optical channel is switched on the millimeter one if the quality of an optical channel is decreased. This can happen for example as a result of external factors such as morning fog, sunlight at noon and continuous or intermittent man-made radio influences, frequency drift, etc. In this case, it is incorrect to average the probability of states over a long period of time, since the noise interference and other influences lead to significant change of the system parameters. Moreover, during the action of periodic noise, the probabilities of the states can differ significantly from similar probabilities in the stationary mode. It means that the reliability, throughput, and queue length in the transient and stationary mode can be different. An inaccurate estimate of these parameters can lead to system overload and a decrease in the quality of information transmission. This influence is especially critical for the transmission of delay-sensitive traffic. So, for example, the signal attenuation in the optical channel is 0–3 dB/km in clear weather, and it is 6–17 dB/km in rainy weather. Mover snow and fog leads to an increase in attenuation up to 6026 dB/km and 50–100 dB/km respectively [11]. In addition, turbulent phenomena in the atmosphere caused by changes in temperature and increased wind can lead to low-frequency flickering fading of the optical channel (0.5 Hz–3 kHz). At the same time, the availability factor of the 1–2 km optical channel is considered equal to 0.99 [12]. Similar phenomena are observed for the millimeter channel in [13]. For example, the attenuation of the millimeter-wave channel during heavy rain is in the order of 30–50 dB/km. In this paper, the approach for analysis of hybrid system transient behavior in the case of one of two channels failure based on the analytical method is proposed. In Sect. 2 the statement of the problem is given. Section 3 presents the translation matrix of the Kolmogorov system (Subsect. 3.1) and the expressions for the probabilities states of the system in normal operation in the absence of jumps of the intensities of transitions (Subsect. 3.2), as well as in the presence of jumps of the transition intensities caused by noise (Subsect. 3.3). Numerical simulation results are presented in Sect. 4.

2   The Statement of the Problem

The considered hybrid system consisting of two data transmission channels is studied in this paper. The basic channel is high-speed optical channel and the millimeter channel is in cold reserve. The case when random interferences lead to a disruption of optical channel operation and the reserve millimeter-wave channel is automatically switched on is considered in this paper. The backup channel is working for a certain period of time until the optical channel is restored or the backup millimeter wave channel is also disrupted due to external factors. The system state graph is shown in Fig. 1. In Fig. 1 the states of the system mean the following: $S_1(t)$ is the state when the optical channel is working, $S_2(t)$ is the state when the system is in the mode of switching from the optical to the millimeter channel, $S_3(t)$ is the state when the millimeter channel is working, $S_4(t)$ is the state when the system is in the mode of switching from the millimeter to the optical channel. The system to be in each of the states during the time which has an exponential distribution. The intensities of transitions from one state to another are λ_{ij} that, in the general case, are piecewise-constant functions, determined by both the climatic conditions of the system operation, its technical characteristics, and the presence or absence of noises. The system state graph is shown in Fig. 1.

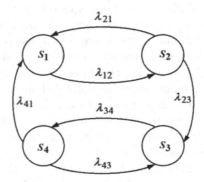

Fig. 1. The state transition diagram.

3   The Method

3.1   Fundamental Matrix of the Kolmogorov System

Let denote by $p_1(t)$, $p_2(t)$, $p_3(t)$, $p_4(t)$ the probabilities of finding the system in the states $S_1(t)$, $S_2(t)$, $S_3(t)$, $S_4(t)$, respectively. Then, using the Δ-method we can write the following system of Kolmogorov equations in matrix form

$$\frac{d}{dt}\mathbf{P}(t) = \mathbf{A}(t)\mathbf{P}(t) \tag{1}$$

where

$$\mathbf{A}(t) = \begin{pmatrix} -\lambda_{12}(t) & \lambda_{21}(t) & 0 & \lambda_{41}(t) \\ \lambda_{12}(t) & -\lambda_{21}(t) - \lambda_{23}(t) & 0 & 0 \\ 0 & \lambda_{23}(t) & -\lambda_{34}(t) & \lambda_{43} \\ 0 & 0 & \lambda_{34}(t) & -\lambda_{41}(t) - \lambda_{43}(t) \end{pmatrix} \tag{2}$$

Here $\mathbf{P}(t) = (p_1(t), p_2(t), p_3(t), p_4(t))^T$ is the state vector of the system, T is the transposition operator, $\mathbf{A}(t)$ is the piecewise constant matrix of equation Kolmogorov coefficients.

The method of the fundamental matrix of the Kolmogorov equations is used to study the behavior of the system in this work. The fundamental matrix of a system of linear homogeneous differential equations allows one to obtain solutions of the system at an arbitrary moment of time for given initial conditions. In this case, this matrix connects the probabilities of the system states at an arbitrary time instant t with these probabilities at the initial time t_0 [17]:

$$\mathbf{P}(t) = \mathbf{M}(t)\mathbf{P}(t_0) \tag{3}$$

If the transition probabilities change in a jump at a certain moment of time, then the fundamental matrix after the jump can be found as the matrix of the first interval at a time value equal to the duration of the interval to the fundamental matrix of the second interval [17]

$$\mathbf{M}(t) = \mathbf{M}_2(t)\mathbf{M}_1(t_1) \tag{4}$$

In the presence of several jumps, the behavior of the system on the ith interval is described by the matrix

$$\mathbf{M}(t) = \mathbf{M}_i\left(t - \sum_{i=1}^{N} t_i\right) \sum_{j=i-1}^{1} \mathbf{M}_j(t_j) \tag{5}$$

The proof of this property will be given below.

Thus, the fundamental matrix makes it possible to describe the behavior of the system with an arbitrary jump-like change transition intensities.

3.2 Probabilities of States for Constant Parameters

In the case when the transition intensities $\lambda_{ij} = const$ $(i, j = \overline{1, 4})$ (in the absence of noises), the system (1) is a system of linear homogeneous differential equations of the fourth order, the solution of which has an exponential form

$$p_i(t) = \sum_{k=1}^{4} \xi_{ik} A_k exp(\gamma_k t) \tag{6}$$

where γ_k are the roots of the characteristic equation of the system (1). And the elements of the fundamental matrix $\mathbf{M}(t)$ of the Kolmogorov equation system at constant transition intensities found in accordance with the method [17] can be written in the form

$$m_{ij}(t) = \sum_{k=1}^{4} \xi_{jk} \frac{\Delta_{ik}}{\det \Xi} \exp(\gamma_k t) = \sum_{k=1}^{4} \chi_{ij}^{(k)} \exp(\gamma_k t) \tag{7}$$

where $\Xi = \| \xi_{ij} \|_4^4$ and Δ_{ik} are the algebraic complements to the elements ξ_{ik} of the matrix Ξ. Then, in accordance with (3), the state probability is determined by the expression

$$p_i = \sum_{j=1}^{4} m_{ij}(t)p_j(0) \tag{8}$$

Here $p_j(0)$ is the initial probability of the j-th state.

The presented method allows one to fully describe the behavior of the considered hybrid system at an arbitrary moment of time t and calculate all parameters of this system. For example, when jumps of the transition intensities are absent the elements of the fundamental matrix have the form (6). In this case as the availability factor equals to the probability of failure-free system operation the formula for it calculation can be written as

$$R(t) = \sum_{j=1}^{4} \left\{ \left[\sum_{i=1}^{4} p_i(t_0) \left(\chi_{1i}^{(j)} + \chi_{3i}^{(j)} \right) \right] \exp\left(\gamma_j t \right) \right\} \tag{9}$$

And the probability of failure

$$F(t) = \sum_{j=1}^{4} \left\{ \left[\sum_{i=1}^{4} p_i(t_0) \left(\chi_{2i}^{(j)} + \chi_{4i}^{(j)} \right) \right] \exp\left(\gamma_j t \right) \right\} \tag{10}$$

Taking into account the probability of failure-free operation of the optical and millimeter communication channels $p_1(t)$ and $p_2(t)$ correspondingly, their throughputs are equal to $Q_{oc}(t) = p_1(t)V_{oc}$ and $Q_{mc}(t) = p_3(t)V_{mc}$. Here V_{oc} and V_{mc} are the maximum throughputs of the optical and millimeter channels, respectively. Then the resulting throughput of the hybrid station is

$$Q_{hs}(t) = p_1(t)V_{oc} + p_3(t)V_{mc} \tag{11}$$

Taking into account (7) and (8), expression (11) can be written as

$$Q_{hs}(t) = \sum_{j=1}^{4} \left\{ \left[\left(\sum_{i=1}^{4} \chi_{1i}^{(j)} \exp\left(\gamma_j t \right) \right) V_{oc} + \left(\sum_{i=1}^{4} \chi_{3i}^{(j)} \exp\left(\gamma_j t \right) \right) V_{mc} \right] p_i(t_0) \right\} \tag{12}$$

3.3   Probabilities of States in the Presence of Jumps of the Transition Intensities Caused by Noises

This section proposes the mathematical model of the hybrid communication system in the presence of jumps of the transition intensities. Such jumps can occur when one of the channels fails because of external weather factors, noises, etc. In the simplest case, the change of the transition intensities can be approximated by a jump function. In this case, the fundamental matrices of the Kolmogorov equation system calculated using the described above method, are different on intervals with normal operation $\mathbf{M}_{no}(t)$ and intervals with violation of the normal mode $\mathbf{M}_{nn}(t)$. And the resulting matrix on the noisy interval is found as the multiplication of the matrices of these intervals. Indeed, from the moment of time when the station is switched on until the effect of the interference, the probabilities of the system states are found as

$$\mathbf{P}(t) = \mathbf{M}_{no}(t)\mathbf{P}(t_0) \tag{13}$$

Then at the moment of the beginning of exposure to noise $t = t_1$ we have

$$\mathbf{P}(t_1) = \mathbf{M}_{no}(t_1)\mathbf{P}(t_0) \tag{14}$$

Taking into account that the state probabilities cannot change abruptly at the moment of noise exposure, for the time interval when the noise affects the system, we write

$$\mathbf{P}(t) = \mathbf{M}_{nn}(t - t_1)\mathbf{P}(t_1) \tag{15}$$

Substituting into (15) the value of $\mathbf{P}(t_1)$ from (13), we obtain for this interval

$$\mathbf{P}(t) = \mathbf{M}_{nn}(t - t_1)\mathbf{M}_{no}(t_1)\mathbf{P}(t_0) \tag{16}$$

Then at the moment of the end of the noise exposure $t = t_1 + t_2$, we have

$$\mathbf{P}(t_1 + t_2) = \mathbf{M}_{nn}(t_2)\mathbf{M}_{no}(t_1)\mathbf{P}(t_0) \tag{17}$$

Taking into account that the probability of the state cannot change abruptly at the moment of the end of the noise, for the time interval when the noise is over, we write

$$\mathbf{P}(t) = \mathbf{M}_{no}(t - t_1 - t_2)\mathbf{P}(t_1 + t_2) \tag{18}$$

Substituting into (18) the value of $P(t_1 + t_2)$ from (17), we obtain an expression describing the state of the system on this interval

$$\mathbf{P}(t) = \mathbf{M}_{no}(t - t_1 - t_2)\mathbf{M}_{nn}(t_2)\mathbf{M}_{no}(t_1)\mathbf{P}(t_0) \tag{19}$$

Further, the procedure is repeated and its duration depends on the amount and duration of the noises. In the general case, the fundamental matrix for the case of N jumps of transition intensities ($N + 1$ intervals with constant parameters) on the N-th interval has the form

$$\mathbf{M}(t) = \mathbf{M}_{N+1}\left(t - \sum_{i=1}^{N} t_i\right) \prod_{i=N}^{1} \mathbf{M}_i(t_i) \tag{20}$$

The obtained result makes it possible to write analytic expressions for different parameters of the system. So, the availability factor at an arbitrary moment of time is equal to the sum of the probabilities that either the optical or the millimeter channel is working $R(t) = p_1(t) + p_3(t)$. Then, taking into account the fundamental matrix (20), we write

$$R(t) = \sum_{i=1}^{4} [M_{1i}(t) + M_{3i}(t)] \, p_i(t_0) \tag{21}$$

Analogously the probability of failure is

$$F(t) = \sum_{i=1}^{4} [M_{2i}(t) + M_{4i}(t)] \, p_i(t_0) \tag{22}$$

Here $M_{ki}(t)$ are the elements of the fundamental matrix (20). In this case, these elements are not presented here because they are very complex. At the same time, they are expressed in the primary parameters of the system in an explicit form. Therefore, one of the main obvious advantages of the presented method is a synthesis of hybrid communication systems for the given availability factor in conditions of noises or abrupt changes of the environment.

The throughputs of the optical and millimeter-wave channels are

$$Q_{oc} = p_1(t)V_{oc} \tag{23}$$

$$Q_{mc} = p_3(t)V_{mc} \tag{24}$$

where V_{oc} is the maximum throughput of the optical channel, and the resulting throughput of the system at an arbitrary moment of time t is equal to

$$Q(t) = Q_{oc}(t) + Q_{mc}(t) = p_1(t)V_{oc} + p_3(t)V_{mc} \tag{25}$$

Taking into account the fundamental matrix (20), expression (25) can be written as

$$Q(t) = \sum_{i=1}^{4} [\mathbf{M}_{1i}V_{oc} + \mathbf{M}_{3i}V_{mc}] \, p_i(t_0) \tag{26}$$

In conclusion of this section, we note that the proposed method can be applied to the case of continuously varying transition intensities. In this case, the continuous function describing these intensities must be represented as a piecewise constant one. Estimating the accuracy of calculations for a system with continuously changing parameters is beyond the scope of this work and it is the subject of a separate paper.

4   Numerical Simulation

4.1   Simulation of a Normal Operation Mode of the System

In this section, numerical calculations of the hybrid communication system (Fig. 1) in the absence and presence of man-made radio and optical noises in the transient mode are carried out. Also three transient modes are considered. In the first mode the station is turned on and in the second and the third modes the optical or millimeter channels are failed respectively. As the calculation of the system transition intensities is a separate problem that requires a large number of experimental tests their values were chosen on the basis of well-known

practical date [14]. It should be noted that the error rates experimentally investigated in [14] strongly depend on season and these regularities are not present
in the paper. Therefore, the average values of transition intensities are taken into
account in this paper. Thus, here we take an average error rate of 0.06 errors
per minute for the optical channel and 0.03 errors per minute for the millimeter
channel. Then the intensity of the transition from the first state to the second
state is $\lambda_{12} = 0.001$, and the intensity of the transition from the third state to the
fourth state is $\lambda_{34} = 0.0005$. Further, $\lambda_{21} = 0.0005$, $\lambda_{23} = 0.9995$, $\lambda_{41} = 0.9995$,
$\lambda_{43} = 0.005$. The results of calculating the probabilities of the system states in
the transient and subsequent stationary modes are shown in Fig. 2a. In Fig. 2a
and Fig. 2b p_1 is the probability of the optical channel operation, p_3 is the probability of the radio channel operation, $p_{switch} = p_2 + p_4$ is the probability that
the system is in the switching mode or there is no connection. The analysis of
transient behavior of the system in the initial moment of time in a larger scale
is presented in Fig. 2. It is seen that when the hybrid communication station is
turned on, the probabilities of the optical and the millimeter-wave channels are
obviously equal to zero and the probability of failure is equal to unity. Then the
probabilities of the channel operation increase and at $t = 4.5$ s we have $p_1 \approx p_3$.
Besides starting from $t = 4.5$ s, the probability p_1 decreases, and the probability
p_3 increases. In general, the transient time is $t_{tm} = 3000$ s. In the stationary
mode, the probability that the optical channel is operating is 0.332, the probability that the millimeter channel is operating is 0.667, and the communication
probability is 0.001. The availability factor is 0.999 and the downtime ratio is
0.001.

We also note that the time of the transient process of such a system is very
long, and a decrease of the transition intensities from the first state to the second
state leads not only to an increase of the probability of failure-free operation,
but also to an increase of the transient time. The throughput of the hybrid
channel is calculated according to (26) is $Q - 398.7$ Mbit/s in the stationary
mode at the maximum throughput of the optical and the millimeter channels
V_{oc}=1 Gbit/s and V_{mc}=0.1 Gbit/s. In this case, the throughput of the optical
channel is $Q_{oc} = 332$ Mbit/s, and the one of the millimeter channel is $Q_{mc} = 66.7$
Mbit/s. In the worst case, for the $\lambda_{12} = 0.00125$, the probability of the optical
channel operating is 0.287, the probability of the millimeter channel operating is
0.71, the probability of failure is 0.003 (Fig. 2c). As it is seen the probability of
failure increases and the probability of system operation in the millimeter-wave
domain increases too. The system availability factor is 0.997. In the best case,
for the $\lambda_{12} = 0.00075$, the probability of the optical channel operation is 0.398,
the probability of the millimeter channel operation is 0.601, the probability of
failure is 0.001 (Fig. 2d). The availability factor is 0.999.

4.2   Modeling the Operation of a Hybrid Station with Channel Faults

The calculating results of the state probabilities of the system under the condition of its normal operation up to $t = 3000$ s (see the previous section) and a

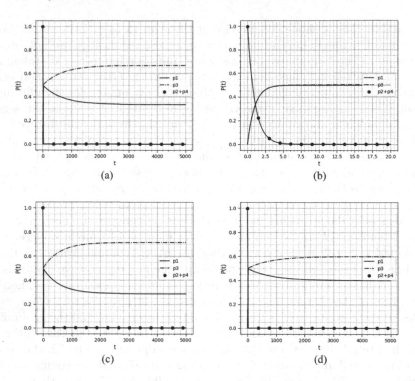

Fig. 2. Dependence of the hybrid system states on time when the station is turned on

failure of one of channels from the moment $t = 3000$ s are presented in Fig. 3 (millimeter channel failure) and in Fig. 4. (optical channel failure). The solid line corresponds to the probability of the optical channel operation and the dash-dot line corresponds to the probability of the millimeter-wave channel operation. The dotted line indicates that the system is in the switching mode. Figure 3b and Fig. 4b show the transient mode in a larger scale. Note that the downtime ratio of the hybrid station sharply increases in the transient mode. In the case of the millimeter channel failure, the maximum value of the downtime ratio is 0.25, and in the case of the optical channel failure, it is 0.125. In the case of the millimeter channel failure (Fig. 3), the probability p_3 is equal to zero and the probability p_1 is equal to 0.999. These probabilities are the system availability factors at the same time. In this case the probability that the system is in the switching mode is 0.001. In the event of the optical channel failure the probability $p_3 = 0.9995$, $p_1 = 0$, $p_{sw} = p_2 + p_4 = 0.0005$ (Fig. 4). Moreover, the indicated values of the probabilities do not depend on the moment of time when the failure has occurred. The duration of the transient mode in the first case is 9s, and in the second one is 8 s.

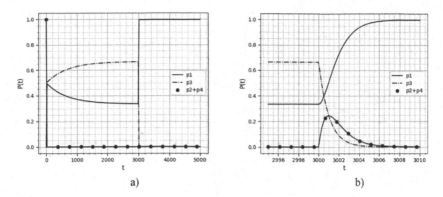

Fig. 3. The probabilities of system states in the event of millimeter channel failure

4.3 Simulation of the Hybrid Station Under the Man-Made Radio Influence

The dependences of the state probabilities are presented in Fig. 5 for the case of man-made radio influence to the optical channel from the moment of time $t = 3000$ s. The noise lasts for 5 s, the effect lasts for 10 s and has five periods. Under the influence of noise $\lambda_{12} = 0.8$, $\lambda_{34} = 0.0005$, $\lambda_{21} = 0.0005$, $\lambda_{23} = 0.9995$, $\lambda_{41} = 0.9995$, $\lambda_{43} = 0.005$. An increase of λ_{12} is obviously in this case, since it increases the probability of communication disruption in the optical channel. Obviously, the availability factor of the optical channel decreases to zero in this case, the availability factor of the millimeter channel increases to 0.999, and the station downtime factor is 0.001. Despite the short-term effect of noises, the restoration of the stationary mode takes about 4000 s. In addition, when the noise is turned on, the station availability factor drops to 0.89, and then increases within 4s to a value of 0.001.

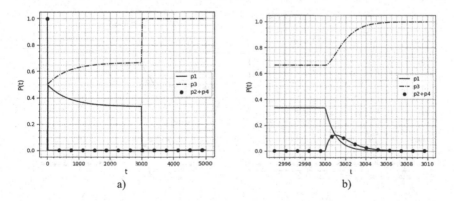

Fig. 4. The probabilities of system states in the event of optical channel failure

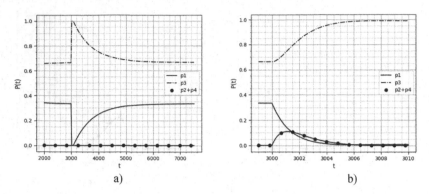

Fig. 5. State probabilities of the system under the man-made radio influence on optical channel

Dependences of state probabilities of the system in the event of the man-made radio influence on millimeter channel are presented on Fig. 6. In this case the noises to the millimeter channel are acting for 5 s, are finishing for 10 s and has five periods. In this case $\lambda_{12} = 0.001$, $\lambda_{34} = 0.8$, $\lambda_{21} = 0.0005$, $\lambda_{23} = 0.9995$, $\lambda_{41} = 0.9995$, $\lambda_{43} = 0.005$. Despite the short-term effect of the noise, in which the availability factor of the millimeter-wave channel is reduced to almost zero, the system recovery time is a long period about 4500 s. It is important that at the beginning of the noise, the system availability decreases to 0.78 within one second, and then increases to 0.001 within 7 s. Figure 7 shows the results of the states probabilities calculation under the man-made radio influence on both channels of the system. Here $\lambda_{12} = 0.8$, $\lambda_{34} = 0.8$, $\lambda_{21} = 0.0005$, $\lambda_{23} = 0.9995$, $\lambda_{41} = 0.9995$, $\lambda_{43} = 0.005$. In Fig. 7a, the dashed line corresponds to the availability of the optical channel; the dash-dot line corresponds to the availability of the millimeter-wave channel, and the solid line corresponds to the availability of the total hybrid system. The dependence of downtime ratio is shown on Fig. 7b. As

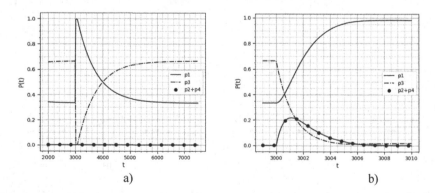

Fig. 6. State probabilities of the system under the man-made radio influence on millimeter channel

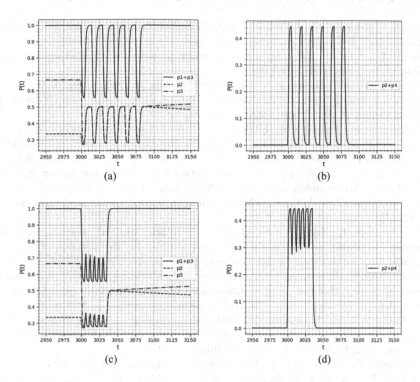

Fig. 7. The state probabilities of the system under the man-made radio influence on both channel

can be seen from the calculations, its maximum value is 0.443 during the time when the noise is acting. With a decrease in the interval between the noises, the system does not have time to restoring. The case when the intervals between the noises are 1s is considered in Fig. 7c and d. As seen, the availability factor does not exceed 0.725, and on average is 0.64.

5   Conclusion

In this paper, the hybrid communication system including optical and millimeter communication channels is considered. The accurate analytical method that allows calculating the state probabilities of the system, and technical parameters of the system such as an availability factor and a throughput, both the stationary and the transient modes are presented. In addition, the cases when one of the channels fails are considered. The influence of noises on the system states changing has been studied.

References

1. Noohani, M.Z., Magsi, K.U.: A review of 5G technology: architecture, security and wide applications. Int. Res. J. Eng. Technol. **7**(5), 34p (2020)

2. Singh, S., Chawla, M.: A review on millimeter wave communication and effects on 5G systems. Int. Adv. Res. J. Sci. Eng. Tecchnol. **4**(7), 28–33 (2017)
3. Singh, C., John, J., Tripathi, K.K.: A review on indoor optical wireless systems. Opt. Quant. Electron. 1–36 (1997)
4. Chowdhury, M.Z., Hasan, M.K., Shahjalal, M., Hossan, M.T., Jang, Y.M.: Optical wireless hybrid networks: trends, opportunities, challenges, and research directions. IEEE Commun. Surv. Tutor. **22**(2), 930–966 (2020)
5. Rahaim, M.B., Vegni, A.M., Little, T.D.C.: A hybrid radio frequency and broadcast visible light communication system. In: Proceedings of IEEE GLOBECOM Workshops (GC Wkshps), Houston, TX, USA, pp. 792—796, December 2011
6. Wu, W., Zhou, F., Yang, Q.: Dynamic network resource optimization in hybrid VLC and radio frequency networks. In: Proceedings of the International Conference Selected Topics Mobile Wireless Networking (MoWNeT), Avignon, France, pp. 1—7 (2017)
7. Pan, G., Lei, H., Ding, Z., Ni, Q.: On 3-D hybrid VLC-RF systems with light energy harvesting and OMA scheme over RF links. In: Proceedings of the IEEE Global Communications Conference, Singapore, pp. 1–6 (2017)
8. Vishnevsky, V.M., Semenova, O.V., Sharov, S.Y.: Modeling and analysis of a hybrid communication channel based on free-space optical and radio-frequency technologies. Autom. Remote Control **72**, 345–352 (2013)
9. Vishnevsky V.M., Semenova O.V. On one performance model of a broadband hybrid communication channel based on laser and radio technologies. Problemy informatiki **2**(6), 43–58. (2010). Novosibirsk (in Russian)
10. Klimenok, V., Vishnevsky, V.: Unreliable queueing system with cold redundancy. In: Gaj, P., Kwiecień, A., Stera, P. (eds.) CN 2015. CCIS, vol. 522, pp. 336–346. Springer, Cham (2015). https://doi.org/10.1007/978-3-319-19419-6_32
11. Zhizhin, V.: The future of broadband radio communications. Novyye tekhnologii **1**, 51–55 (2017). (in Russian)
12. Lebedev A.: Choosing a wireless technology for an industrial application. Elektronnyye komponenty **4** (2012). (in Russian)
13. Bashilov G. Wireless millimeter lines. Zhurnal setevykh resheniy/LAN (2010) (in Russian)
14. Belov, V.V., Tarasenkov, M.V., Abramochkin, V.N.: Bistatic atmospheric optoelectronic communication systems (field experiments). Pis'ma v ZhTF **40**(19), 89–95 (2014). (in Russian)
15. Rubino, G.: Transient analysis of Markovian queueing systems: a survey with focus on closed forms and uniformization. Queueing Theory 2. Advanced Trends, pp. 269–307 (2021)
16. Gantmacher, F.R.: The Theory of Matrices. vols. 1, 2, P.375, P.270 (1959)
17. Vytovtov, K., Barabanova, E., Vishnevskiy, V.M.: Accurate mathematical model of two-dimensional parametric systems based on 2 × 2 matrix. In: Vishnevskiy, V.M., Samouylov, K.E., Kozyrev, D.V. (eds.) DCCN 2019. CCIS, vol. 1141, pp. 199–211. Springer, Cham (2019). https://doi.org/10.1007/978-3-030-36625-4_17

The Overflow Probability Asymptotics in a Single-Class Retrial System with General Retrieve Time

Evsey Morozov[1,2,3] and Ksenia Zhukova[1,2(✉)]

[1] Institute of Applied Mathematical Research, Karelian Research Centre
of the Russian Academy of Sciences, Petrozavodsk, Russia
emorozov@karelia.ru
[2] Petrozavodsk State University, Petrozavodsk, Russia
[3] Moscow Center for Fundamental and Applied Mathematics,
Moscow State University, 119991 Moscow, Russia

Abstract. We consider the logarithmic asymptotics of the large deviation probability in a single-server queue with Poisson input, where server, after completion of service, seeks a customer in a virtual orbit (retrial customer) for the next service, unless new arrival captures server. This system is described by a regenerative process, and under stability assumption, the logarithmic asymptotics of the stationary probability that the number of the customers in the system reaches a high threshold within a regeneration cycle is found. Some examples are given for particular retrieval time distributions.

Keywords: Retrial queue · Large deviations · Overflow probability · Logarithmic asymptotics · Stability conditions · Regenerative approach

1 Introduction

Many modern communication, computer and service systems are described and studied by means of the queueing models with orbits called retrial queues [2,6]. In such systems, if an arrival finds server busy, he joins infinite-capacity virtual queue (orbit). To get the service, this orbital customer makes the attempts (retrials) to capture the server. For instance, a customer that does not meet an idle manager at a call center upon arrival, may try to call back later. In other cases an orbital customer may wait until the server selects him for the next service. This may happen for example, when a customer leaves a contact information and waits for a calling back. An important quality of service parameter in such systems is the probability that the number of potential customers waiting service exceeds a given large threshold, and the probability of this event is called the *overflow probability*. This probability can be evaluated by means of the *large*

The research is supported by the Russian Foundation for Basic Research, project No. 19-07-00303.

deviation analysis [4]. In particular, an asymptotic analysis of the logarithm of the overflow probability, when the overflow occurs during a *regeneration cycle*, as the thresholds increases, has been developed in the paper [12] for the classic multi-server buffered queueing system. This approach then has been adapted for the asymptotic analysis of the overflow probability in the tandem queues in [3], for the multi-server single-class retrial queue in [9] and for the multi-class retrial queue in [14].

In this paper we apply the technique developed in [9] to a single-server single-class system with Poisson input in which server, when becomes idle after service completion, needs some random time to take a customer from the orbit for the next service. We assume that such *seeking (or retrieve)* times are independent and identically distributed (iid) with a general distribution. This system has a lot of common with the system with constant retrial rate [7].

The main contribution of this work is that we now construct the *exact logarithmic asymptotics* of the overflow probability, unlike our previous works [9,14] where different upper and lower bounds have been obtained. The main idea of the approach is to include the retrieve time in the 'generalized' service time of the customer which will be served next. At that we study the system on the regeneration period where orbit can not be idle, and by this reason the generalized service time of each customer includes, besides his service time, also a random variable which is the minimum between the interarrival time and retrieve time. Note that in this analysis the memoryless property of the exponential distribution of the interarrival time plays a crucial role.

The paper is organized as follows. In Sect. 2 we describe the model in detail. In particular, we emphasize the regenerative structure of the processes describing the model. This property is crucial for the analysis we develop in the paper. In Sect. 3 we construct an equivalent *buffered system* in which each customer within a regeneration period has the generalized service time. Section 4 contains stability analysis of the system, which is based on the regenerative approach. In this section we also apply a coupling approach assigning the service time to customers not in the order of the arrival but when each customers enter the server. The main contribution of this section is the regenerative stability analysis. We obtain the sufficient stability condition considering an embedded Markov chain obtained by the observation of the number of customers in the system at the departure instances. Then we deduce the necessary stability condition based on regenerative approach and a balance equation connecting input and output data. Finally, in Sect. 5 we give the main result concerning the logarithmic asymptotics of the overflow probability, and also consider a few examples in which these asymptotics are obtained in the explicit form.

2   Description of the Model

We consider a single-server system with an infinite-capacity virtual orbit, in which a new customer joins the orbit if finds server busy upon arrival. After service completion, server begins to seek a customer from the orbit to be served next according to the First-In-First-Out (FIFO) service discipline.

If the seeking (retrieval) time is completed before the next arrival, the server begins to serve the oldest customer from the orbit. If a new customer arrives during the retrieval time, then the server stops seeking process and immediately begins to serve this new customer. The described model is well-motivated by practical applications, see for instance, [6,9]. Note that even more general variants of such a system with various state dependencies are considered, for example, in the papers [2,11].

We assume a Poisson input of customers arriving at the instants $\{t_n, n \geq 1\}$, with (exponential) interarrival times $\tau_n = t_{n+1} - t_n$, $t_1 = 0$, and with rate $\lambda = 1/E\tau$. (We omit serial index to denote generic element of an iid sequence.) The service times $\{S_n\}$ are assumed to be iid. Moreover, the retrieval times $\{A_n\}$ are assumed to be iid as well, where A_n is the seeking time of the server after the $(n-1)$-th departure from the system, and we put $A_1 := 0$. Denote the service rate by $\mu = 1/ES$ and the retrieval rate by $\gamma = 1/EA$.

The basic processes describing the dynamics of the system (such as the number of customers and the remaining work) are regenerative, and regeneration instances are generated by the customers which arrive in an *empty system*. More exactly, denote by $\nu(t)$ the number of customers in the system at instant t^-, and let $\nu(t_n) = \nu_n$, $n \geq 1$. Then the regeneration instances $\{T_n\}$ of the process $\{\nu(t), t \geq 0\}$ are defined as (see, for instance, [8]):

$$T_{n+1} = \inf(t_k > T_n : \nu_k = 0), \; n \geq 0, T_0 := 0.$$

Then the random variable $T_{n+1} - T_n$ is called the *regeneration length* of the nth *regeneration cycle* defined by

$$G_n = \{\nu(t), T_n \leq t < T_{n+1}\}, n \geq 0.$$

It is easy to show that the regeneration cycles are the iid random elements (of the space of the trajectories of the process) and regeneration cycle lengths are iid random variables. Thus, regeneration cycle contains two parts: the busy period, when there is at least one customer in the system, and the idle period which starts when a customer leaves server while the orbit is idle. We will study the overflow probability during one busy period only, and in this case the orbit has always a customer to be served. This observation is important from the point of view of further analysis.

Our purpose is to find the logarithmic asymptotics of the stationary overflow probability P_N that the number of customers in the described system exceeds a fixed level N during regeneration cycle, as $N \to \infty$. To be more exact, for each value N, P_N is the stationary probability that the number of customers during a regeneration cycle exceeds the threshold N. Then we consider a series of such stationary systems corresponding an increasing series of the thresholds N. This analysis assumes that we know stability (stationarity) conditions of the system we analyze, and indeed we develop such an analysis based on the regenerative approach.

Our further analysis is based on the adapting the results from the work [12] obtained for classical m-server system $GI/GI/m$ with a renewal input process and iid service times having general distribution. A similar approach has been previously applied in the works [9,14] to construct the lower and upper bounds for the logarithmic asymptotics of the large deviation probability in the retrials queue with constant and classic retrial policies. We emphasize that the asymptotics obtained in these works are different and in general are not tight, unlike the present work, where found lower and upper bounds coincide.

3   An Equivalent Buffered System

The analysis is based on a comparison of the original system, denoted Σ, with an associated system $\hat{\Sigma}$ with infinite capacity buffer for the waiting customers. Now we describe the new system which in fact is a modified original system. In this modified system we remain the number of customers unchanged because we assume that each new customer arriving in the system joins the orbit *regardless of the state of server*. At that, if a new customer arrives during a seeking time, then the server starts to serve the oldest orbital customer immediately instead of the new arrival, which in turn joins the 'end of queue' in the virtual orbit. In this regard we recall that FIFO discipline in orbit is assumed. If the seeking time of the server is not interrupted by the newly arrived customer, then the behaviour of the system remains as in the system Σ: the server takes the oldest customer from the orbit. When the served customer leaves the system, server starts to seek the next orbital customer, and the process repeats. We stress that a possible change of the order of the customers in the modified system $\hat{\Sigma}$ does not change distribution of the total number of customers in this system because of the stochastic equivalence of the service times of the customers. To use this equivalence, the service time of a customer is assigned not at the arrival instant but at the instant when the server starts his service. On the other hand, the described order of the customers is convenient for the analysis, because we keep the order of the service the same as the order of the arrivals, preserving FIFO discipline. Thus this modification allows us to consider the retrial system $\hat{\Sigma}$ with the orbit as a classic *buffered system* and apply the large deviation analysis developed for such systems in [12]. (For more detail see [9].)

Denote now by η_n the idle time of server before to start service of customer $n \geq 1$. In what follows we will interpret the idle time η_n as a *part of the service time* of customer n. Define now the nth service time in the modified system as

$$\hat{S}_n = S_n + \eta_n, \ n \geq 1.$$

At that, it follows that $\eta_n = A_n$ if the seeking time is not interrupted by a new arrival, and $\eta_n = \tau_n$ otherwise because the input flow is Poisson and, by the memoryless property, the remaining interarrival time is distributed as τ_n with given exponential distribution with parameter λ. Then constructed sequence of the service times $\{\hat{S}_n\}$, provided the system is considered over a regeneration cycle, is iid and can be written as

$$\hat{S}_n = S_n + \min(A_n, \tau_n), \tag{1}$$

with generic variable $\hat{S} = S + \min(A, \tau)$ because $\{\min(\tau_n, A_n), n \geq 1\}$ is an iid sequence with generic element $\min(A, \tau)$.

Remark 1. It is worth mentioning that when a customer leaves the system empty, then a new regeneration period is initiated by the next arrival, and in this case the idle time of the server is distributed as τ. In this case the equality (1) is violated. However, we do not consider such empty periods because our analysis is concentrated on the busy periods only.

Due to the mentioned stochastic equivalence, the number of customers in the original system Σ and in the modified (buffered) system $\hat{\Sigma}$ with service time $\hat{S}$ are equal [10]. Based on this fact, in Sect. 5 we obtain the logarithmic asymptotics of the overflow probability P_N as $N \to \infty$ applying the large deviation analysis developed for the classic (multi-server) system in [12].

4   Stability Analysis

First of all, we recall that the sufficient stability condition of a regenerative queueing system, as a rule, is formulated provided the system approaches a 'saturated' regime: the orbit size increases unlimitedly in probability. Intuitively, it means that the fraction of the idle periods separating adjacent regeneration cycles becomes negligible and the service times in the modified system approaches $\hat{S}$. On the other hand, the well-known stability condition of classic single-server (buffered) system is $\lambda\hat{S} < 1$ [1,8]. Indeed, as we show below, it is the case.

First of all we note that stability of a regenerative process means its *positive recurrence* [1]. In other words, the mean regeneration cycle length is finite, $\mathsf{E}T < \infty$, where T denotes the generic cycle length. In particular, if the process $\{\nu(t)\}$ is positive recurrent then the weak limit $\nu(t) \Rightarrow \nu$, as $t \to \infty$, exists, where ν represents the stationary number of customers in the system. (The same holds for the embedded process ν_n as $n \to \infty$.) We call the system positive recurrent (stable) if the basic regenerative process $\{\nu(t)\}$ is so. Also the system is initially empty if $\nu_1 = 0$. Denote $\rho = \lambda\mathsf{E}S = \lambda/\mu$ and prove the following statement containing sufficient stability condition of the system.

Theorem 1. *If the condition*

$$\rho + \lambda\mathsf{E}[\min(A, \tau)] < 1, \tag{2}$$

holds, then the initially empty retrial system under consideration is positive recurrent.

Proof. To establish stability, we to consider the Markov chain $\{\nu_n^*\}$ obtained as the embedded process at the departure instants $\{t_n^*\}$ of the customers leaving the system. More precisely, consider $\nu(t_n^* + 0) =: \nu_n^*$, that is the number of customers in the system just after the nth departure, $n \geq 1$. Because the input process is

Poisson, then it is easy to verify that the sequence $\{\nu_n^*, n \geq 1\}$ constitutes an aperiodic and irreducible Markov chain. This chain is defined by the following recursion (if $\nu_n^* > 0$):

$$\nu_{n+1}^* = \nu_n^* + \Delta_n - 1, \tag{3}$$

where Δ_n is the number of new customers who arrive in the system in the interval of time $(t_n^*, t_{n+1}^*]$ between the nth and the $(n+1)$th departures. Note that the sequence $\{\Delta_n\}$ is iid, and we denote by Δ a generic element. Also it is well known that the Markov chain is ergodic if the conditional mean increment

$$\mathsf{E}[\nu_{n+1}^* - \nu_n^* | \nu_n^* = i] = \mathsf{E}\Delta - 1 < 0, \tag{4}$$

for all (large) values i, see for instance, [1]. In our case, it is enough that this condition holds for all $i \geq 1$. A simple calculation gives, provided the orbit is not idle, that the mean increment satisfies the equality

$$\mathsf{E}\Delta = \lambda \mathsf{E}[S + \min(A, \tau)].$$

Then it is seen that requirement (4) coincides with condition (2), and the proof of ergodicity it completed. On the other hand, it is well known that the ergodic Markov chain is positive recurrent regenerative process (for instance, see [1]), and, by the memoryless property of the interarrival time, the arrival instants $\{T_n\}$ in particular are regeneration epochs of the Markov chain $\{\nu_n^*, n \geq 1\}$. Thus the proof is completed.

We note that because τ is exponential, we can rewrite, after some algebra, the sufficient stability condition (2) in the following form:

$$\rho < \mathsf{E}e^{-\lambda A}, \tag{5}$$

where the right-hand side is the Laplace-Stieltjes transform of the (generic) seeking time A. To calculate (5), we use the independence between τ_n, A_n, and also the integration by parts in the representation

$$\mathsf{E}[\min((A, \tau)] = \int_0^\infty \mathsf{P}(A \geq x)\mathsf{P}(\tau \geq x)dx = \int_0^\infty e^{-\lambda x}G(dx),$$

where G is the distribution function of the seeking time A. Now we denote

$$V(t) = \sum_{n=1}^{A(t)} S_n,$$

the total work arrived in the system in the time interval $[0, t]$, where $A(t)$ is the number of arrivals. Also denote by $B(t)$ the busy time and by $I(t)$ the idle time of the server, so $t = B(t) + I(t)$, in the time interval $[0, t]$, $t \geq 0$. Note that $B(t)$ is also the amount of the processed work by server in the time interval $[0, t]$. Moreover, we denote by $W(t)$ the remaining work in the system at instant t^-, and then $W(0)$ is the initial work in the system. Also we denote by $J(t)$ the idle time of server in $[0, t]$ composed by the idle periods *preceding regeneration instants*. Note that each such a period is distributed as the interarrival time τ. Finally, we denote by $R(t)$ the total time, in the interval $[0, t]$, when the server is seeking a new customer to be served next. In other words, in this time the server is free but the system is not empty because the orbit is not empty. We note that it is assumed in the model that the server 'knows' when the orbit is free, and by this reason $R(t)$ does not include the idle time $J(t)$. Then it follows that $I(t) = R(t) + J(t)$, $t \geq 0$.

To find the necessary stability condition, we assume that the system is positive recurrent, that is $ET < \infty$, and write the following balance equation

$$V(t) + W(0) = W(t) + B(t) = W(t) + t - I(t)$$
$$= W(t) + t - J(t) - R(t), \ t \geq 0. \tag{6}$$

Let the indicator function $1_n = 1$ if the nth arrival meets server busy, and $1_n = 0$ otherwise. Then we obtain representation

$$R(t) = \sum_{n=1}^{A(t)} 1_n \min(\tau_n, A_n), \ t \geq 0. \tag{7}$$

It follows from theory of regenerative processes (see for instance, [8,13]) that in the positive recurrent case the following limits exist with probability 1 (w.p.1):

$$\lim_{t \to \infty} \frac{V(t)}{t} = \lim_{t \to \infty} \frac{\sum_{n=1}^{A(t)} S_n}{A(t)} \frac{A(t)}{t} = \lambda ES = \rho =: P_B;$$

$$\lim_{t \to \infty} \frac{W(t)}{t} = 0; \ \lim_{t \to \infty} \frac{I(t)}{t} = P_0, \tag{8}$$

where we apply the Strong Law of Large Numbers (SLLN) and, from the 1st equality in (6), obtain the expression $P_B = \rho$ for the stationary busy probability of the server. Also $P_0 = 1 - P_B = 1 - \rho$ denotes the stationary idle probability of the server. Again, invoking SLLN and a regenerative argument we obtain from (7) that

$$\lim_{t \to \infty} \frac{R(t)}{t} = \lim_{t \to \infty} \frac{A(t)}{t} \frac{\sum_{n=1}^{A(t)} 1_n \min(\tau_n, A_n)}{A(t)} = \lambda P_B E[\min(\tau, A)], \tag{9}$$

where we use independence between indicator 1_n and $\min(\tau_n, A_n)$ and also invoke the property PASTA allowing to equate the probability P_B from (8) and the the mean value of the weak limit of indicator $1_n \Rightarrow 1$, that is $E1 = P_B$. It remains to mention that, again by a regenerative argument, there exists the limit (w.p.1)

$$\lim_{t \to \infty} \frac{J(t)}{t} = \frac{E\tau}{ET} =: P_o,$$

where we use the fact that the idle period between regeneration cycles is distributed as τ. By positive recurrence, $ET < \infty$, and then $P_o > 0$. Now, collecting (9) and (8) we obtain from (6), dividing by t and letting $t \to \infty$, that

$$\rho + \lambda\rho E[\min(\tau, A)] = 1 - P_o < 1. \tag{10}$$

Now we are ready to formulate the obtained necessary stability condition as the following statement.

Theorem 2. *If the system with arbitrary initial state is positive recurrent then the following condition holds:*

$$\rho(1 + \lambda E[\min(\tau, A)]) < 1. \tag{11}$$

Remark 2. Comparing the sufficient stability condition (2) and the necessary condition (11), we see that, as required, condition (2) implies (11), because then $\rho < 1$. On the other hand, the 'gap' between these conditions,

$$\lambda E[\min(\tau, A)](1 - \rho) \to 0 \ \text{ as } \rho \to 1.$$

5   Large Deviation Results with Examples

In this section, we formulate the main large deviation results and give some examples related to some distributions of service time and retrieval time.

Before to formulate the main result, we denote by

$$\Lambda_X(\theta) = \log E e^{\theta X}, \ \theta > 0,$$

the *logarithmic moment generating function* of a random variable X, and assume that it is finite for some value of parameter $\theta > 0$. Moreover we define the *maximal moment indexes*

$$\theta_1 = \sup(\theta > 0 : E e^{\theta S} < \infty) > 0, \tag{12}$$
$$\theta_2 = \sup(\theta > 0 : E e^{\theta \min(A, \tau)} < \infty) > 0. \tag{13}$$

The following exact logarithmic asymptotics for the P_N is proved as in [9] (because the proof in [9] is applied equally to both lim sup and lim inf).

Theorem 3. *Assume that condition (2) holds true. Then*

$$\lim_{N \to \infty} \frac{1}{N} \log P_N = \Lambda_\tau(-\theta_*), \tag{14}$$

where parameter θ_* of the moment generating function Λ_τ of the interarrival time τ is defined as

$$\theta_* = \sup_\theta \left(0 < \theta < \min(\theta_1, \theta_2) : \Lambda_\tau(-\theta) + \Lambda_S(\theta) + \Lambda_{\min(A, \tau)}(\theta) \leq 0\right). \tag{15}$$

Now we give a few examples of distributions of the service and retrieval times for which function $\Lambda(\theta)$ and parameter θ_* are available in an explicit form.

Example 1. The interval τ is exponential with parameter λ, service time is exponential with parameter μ and retrieval time A is exponential with parameter γ. In this case, as it is easy to calculate, $\theta_1 = \mu$, $\theta_2 = \lambda + \gamma$ and

$$\Lambda_\tau(-\theta) = \log \frac{\lambda}{\lambda + \theta}, \quad \Lambda_S(\theta) = \log \frac{\mu}{\mu - \theta}, \quad \Lambda_{\min(A,\tau)}(\theta) = \log \frac{\gamma + \lambda}{\gamma + \lambda - \theta}.$$

Then parameter (15) satisfies

$$\theta_* = \sup\left(\theta \in (0, \min(\gamma + \lambda, \mu)) : \frac{\lambda\mu(\gamma + \lambda)}{(\lambda + \theta)((\gamma + \lambda) - \theta)(\mu - \theta)} \leq 1\right).$$

Stability condition (2) in this case can be written as

$$\rho + \frac{\lambda}{\gamma + \lambda} < 1.$$

Example 2. Consider a queueing system with non-exponential retrieval time. The time τ and S remain exponential with parameters λ and μ, respectively, and retrieval time A has *Weibull* distribution

$$F(x) = 1 - e^{-(x/b)^a}, \quad x \geq 0, \ b > 0, \ a \geq 1.$$

We take $a = 2$, $b = 1$, and, after some algebra, obtain the distribution function $F_{\min(A,\tau)}(x) = 1 - e^{-\lambda x - x^2}$, and hence

$$\mathsf{E}e^{\theta \min(A,\tau)} = \int_0^\infty (\lambda + 2x)e^{-x^2 + (\theta - \lambda)x}dx. \tag{16}$$

Then we derive (see [5])

$$\Lambda_{\min(A,\tau)}(\theta) = \log \int_0^\infty (\lambda + 2x)e^{-x^2 + (\theta - \lambda)x}dx$$

$$= \log\left[1 + \frac{\theta}{2}\sqrt{\pi}e^{\frac{(\lambda - \theta)^2}{4}}\left(1 - \Phi\left(\frac{\lambda - \theta}{2}\right)\right)\right],$$

where function

$$\Phi(x) = \frac{2}{\sqrt{\pi}}\int_0^x e^{-t^2}dt.$$

Note that Λ_τ, Λ_S and θ_1 remain the same as in the previous example, θ_2 can be calculated numerically using (13) and (16). To find θ_*, we must solve equation

$$\log \frac{\lambda}{\lambda + \theta} + \log \frac{\mu}{\mu - \theta} + \log\left[1 + \frac{\theta}{2}\sqrt{\pi}e^{\frac{(\lambda - \theta)^2}{4}}\left(1 - \Phi\left(\frac{\lambda - \theta}{2}\right)\right)\right] = 0. \tag{17}$$

For example, if we take $\lambda = 1$, $\mu = 4$, then the numerical solution to Eq. (17) is $\theta_* = 0.398$, implying

$$\Lambda_\tau(-\theta_*) = \log \frac{1}{1+\theta_*} = -0.146.$$

Figure 1 demonstrates a remarkable consistency between the theoretical asymptotics of the overflow probability and the estimate of this probability in this system (as functions of the increasing threshold N).

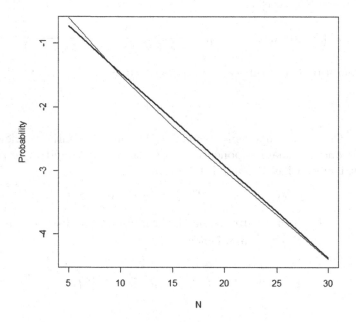

Fig. 1. The estimated overflow probability (grey) and theoretical asymptotics (black) vs. overflow level N; exponential retrieval times, logarithmic scale.

Example 3. Again, we consider the queueing system with non-exponential retrieval time. The variables τ and S remain exponential with parameters λ and μ, respectively, and retrieval time A has *Pareto* distribution

$$F(x) = 1 - \frac{1}{x^2}, \ x \geq 1 \ (F(x) = 0, \ x \leq 1).$$

After some algebra, we obtain the following distribution function of $\min(A, \tau)$:

$$F_{\min(A, \tau)}(x) = \begin{cases} 1 - e^{-\lambda x}, & \text{if } x \in [0, 1), \\ 1 - e^{-\lambda x}x^{-2}, & \text{if } x \geq 1. \end{cases}$$

This implies the following expression for the moment generating function (where we assume that $\theta < \lambda$):

$$\mathsf{E}e^{\theta \min(A, \tau)} = \lambda \int_0^1 e^{(\theta - \lambda)x} dx + \int_1^\infty e^{-(\lambda - \theta)x} \left(\frac{\lambda}{x^2} + \frac{2}{x^3} \right) dx. \qquad (18)$$

In this case

$$\Lambda_{\min(A, \tau)}(\theta) = \log \left[\frac{\lambda}{\theta - \lambda} (e^{\theta - \lambda} - 1) + \int_1^\infty e^{-(\lambda - \theta)x} \left(\frac{\lambda}{x^2} + \frac{2}{x^3} \right) dx \right].$$

Then we derive, again using [5], that

$$\lambda \int_1^\infty e^{(\theta - \lambda)x} x^{-2} \, dx = \lambda(\lambda - \theta) E_i(\lambda - \theta),$$

and

$$2 \int_1^\infty e^{(\theta - \lambda)x} x^{-3} \, dx = e^{\theta - \lambda} + (\theta - \lambda)e^{\theta - \lambda} - (\theta - \lambda)^2 E_i(\lambda - \theta),$$

where $E_i(x)$ is an *exponential integral* [5]. Therefore, we obtain the following expression for the logarithmic moment generating function

$$\Lambda_{\min(A, \tau)}(\theta) = \log \left[\frac{\lambda e^{\theta - \lambda} - \lambda}{\theta - \lambda} + \lambda(\lambda - \theta) E_i(\lambda - \theta) + e^{\theta - \lambda} \right.$$
$$\left. + (\theta - \lambda)e^{\theta - \lambda} - (\theta - \lambda)^2 E_i(\lambda - \theta) \right]. \qquad (19)$$

Note that Λ_τ, Λ_S and θ_1 remain the same as in previous experiment, while θ_2 can be calculated numerically using representation (16). Finally, to find parameter θ_*, we must solve equation

$$\log \frac{\lambda}{\lambda + \theta} + \log \frac{\mu}{\mu - \theta} + \log \left[\frac{\lambda e^{\theta - \lambda} - \lambda}{\theta - \lambda} + \lambda(\lambda - \theta) E_i(\lambda - \theta) \right.$$
$$\left. + e^{\theta - \lambda} + (\theta - \lambda)e^{\theta - \lambda} - (\theta - \lambda)^2 E_i(\lambda - \theta) \right] = 0, \qquad (20)$$

that can be done numerically as well.

6   Conclusion

In this paper we consider a single-server single-class retrial queueing system with one orbit, in which the server, when it finishes service of a customer, takes a random time to select an orbital customer for the next service unless new arrival interrupts this seeking time. The input process is Poisson while the distributions of the retrieval time and service time are assumed to be general. First we establish the stability conditions of this system using a modified system with an infinite buffer for the waiting customers, in which the queue-size process is distributed as this process in the original system. The system has a regenerative structure, and we study the logarithmic asymptotics of the stationary overflow probability that the number of customers in the system reaches a (high) level N within a regeneration cycle. We show that, unlike previously considered systems, in this system the upper and lower bounds of the asymptotics coincide. Some analytical examples are included as well in which the parameters corresponding to the asymptotics are available in an explicit form or can be easily found numerically.

References

1. Asmussen, S.: Applied Probability and Queues, 2nd edn. Springer-Verlag, New York (2003). https://doi.org/10.1007/b97236
2. Baron, O., Economou, A., Manou, A.: The state-dependent M/G/1 queue with orbit. Queueing Syst. **90**, 89–123 (2018). https://doi.org/10.1007/s11134-018-9582-1
3. Buijsrogge, A., de Boer, P.-T., Rosen, K., Scheinhardt, W.: Large deviations for the total queue size in non-Markovian tandem queues. Queueing Syst. **85**, 305–312 (2017). https://doi.org/10.1007/s11134-016-9512-z
4. Ganesh, A., O'Connell, N., Wischik, D.: Big Queues. LNM, vol. 1838. Springer, Heidelberg (2004). https://doi.org/10.1007/978-3-540-39889-9
5. Gradshteyn, I.S., Ryzhik, I.M.: Table of Integrals, Series and Products. Elsevier, Boston (2007)
6. Kim, J., Kim, B.: Tail asymptotics for the queue size distribution in the MAP/G/1 retrial queue. Queueing Syst. **66**, 79–94 (2010). https://doi.org/10.1007/s11134-010-9179-9
7. Morozov, E., Avrachenkov, K., Steyaert, B.: Sufficient stability conditions for multiclass constant retrial rate systems. Queueing Syst. **82**, 149–171 (2016)
8. Morozov, E., Delgado, R.: Stability analysis of regenerative queues. Autom. Remote Control **70**, 1977–1991 (2009)
9. Morozov, E., Zhukova, K.: A large deviation analysis of retrial models with constant and classic retrial rates. Perform. Eval. **135**, 102021 (2019)
10. Muller, A., Stoyan, D.: Comparison Methods for Stochastic Models and Risks. Wiley, New York (2002)
11. Phung-Duc, T., Rogiest, W., Wittevrongel, S.: Single server retrial queues with speed scaling: analysis and performance evaluation. J. Ind. Manage. Optim. **13**, 1927 (2017)
12. Sadowsky, J.S.: Large deviations theory and efficient simulation of excessive backlogs in a GI/GI/m queue. IEEE Trans. Autom. Control **36**(12), 1383–1394 (1991)
13. Serfozo, R.F.: Basics of Applied Stochastic Processes. Probability and Its Applications, Springer-Verlag, Heidelberg (2009). https://doi.org/10.1007/978-3-540-89332-5
14. Morozov, E., Zhukova, K.: The overflow probability asymptotics in a multiclass single-server retrial system. In: Vishnevskiy, V.M., Samouylov, K.E., Kozyrev, D.V. (eds.) DCCN 2020. CCIS, vol. 1337, pp. 394–406. Springer, Cham (2020). https://doi.org/10.1007/978-3-030-66242-4_31

Local Hybrid Navigation System
of Tethered High-Altitude Platform

Vladimir Vishnevsky⬤, Konstantin Vytovtov(✉)⬤, Elizaveta Barabanova⬤,
V. E. Buzdin⬤, and S. A. Frolov

V.A. Trapeznikov Institute of Control Sciences RAS, Profsoyuznaya 65 Street,
Moscow, Russia

Abstract. This paper describes the principles of constricting the local
navigation system for a tethered high-altitude platform. In particular,
the navigation system for the platform "Albatross" developed by the
scientists of the Institute of Control Sciences of the Russian Academy of
Sciences is considered here. The hybrid navigation system is proposed
here. The system includes the millimeter and optical subsystems. The
millimeter system effectively functions at altitudes above 5–10, and the
optical system effectively functions at altitudes below 5–10. Moreover,
each of them can be used as a backup in the corresponding range.

Keywords: Navigation system · Millimeter domain · Optical domain

1 Introduction

Development of autonomous navigation systems for unmanned aerial vehicles is
a very urgent problem [1–8]. First of all, these systems are divided according to
the wavelength that are used by the navigation system [1]. The acoustic domain
is the lowest one. A total spectrum of acoustic radiation from UAVs is due to
harmonic and broadband components [1–5]. The optical domain is most widely
used in UAV navigation. The systems of this domain have a high accuracy in
determining coordinates; they can measure speed using both the longitudinal
and transverse Doppler effect, and effectively measure the range. However UAV
optical tracking is strongly dependent on environmental conditions [1,6]. Special
application systems also use the infrared domain. In this case, the possibility of
detecting and tracking a UAV is determined by its emissivity, contrast and radiation area [1,6]. UAV navigation in radiofrequency domain is quite productive,
since they have a relatively large pulse search volume and a significant range
[1,8]. However radiofrequency methods of UAV navigation can be acceptable
only when there are no requirements for the secrecy of the operation or mobility
of the system [1,7].

Computer vision solutions have been often utilized for UAV navigation
problem also. Modern computer vision solutions include several main problems

The reported study was funded by RFBR, project numbers 19-29-06043, 20-37-70059.

V. M. Vishneviy et al. (Eds.): DCCN 2021, LNCS 13144, pp. 67–79, 2021.
https://doi.org/10.1007/978-3-030-92507-9_7

[9–15]: 1. Development of the physical concept of the photo detector; 2. Developing the algorithms of optical flow and recognition of special points (lines, objects, angles, etc.) and SLAM. Depth or stereo cameras can be used as the photo detectors [9–11]. For the example, the work [9] demonstrates the development of modern stereo cameras for both computer vision and photogrammetry. In [10] the authors developed the object detection algorithm based on the point cloud produced by the device. There are also event-based cameras [11] and machine learning methods.

Fig. 1. The platform "Albatross" developed in the Institute of Control Sciences of the Russian Academy of Sciences.

Next, it must be consider optical flow algorithms and singular point recognition problem [12–16]. In the work [12] the method of constructing a dense cloud of points for creating a map has been given. The author has obtained a fairly dense cloud by using the Epic-Flow algorithm. Applying neural networks is one of the promising areas in the process of recognizing objects or creating a three-dimensional map. In [13] the algorithm of building a high-precision map has been described. The work [14] allows us to realizing SLAM resistant to monotextile surfaces. The authors have developed the mathematical apparatus for working with such images. There are many other geometric image processing algorithms such as ORB-SLAM [15] and DSO-SLAM [16]. However it is necessary to note that all these algorithm have many limitations.

In order to exclude the noted above disadvantage the hybrid navigation system for high-altitude platforms which includes the radio frequency [17] and optical subsystems is presented in this paper. The first of the subsystem works effectively at an altitude above 5 m, and the second one has the best quality indicators at an altitude below 10 m. The navigation system provides stabilization of the drone in height, angular coordinates and speed both in the case of a stationary state and during the movement of the system. As example we consider the platform "Albatross" developed in the Institute of Control Sciences of the Russian Academy of Sciences (Fig. 1).

2 The Statement of the Problem

In this paper, the authors propose the local hybrid navigation system for the high-altitude platform based on the drone with the characteristics: the carrying capacity 32 kg, the maximal flight altitude 100 m and the landing speed 0.4–1 m/s that has been made in the Institute of Control Sciences. The proposed navigation system is corresponded to the following requirements:

- The system is completely autonomous and it provides automatic tracking and control of the drone attitude in the absence of GPS/GLONAS signals.
- When the drone "leaves" the system's visibility zone, it must scan the space and search for it.
- The system provides the stable control of a drone's position in the presence of interference.
- The accuracy of determining the angular coordinates is 30'.
- The accuracy of determining the range is 0.1 m at an altitude above 10 m, while landing 0.05 m at an altitude below 10 m.
- Accuracy of speed determination 0.1 m/s.

Thus the authors propose the hybrid navigation system, which includes the millimeter and optical subsystems.

3 Hybrid Navigation System Structure

To detect a position location the difference signal method in the millimeter domain is used at an altitude more than 10 m, the computer vision methods are used at a altitude less than 5 m, both these methods are used together in the range from 5 m to 10 m. Ranging is based on the measurement of a delay time of a reflected signal at altitude more than 10 m, and it is based on the computer vision methods if an altitude less than 5–10 m. The speed and acceleration of the drone must be determined using computer vision techniques at any altitude.

The frequency domain of our millimeter subsystem is 60 GHz as there is a strong signal attenuation in this one, that provides reliable noise protection along a side direction on the one hand and reliable signal reception at a distance of up to 100 m along an antenna radiation direction on the other hand. Additionally, a phase-shift keyed signal is also used in this subsystem for better noise immunity.

When developing this navigation subsystem, there are a number of the problems that is solved in this work. First of all, the base platform on which the system is located is the metal square with the area 1×1 m. Thus, the base between the antennas in each plane is $b = \sqrt{2}$ when using the equal-signal direction method or the comparison method. Its increase is impossible due to restrictions on the dimensions and weight of the car. In this regard, the use of antennas with a very narrow radiation pattern is impractical. However, antennas with a wide radiation pattern have a low steepness of this pattern, that reduces the accuracy of the drone's direction finding. Moreover, antennas with a wide radiation pattern are more susceptible to external influences, in particular, deliberate interference. Next important problem is the fact that the effective scattering surface of the drone is very small indicate the area (this surface is comparable to the one of large birds: eagle, raven, etc.) Thus, the use of the semi-active method when equipment is installed in the ground part of the system is also impractical in this case. Moreover the radial speed of the drone in an ideal situation is equal to zero. And it is also close to zero in the case when the drone is displaced due to external factors. This fact makes it impossible to use the Doppler effect to determine the displacement velocity. Additionally, the elevation angle of a drone is strongly dependent on its height (Fig. 2). As we can see when the drone is in an equisignal direction, it changes from 22′48″ at an altitude of 100 m to 90° at an altitude of 0 m. In other words, the classical approach to solving this problem has maximum and minimum height restrictions.

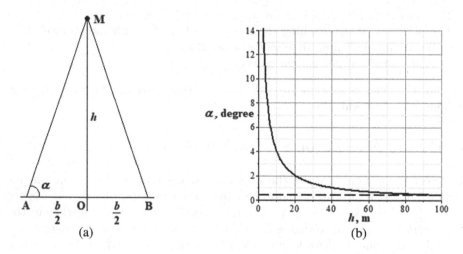

Fig. 2. Illustration of the dependence of the elevation angle on the height: a) the geometry of the problem; b) the dependence of the elevation angle on the altitude.

The millimeter subsystem is functioning at 60 GHz. The choice of this domain is due to the fact that the signal strength is sufficient to ensure stable navigation at a distance of up to 100 m, but due to the strong attenuation, the range provides high noise immunity. Here, the equal-signal method of drone direction finding is proposed. The transmitting device is located on the drone. In this case, the turnstile antenna of circular polarization has an omnidirectional pattern. The receiving device is located on the ground. In terrestrial receivers, it is proposed to use helical antennas with a sufficiently wide pattern.

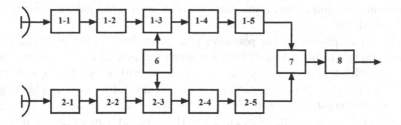

Fig. 3. Block diagram of the millimeter-wave receiver.

The receiver block-scheme of the millimeter subsystem is shown in Fig. 3. Here the receiver has two identical channels. In Fig. 3: 1-1, 2-1 are the input filters, 1-2, 2-2 are the high frequency amplifiers, 1-3, 2-3 are the mixers, 1-4, 2-4 are the intermediate frequency amplifiers, 1-5, 2-5 are the amplitude detectors, 6 is the heterodyne, 7 is the subtractor, 8 is the analog-to-digital converter. The SDR transceiver AD9361 of Analog Device Company is used here. Additionally, the receiver uses a 60 GHz to 5.1 GHz down converter, the transmitter uses a 5.1 GHz to 60 GHz up-converter.

The main purpose of optical navigation is platform detection in various weather conditions like light fog and rain. It also must be taken into account that the platform can move. Obviously, this requirement limits the list of possible algorithms. Many of them are focused on detecting objects in a uniform, non-changing background with the accuracy necessary for landing a drone. Note, that the optical system based on computer vision has a high accuracy in determining the angular coordinates, range and speed. However, it is not applicable at long range in heavy smoke, heavy fog or rain. Moreover, the accuracy of determining coordinates by such a system decreases at heights above 10–20 m. The optical system consists of an Nvidia Jetson Xavier NX on-board computer. To minimize the delay and eliminate the rolling shutter effect, the camera based on an OV9281 matrix, the pixhawk based the flight controller, and the ArduPilot control system are used. Flight simulation and testing is provided by using the AirSim.

The range of drone angles (roll pitch) at which the optical system normally functions depends on the focal length of the lens, if the camera is rigidly attached to the vehicle body. But the larger the viewing angle, the lower the height the system can work at. In our case, a wide-angle lens is used, since the roll and pitch of the device are limited from −30 to 30°. An alternative solution may be a gyro-stabilization system for the camera.

First of all, let us consider the microwave subsystem and then we describe the optical subsystem.

3.1   Radio-Frequency Subsystem

Determination of Angular Coordinates. In the proposed subsystem the transmitter is located on the drone. Ideally the pattern of the transmitting antenna must be omnidirectional. Here we offer to use the helical antenna with pattern width 70°.

The receiver placed on car platform is used to determine the angular coordinates at heights above 5 m. If four antennas are placed in the corners of the square site with the area of 1×1 m (it is note above), then the angle between vertical and direction to the drone $\alpha = 4°$ at a height of $h = 10$ m. This angle is $\alpha = 8°$ at the height of $h = 8$ m. Further, the angle tends to $\alpha_{max} = 90°$ as the drone descends to $h = 0$ m (Fig. 2). Thus the minimum altitude for using the radar system is approximately $h = 3 - 5$ m.

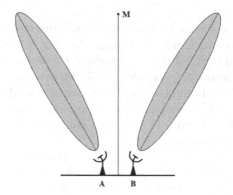

Fig. 4. Illustration of the location of the millimeter-wave receiving antennas.

To use an interval of the pattern with a greater steepness at high altitudes, antennas should be directed not perpendicular to the base surface, but inclined relative to each other in opposite directions at an angle α_0 (Fig. 4). Note also that, on the other hand, this increases the minimum allowable tracking altitude for the drone. Today horn, dielectric, helical antennas and other are used in the considered frequency domain. Here we suggest using helical antennas because their radiation is circularly polarized.

Figure 5 shows the radiation patterns of two helical antennas for angular detection in one of the planes that inclined in different directions relative to a vertical to platform. Here the antenna parameters: the frequency is $f = 60$ GHz, the number of turns is $n = 5$, the wavelength is $l = 0.005$ m, the winding pitch is $h = 0.0012$ m, the wire diameter is $d = 0.0001$ m the antenna input impedance is 140Ω, the inclination angle of the antennas is $\alpha_0 = 60°$. As we can see the pattern width is 60°.

However, the accuracy of determining angular coordinates depends not so much on the radiation pattern width, but on the derivative of a function describing this pattern. Figure 6 shows the dependence of the derivative of the function describing the radiation patterns of these antennas on the angle. This derivative is zero at $\alpha = \pm 21°$ and $\alpha = \pm 60°$.

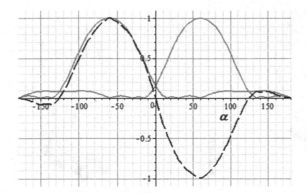

Fig. 5. Receiving antenna patterns.

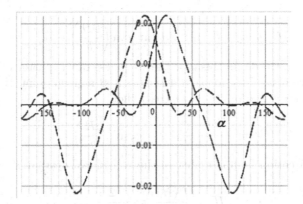

Fig. 6. Dependence of the derivatives of the directional diagrams of the receiving antennas on the azimuth.

It is important that the receiver antennas are located at the corners of the platform, which means that their patterns are not symmetrical about the antenna axis. This leads to distortion of the patterns. However, the entire structure is entirely symmetrical. Therefore, all patterns are symmetrical about the center of the platform. Figure 7 shows the ideal and distorted pattern of the antenna. The calculations are carried out for the antenna parameters written above.

Now let's calculate the measure error of the angular coordinates of the antenna system. Taking into account the inclination angle α_0 (Fig. 2, Fig. 4), the signal at the output of the receiving device is proportional to

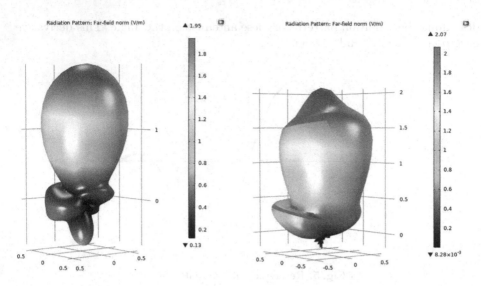

Fig. 7. Three-dimensional radiation patterns of receiving antennas.

$$f(h) = f(\alpha + \alpha_0) = f\left[\arctan\left(\frac{2H}{b}\right) + \alpha_0\right] \tag{1}$$

Here H is an altitude. In this case the steepness of the pattern is determined by the derivative of the function that describes it at a given point. If, for example, we accept the simplification $f(\alpha) = n\cos\alpha$ then

$$f'(h) = -\frac{2n}{b + 2H}\sin\left[\arctan\left(\frac{2H}{b}\right) + \alpha_0\right] \tag{2}$$

Then the direction finding ability of the considered system is

$$\Pi = \left|\frac{f'(H)}{f(H)}\right| = \frac{2}{b + 2h} \tag{3}$$

It is obvious that an increase in the height or base of the system leads to a decrease in its direction finding ability. In this case, the measure error of the angle is

$$\Delta\alpha = m\frac{b + 2H}{2} \tag{4}$$

here m is the amplitude modulation factor caused by the displacement of the direction finding characteristics.

To increase the steepness of pattern of the receiver antenna we also offer to use the array of the four helical antennas. The pattern for the considered case is presented in Fig. 8. The calculation results show that the pattern width of the antenna array decreases by three times. However the side lobes of the array are significantly more than ones for single antenna.

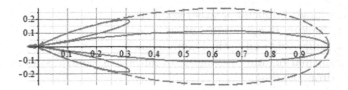

Fig. 8. Antenna array radiation pattern.

Determination of Range. Amplitude, frequency or phase methods exist for determining the range to a suspended mobile platform in radar. The use of the phase method in this case is inappropriate since it either is necessary to use the meter wavelength domain in this case, or use the vernier (multifrequency) approach. At the same time, the transmitting and receiving equipment must be placed on the drone. Placing this equipment on the ground side of the system is impractical because the effective dispersing surface of the drone is very small. However, placing it on the drone will result in significant weight gain. The frequency method is preferable from the point of view of noise immunity and ranging accuracy, but its use will lead to complications of the on-board equipment. The most simple and preferable in this case seems to be the amplitude method, although it is susceptible to interference. Therefore, in this case, it is advisable to use manipulated signals.

Thus, to determine the range, it is advisable to install the transceiver on the drone. It should be borne in mind that the measured range varies from approximately 0 to 100 m.

Then, for a range of 100 m, we have the maximum signal delay time $t_{max} = 2L/c = 0.6(7) \cdot 10^{-6}$ s. Here L is the drone range, c is the light speed. In addition, the range resolution is determined by the duration of the radio pulse. So, for a pulse of duration $\tau = 10^{-9}$ s, the resolution is 0.3 m. Thus, the period of high-frequency oscillations should be less than $T_{max} = 10^{-10}$ s. Those the operating frequency of the on-board radar navigation system must be greater than $f_{min} = 10\,\text{GHz}$ ($\lambda < 3\,\text{cm}$).

3.2 The Optical Subsystem

The numerical calculations show that the radar navigation subsystem can be effectively used at altitudes above 5–10 m. At altitudes from 0 m to 10 m it is advisable to use optical navigation subsystems. Moreover, it can be used as a backup at high altitudes. Here, fiducial points and correlation methods for identifying point images are used.

It should also be noted that in this paper we, first of all, are considering local navigation problem. It is also assumed that the platform will be in line of sight most of the time. Based on these facts, the resulting technique is developed here. This technique should allow us to solve the following problems:

- Object tracking in difficult weather conditions
- Restoring tracking in case of loss of an object
- Choice of computing and optical equipment
- Data transformation and filtering in an automatic control system
- Precision drone landing

In order to develop an optical subsystem, the experiment was carried out with a moving platform, since this is one of the main and most complex requirements for classical algorithms. For experimental verification, we chose three main algorithms [18–20] and carried out the calculations on NVIDIA JETSON XAVIER AGX. The results of experiment are presented in Fig. 9. The left column of Fig. 9 shows the initial frame and the region of interest, which in the experiment was the same for the three algorithms. The right column shows the calculated tracking result of the algorithm. As it can be seen, the KCF algorithm lost the area of interest and focused on the trees. This trend is observed in all videos with a lot of noise. But DaSiamRPN performs much better tracking than other algorithms.

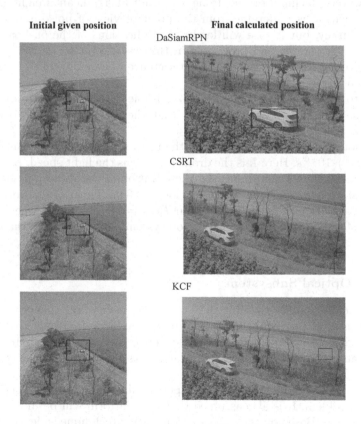

Fig. 9. The results of comparative calculations.

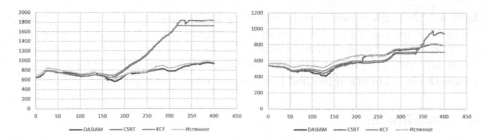

Fig. 10. The platform tracking errors.

The graphs (Fig. 10) show one of the tracking experiments. On the X-axis is the frame number, on the Y-axis is the ordinal number of the pixel. The first graph shows corresponds to the horizontal axis of the frame. The second one is vertical.

The graph shows the platform tracking error for the next frame. As can be seen from the graph and images above. The KCF and CSRT methods quickly lost the object of interest. For our work, this is unacceptable because information is lost on the basis of which the ability to correctly calculate the position of the device is lost.

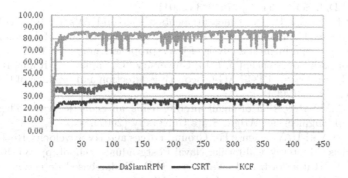

Fig. 11. The frame rate graph.

Each algorithm has a different complexity and requires different computing power. In the graph (Fig. 11), you can see the frame rate obtained in a practical experiment for each algorithm on NVIDIA Jetson Xavier AGX. Despite the fact that KCF works more than twice as fast, the quality of tracking is unsatisfactory. We have chosen the DaSiamRPM algorithm that gives us a frequency of more 30 Hz.

Based on experiments, the DaSiamRPN method has shown the greatest efficiency. It provides the longest possible retention of the object of interest. Moreover the calculation results show that DaSiamRPN is the most resistant to noise. Nevertheless, when working on a single-threaded CPU, the CSRT algorithm will have a higher frame rate, which is preferable in the case of working on weak computing devices. So at this moment we are using the DaSiamRPN algorithm, but in the future we plan to add auxiliary methods for error correction and precision landing of the drone. Thus, in this navigation system, it will be advisable to use the hybrid optical method based on all described algorithm.

4 Conclusions

This paper describes the principles of building the local navigation system for a tethered high-altitude platform "Albatross", developed by the scientists of the Institute of Control Sciences of the Russian Academy of Sciences. It is offered the hybrid navigation system containing millimeter subsystem and computer vision subsystem.

References

1. Kaktashov, V.M., Oleinikov, V.N., Sheiko, S.A., Babkin, S.I., Korytsev, I.V., Zubkov, O.V.: Features of detection and recognition of small unmanned aerial vehicles. Radioengineering. **195**, 235–243 (2018). (in Russian)
2. Sadasivan, S., Gurubasavaraj, M., Sekar, S.R.: Acoustic signature of an unmanned air vehicle - exploitation for aircraft localisation and parameter estimation. Eronautical Def. Sci. J. **51**(3), 279–283 (2001)
3. Massey, K., Gaeta, R.: Noise measurements of tactical UAVs. In: 16th AIAA/CEAS Aeroacoustics Conference, pp. 1–16 (2010). Georgia Institute of Technology
4. Marino, L.: Experimental analysis of UAV-propellers noise. In: 16th AIAA/CEAS Aeroacoustics Conference, pp. 1–14. University "La Sapienza", Rome, Italy (2010)
5. Pham, T., Srour, N.: TTCP AG-6: acousting detection and tracking of UAVs. Proc. SPIE **54**, 24–29 (2004)
6. Zelnio, A.M.: Detection of small aircraft using an acoustic array. Thesis, B.S. - Electrical Engineering, Wright State University (2007). 55p
7. Beel, J.J.: Anti-UAV Defense For Ground Forces and Hypervelocity Rocket Lethality Models. Monterey, California: Naval Postgraduate School, pp. 36–46 (1992)
8. Moses, A., Rutherford, M.J., Valavanis, K.P.: Radar-based detection and identification for miniature air vehicles. In: IEEE International Conference on Control Applications, Denver, CO, USA, pp. 933–940 (2011)
9. Kahmen, O., Rofallski, R., Luhmann, T.: Impact of stereo camera calibration to object accuracy in multimedia photogrammetry. Remote Sens. **12**, 2057 (2020)
10. Isobe, Y., Masuyama, G., Umeda, K.: Occlusion handling for a target-tracking robot with a stereo camera. ROBOMECH J. **5**(1), 1–13 (2018). https://doi.org/10.1186/s40648-018-0101-2
11. Mueggler, E., Rebecq, H., Scaramuzza, D.: The event-camera dataset and simulator: event-based data for pose estimation, visual odometry, and SLAM. Int. J. Robot. Res. **36**(49), 142–149 (2016)
12. Chen, Q., Poullis, Ch.: Single-Shot Dense Reconstruction With Epic-Flow. 3DTV-Conference: The True Vision - Capture, Transmission and Display of 3D Video (2018)
13. Vijayanarasimhan, S., Ricco, S., Schmid, C., Sukthankar, R., Fragkiadaki, K.: Learning of structure and motion from video (2017)

14. Engel, J., Schöps, T., Cremers, D.: LSD-SLAM: large-scale direct monocular SLAM. In: Fleet, D., Pajdla, T., Schiele, B., Tuytelaars, T. (eds.) ECCV 2014. LNCS, vol. 8690, pp. 834–849. Springer, Cham (2014). https://doi.org/10.1007/978-3-319-10605-2_54

15. Mur-Artal, R., Montiel, J.M., Tardos, J.D.: ORB-SLAM: a versatile and accurate monocular SLAM system. IEEE Trans. Robot. **31**(5), 1147–1163 (2015)

16. Yamada, K., Kimura, A.A.: Performance evaluation of keypoints detection methods SIFT and AKAZE for 3D reconstruction. In: 2018 International Workshop on Advanced Image Technology, Chiang Mai, Thailand (2018)

17. Vishnevsky, V.M., Vytovtov, K.A., Barabanova, E.A.: Model of navigation and control system of an airborne mobile station. In: Vishnevskiy, V.M., Samouylov, K.E., Kozyrev, D.V. (eds.) DCCN 2020. LNCS, vol. 12563, pp. 643–657. Springer, Cham (2020). https://doi.org/10.1007/978-3-030-66471-8_49

18. Henriques, J.F., Caseiro, R., Martins, P., Batista, J.: High-speed tracking with kernelized correlation filters. IEEE Trans. Pattern Anal. Mach. Intell. **37**, 583–596 (2014)

19. Lukezic, A., Voj'ir, T., Zajc, L.C., Matas, J., Kristan, M.: Discriminative correlation filter tracker with channel and spatial reliability. Int. J. Comput. Vis. **126**, 671–688 (2018)

20. Zhu, Z., Wang, Q., Li, B., Wu, W., Yan, J., Hu, W.: Distractor-aware siamese networks for visual object tracking. In: Ferrari, V., Hebert, M., Sminchisescu, C., Weiss, Y. (eds.) ECCV 2018. LNCS, vol. 11213, pp. 103–119. Springer, Cham (2018). https://doi.org/10.1007/978-3-030-01240-3_7

The Probabilistic Measures Approximation of a Resource Queuing System with Signals

Kirill Ageev[1] , Eduard Sopin[1,2](✉) , Sergey Shorgin[2] ,
and Alexander Chursin[1]

[1] Peoples Friendship University of Russia (RUDN University),
6 Miklukho-Maklaya Street, Moscow 117198, Russian Federation
{ageev-ka,sopin-es,chursin-aa}@rudn.ru
[2] Institute of Informatics Problems, FRC CSC RAS, 44-2 Vavilova Street,
Moscow 119333, Russian Federation
shorgin-sy@ipi.ran.ru

Abstract. Modern wireless networks are characterized by high user mobility. This factor can lead to changes in the quality of the channel during the lifetime of the interaction session. To take into account the fact of user movement in a resource queuing system with signals are introduced. Signal arrivals trigger resource reallocation of customers in the system. In this paper, we introduce an approximate model that replaces the flow of resource reallocations with the additional flow of customers. The convergence of the approximate iterative evaluation method is proved. Besides, we provide a case study for the accuracy assessment of the approximate method.

Keywords: Resource queuing system · Random requirements ·
Performance analysis · Approximate method convergence

1 Introduction

Resource queuing systems with signals can be applied for the analysis of the performance metrics of modern wireless networks [1,2]. Upon the arrival of a signal, a customer leaves the system and immediately comes back with a new resource requirement, which is determined according to the same probability distribution. Signal arrival indicates that a different amount of resources is required for the customer.

Analytical calculations of probabilistic measures of resource queuing systems with signals under Poisson arrivals are presented in [3–5]. The application of such analytical formulas to calculate the stationary characteristics is rather complicated since it implies the calculation of multiple convolutions of resource requirements distribution, and the stationary distribution is obtained as the solution

The reported study has been funded by RFBR, projects no. 19-07-00933 and 20-07-01052.

of the system of equilibrium equations. Therefore, in [6] we have developed the event-driven simulation tool for resource queuing systems with signals. In [7] we proposed an approximate method for evaluation of the stationary characteristics of the model. We provided a comparison of calculation accuracy with the methods proposed in [3, 4].

In this paper, we prove the convergence of the iterative method for the approximate model. The rest of the paper is organized as follows. Section 2 describes a mathematical model of the resource queuing system with signals, Sect. 3 introduces an approximation method for the considered model and proves the convergence of the method. In Sect. 4, the numerical experiment of the approximate method is performed, the results are compared with simulations. Section 5 concludes the paper.

2 Resource Queuing System with Signals

We consider a multiserver queuing system with N servers and R number of resources. An arriving customer occupies a server and some volume of limited resources. Volumes of customers' resource requirements are independent identically distributed random variables with discrete CDF $F(x)$, which is represented by probability distribution $\{p_i\}, i \geq 0$. Customers arrive according to the Poisson process with intensity λ, and the service times have exponential distribution with the rate μ. Each customer in the system produces a flow of signals. Signals arrive according to the Poisson distribution with intensity γ. When a signal arrives, the customer releases the server and occupied resources and returns to the system again with new resource requirements.

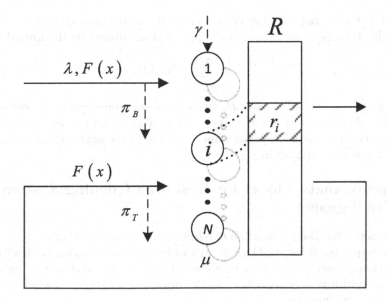

Fig. 1. Schema of resource queuing system with signals

Let $\xi(t)$ be the number of customers in the system at moment $t > 0$ and $\gamma(t) = \gamma_1(t) + \ldots + \gamma_{\xi(t)}(t)$ - the sum of resource volumes occupied by all customers. The system behavior can be approximately described by the stochastic process $X(t) = (\xi(t), \gamma(t))$ over the set of states $S = \{(n,r) : 0 \le n \le N, 0 \le r \le R, p_r^{(n)} > 0\}$, where $\{p_r^{(n)}\}, r \ge 0$ is n-fold convolution of distribution $\{p_i\}$ [6]. Figure 1 shows the scheme of the queuing system.

The stationary probabilities are defined as follows:

$$q_n(r) = \lim_{t \to \infty} \{\xi(t) = n, \gamma(t) = r\}, (n,r) \in S. \tag{1}$$

In [4], system of equilibrium equations is derived. The equilibrium equations allow evaluating the stationary probabilities of the system. Then, a number of performance measures are deduced. The blocking probability on arrival π_b has the following form:

$$\pi_B = 1 - \sum_{(n,r) \in S, n < N} q_n(r) \sum_{j=0}^{R-r} p_j. \tag{2}$$

The probability π_s that an accepted customer is blocked on a signal arrival can be written in the following form:

$$\pi_T = \sum_{(n,r) \in S, n > 0} q_n(r) \sum_{j=0}^{r} \frac{p_j p_{j-r}^{(n-1)}}{p_r^n} \left(1 - \sum_{i=0}^{R-r+j} p_i\right). \tag{3}$$

The probability π_t that an accepted customers is terminated is given by

$$\pi_S = \frac{\gamma \widetilde{N} \pi_T}{\lambda(1 - \pi_B)}, \tag{4}$$

where $\widetilde{N}$ is the average number of customers in the queuing system.

Finally, the resource utilization coefficient is calculated by the formula

$$\delta = \frac{1}{R} \sum_{(n,r) \in S} r q_n(r). \tag{5}$$

With the increase of N and R, the size of the state space, as well as the dimension of the generator matrix, increases very fast, and this leads to complex numerical evaluations of the performance metrics. In the next section, we present our approximation approach.

3   Approximate Model for Resource Queuing System with Signals

One can note, that the reallocation of resources for a customer on a signal arrival can be interpreted as leaving the system and immediately coming back with new resource requirements. It follows that customers leave the system with intensity $\mu + \gamma$, and return with probability $\frac{\gamma}{\mu + \gamma}$. So, for the approximate model, we make the following assumption.

Assumption 1. Rearriving customers form a new type of customer flow with intensity $\tilde{N}\gamma$, where $\tilde{N}$ is an average number of customers in the system.

So, we further refer to the initially arriving customers as the first type, and to the rearriving as the second type customers. We assume here that resource requirements are independent of arrival and serving processes.

Figure 2 shows the scheme of the approximate model.

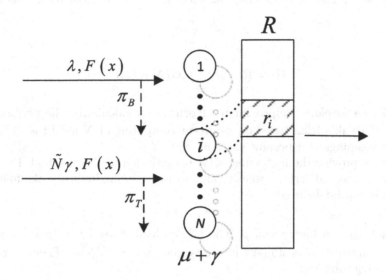

Fig. 2. Schema of approximate model

Denote $p_{l,r}$ the probability that the customer of type l will require r resources. Then $p_{l,r}^{(k)}$ - the probability that k customers of type l will require r resources, where $p_{l,r}^{(k)}$ k-fold convolution of probabilities $p_{l,r}$. Let $\rho_1 = \frac{\lambda}{\mu+\gamma}$ be the offered load of the first type customers and $\rho_2 = \frac{\tilde{N}\gamma}{\mu+\gamma}$ for the second type.

According to [5], we can replace both customer flows by one aggregated flow with resource requirements distribution $p_r = \sum_{l=1}^{L} \frac{\rho_l}{\rho} p_{l,r}$, where $\rho = \sum_{l=1}^{L} \rho_l$, and calculate stationary probabilities according to results for the resource queuing systems without signals:

$$q_n(r) = q_0 \frac{\rho^k}{k!} p_r^{(k)}, \tag{6}$$

$$q_0 = \left(\sum_{k=0}^{N} \sum_{j=0}^{R} \frac{\rho^k}{k} p_j^{(k)} \right)^{-1}, \tag{7}$$

In [5], the convolution algorithm is developed for the evaluation of normalization constant $G(N, R) = q_0^{-1}$:

$$G(n, r) = G(n-1, r) + \frac{\rho}{n} \sum_{i=0}^{r} p_i(G(n-1, r-i) - G(n-2, r-i)). \quad (8)$$

In the approximate model, the arrival intensity of second type customers depends on $\tilde{N}$. The latter can be evaluated using the normalization constant.

$$\tilde{N} = \sum_{n=1}^{N} \sum_{r=0}^{R} n q_{n,r}, \quad (9)$$

$$\tilde{N} = G(N, R)^{-1} \rho \sum_{j=0}^{R} p_j G(N-1, R-j). \quad (10)$$

In [7], we developed approximate algorithm for calculating the performance metrics. This algorithm based on recurrent computing of $\tilde{N}$ until the difference between meanings of it become small enough.

Here we provide the mathematical substantiation for the method. From the equality of accepted arrival intensity and average serving intensity, the following expression can be deduced:

$$\tilde{N} = \rho(1 - \pi_B). \quad (11)$$

Note that π_B is a function of ρ. The offered load ρ itself is a function of the average number $\tilde{N}$ of customers in the system, i.e. $\rho = \frac{\lambda + \tilde{N}\gamma}{\mu + \gamma}$. Then, introduce the following function:

$$Z(x) = \frac{\lambda + \gamma x}{\gamma + \mu} \left(1 - \pi_B \left(\frac{\lambda + \gamma x}{\gamma + \mu}\right)\right), \quad (12)$$

where

$$\pi_B(\rho) = 1 - \left(1 + \sum_{n=1}^{N} \frac{\rho^n}{n!} F^{(n)}(R)\right)^{-1} \sum_{n=0}^{N-1} \frac{\rho^n}{n!} F^{(n+1)}(R), \quad (13)$$

Lemma 1. *The equation $x = Z(x)$ has a single solution on $(0; +\infty)$, and it can be found by fixed-point iteration.*

Proof. 1. Consider the case, in which the resource requirements after a signal arrival have the same distribution as the first type customer. Then in the approximate model:

$$(\lambda + \gamma\tilde{N})(1 - \pi_B(\rho)) = \tilde{N}(\gamma + \mu), \quad (14)$$

where $\rho = \frac{\lambda + \gamma\tilde{N}}{\gamma + \mu}$.

From here follows:

$$\tilde{N} = \rho(1 - \pi_B(\rho)), \quad (15)$$

Thus, the average number of customers in the system is subject to the expression

$$\widetilde{N} = Z(\widetilde{N}). \tag{16}$$

According to the principle of contraction, if $Z(x)$ is a contraction for $x \geq 0$, then it has a single fixed point on the semi-line $x \geq 0$, which is the solution of the equation $Z(x) = x$.

Calculate the derivative of $Z(x)$:

$$Z'(x) = \frac{\gamma}{\mu + \gamma} \left(1 - \pi_B \left(\frac{\lambda + \gamma x}{\gamma + \mu} \right) \right) - \frac{\lambda + \gamma x}{\gamma + \mu} \frac{\gamma}{\mu + \gamma} \pi'_B \left(\frac{\lambda + \gamma x}{\gamma + \mu} \right). \tag{17}$$

Take into account that $\rho = \frac{\lambda + \gamma x}{\gamma + \mu}$

$$Z'(x) = \frac{\gamma}{\mu + \gamma} (1 - \pi_B(\rho) - \rho \pi'_B(\rho)). \tag{18}$$

Let us now estimate the derivative of $\pi'_B(\rho)$. From physical considerations, we can conclude that $\pi'_B(\rho) \geq 0$, because with increasing offered load blocking probability couldn't decrease. We introduce an additional notation:

$$G(\rho) = 1 + \sum_{n=1}^{N} \frac{\rho^n}{n!} F^{(n)}(R), \tag{19}$$

where $G(\rho)$ is a normalization constant. It is easy to see that the expression for the blocking probability $\pi_B(\rho)$ looks like

$$\pi_B(\rho) = 1 - \frac{G'(\rho)}{G(\rho)}, \tag{20}$$

and its derivative

$$\pi'_B(\rho) - \left(\frac{G'(\rho)}{G(\rho)} \right)^2 - \frac{G''(\rho)}{G(\rho)}. \tag{21}$$

Then we estimate each summand of $\pi'_B(\rho)$.

$$\frac{G'(\rho)}{G(\rho)} = G^{-1}(\rho) \sum_{n=0}^{N-1} \frac{\rho^n}{n!} F^{(n+1)}(R)$$

$$= \frac{1}{\rho} G^{-1}(\rho) \sum_{n=0}^{N-1} (n+1) \frac{\rho^{n+1}}{(n+1)!} F^{(n+1)}(R) \tag{22}$$

$$= \frac{1}{\rho} G^{-1}(\rho) \sum_{n=1}^{N} n \frac{\rho^n}{n!} F^{(n)}(R).$$

It is easy to see that the expression on the right represents the mathematical expectation of the number of applications in the resource queuing system:

$$\frac{G'(\rho)}{G(\rho)} = \frac{\widetilde{N}}{\rho}. \tag{23}$$

Then turn to the second term on the right-hand side of (21).

$$\frac{G''(\rho)}{G(\rho)} = G^{-1}(\rho) \sum_{n=2}^{N} \frac{\rho^{n-2}}{(n-2)!} F^{(n)}(R)$$

$$= \frac{1}{\rho^2} G^{-1}(\rho) \sum_{n=2}^{N} n(n-1) \frac{\rho^n}{n!} F^{(n)}(R) \tag{24}$$

$$= \frac{1}{\rho^2} (G^{-1}(\rho) \sum_{n=2}^{N} n^2 \frac{\rho^n}{n!} F^{(n)}(R) - G^{-1}(\rho) \sum_{n=2}^{N} n \frac{\rho^n}{n!} F^{(n)}(R)).$$

where $\widetilde{N}^{(2)}$ is the second moment of the number of customers in the system. Taking into account the obtained relations, the expression for $\pi_B(\rho)$ takes the following form:

$$\pi'_B(\rho) = \frac{\widetilde{N}^2 - \widetilde{N}^{(2)} + \widetilde{N}}{\rho^2} = \frac{\widetilde{N} - (\widetilde{N}^{(2)} - \widetilde{N}^2)}{\rho^2} \tag{25}$$

The expression in parentheses represents the positive variance of the number of customers in the system, which means

$$\pi'_B(\rho) < \frac{\widetilde{N}}{\rho^2} = \frac{(1 - \pi_B(\rho))}{\rho} < \frac{1}{\rho}. \tag{26}$$

Substituting the resulting estimate in the expression for $Z'(x)$, we get

$$|Z'(x)| < \frac{\gamma}{\mu + \gamma} < 1. \tag{27}$$

which is a sufficient condition for a contraction - which was required to be proved. This means that the average number of customers in the system can be found using the fixed-point iteration method.

3.1   Approximate Algorithm for Calculating Performance Metrics

Figure 3 shows the approximation algorithm for calculating performance metrics. Before start we must define min_dif - the accuracy threshold for average number of customers $\widetilde{N}$. At the beginning of approximation, we have only first type of customers. We calculate $\widetilde{N}$ with first type of customer and denote it as $prev\_N$. Having $\widetilde{N}$, we can calculate input parameters for second type of customers, and calculate $\widetilde{N}$ for two flows of customers with intensities λ and $\gamma\widetilde{N}$. On the next step we compare the difference between $\widetilde{N}$ and $prev\_N$ according to the accuracy threshold min_dif. If the difference less then min_dif we calculate blocking probability of newly arriving customers π_B, blocking probability of second-type customers π_T and average amount of occupied resources δ, else return to denoting $\widetilde{N}$, and calculate it with new intensity of the second flow.

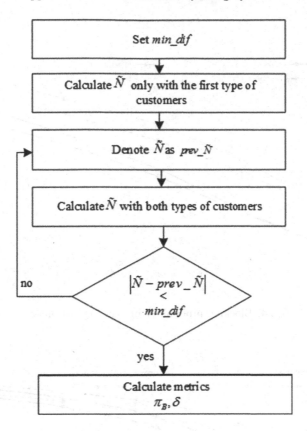

Fig. 3. Block diagram of the approximate algorithm

4   Case Study

The results of the approximation method are compared with the results obtained using the simulation modeling for two scenarios: with different parameters for the distribution function of the first and the second type of customers, and with the same parameters.

Assume that the system has $N = 100$ servers and $R = 200$ units of resources, arrival intensity of customers $\lambda \in [20; 60]$ with constant arrival intensity of signals $\gamma = 5$, serving intensity $\mu = 1$. The distribution of resource requirements is determined by a geometric distribution with parameters for scenario 1: $q_1 = 0.75$, $q_2 = 0.5$, and for scenario 2: $q_1 = q_2 = 0.75$, $p_{l,r} = (1 - q_l)q_l^{r-1}, l = \{1, 2\}, r \geq 1$.

Figure 4 shows the blocking probability π_B of the newly arriving applications. With an increase in the arrival intensity, the blocking probability also increases.

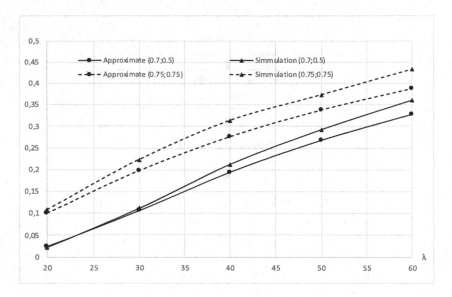

Fig. 4. Blocking probability of arriving customers

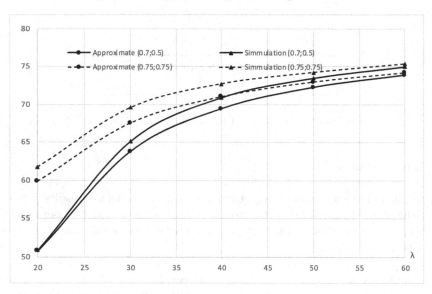

Fig. 5. Average number of occupied resources

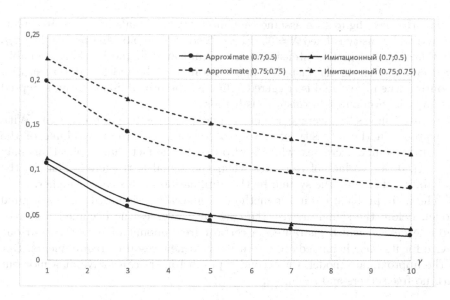

Fig. 6. Blocking probability of arriving customers

Figure 5 shows the average number of occupied resource in the system. As on the previous figure, an increase in the arrival intensity, the average volume of the occupied resource grows to the maximum value, that is, up to 75 in our case.

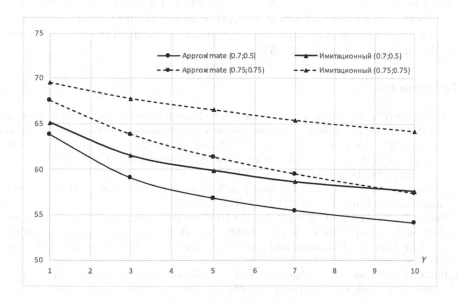

Fig. 7. Average number of occupied resources

For the next figures we assume that intensity of signals is $\gamma \in [1; 10]$, while the arrival intensity is fixed $\lambda = 50$. Figure 6 shows the blocking probability π_B. With an increase in the intensity of the signal arrival, the blocking probability decreases, due to the fact that applications may be blocked due to lack of resources when the signal is triggered. That is, the probability that an accepted customer is terminated in this case is higher.

Figure 7 shows the average amount of occupied resource in the system. With an increase in the intensity of the signal receipt, the average volume of the occupied resource decreases, which is also due to the fact that applications may be blocked due to lack of resources when the signal is triggered. That is, the probability of leaving the system by the application in this case is higher.

The main problem lies in the method of generating signals. If in the original model, before the signal arrives, the customer releases the occupied resources and one device, which guarantees a certain free amount of resources and one device for it, since it immediately returns with new resource requirements. But in the approximate model, the second type customers can arrive at a moment with no free servers and resources.

5   Conclusion

In the paper we described an approximation approach for analysis of stationary characteristics in the resource queuing system with signals. The method can be used to evaluate performance metrics in contemporary wireless network.

The mathematical justification of the application of the approximate model is given. Based on the numerical experiment in comparison with the results of simulation modeling, it is concluded that the algorithm can be used to calculate the blocking probability of newly arriving customers, the average number of customers in the system and the average amount of occupied resource, since the calculation error is from 5% to 10%.

References

1. Lu, X., Dohler, M., Sopin, E., et al.: Integrated use of licensed- and unlicensed-band mmWave radio technology in 5G and beyond. IEEE Access **7**, 24376–24391 (2019)
2. Galinina, O., Andreev, S., Turlikov, A., Koucheryavy, Y.: Optimizing energy efficiency of a multi-radio mobile device in heterogeneous beyond-4G networks. Perform. Eval. **78**, 18–41 (2014)
3. Tikhonenko, O.: Generalized Erlang problem for service systems with finite total capacity. Probl. Inf. Transm. **41**(3), 243–253 (2005)
4. Sopin, E., Vikhrova, O., Samouylov, K.: LTE network model with signals and random resource requirements. In: Proceedings of 9th International Congress on Ultra Modern Telecommunications and Control Systems and Workshops (ICUMT), pp. 101–106 (2017)
5. Sopin, E.S., Ageev, K.A., Markova, E.V., Vikhrova, O.G., Gaidamaka, Y.V.: Performance analysis of M2M traffic in LTE network using queuing systems with random resource requirements. Autom. Control. Comput. Sci. **52**(5), 345–353 (2018). https://doi.org/10.3103/S0146411618050127

6. Sopin, E., Ageev, K., Shorgin, S.: Simulation of the limited resources queuing system for performance analysis of wireless networks. In: Proceedings 32st European Conference on Modelling and Simulation, pp. 505–509 (2018)
7. Sopin E.S., Ageev K.A., Samouylov K.E.: Approximate analysis of the limited resources queuing system with signals. In: 33rd International ECMS Conference on Modelling and Simulation, ECMS 2019 Caserta, Italy, 11–14 June 2019, Proceedings, pp. 462–465. ECMS proceedings (2019)

LoRa Link Quality Estimation Based on Support Vector Machine

Van Dai Pham[1], Phuc Hao Do[2], Duc Tran Le[3(✉)], and Ruslan Kirichek[1]

[1] Bonch-Bruevich Saint-Petersburg State University of Telecommunications, Saint-Petersburg, Russian Federation
fam.vd@spbgut.ru, kirichek@sut.ru
[2] Danang Architecture University, Danang, Vietnam
haodp@dau.edu.vn
[3] The University of Danang – University of Science and Technology, Danang, Vietnam
letranduc@dut.udn.vn

Abstract. Nowadays, the LoRa technology is one of the promising technologies used for the Internet of Things (IoT) networks. Over the LoRa transmission link, two devices can communicate with each other over a long distance. As perspective research on LoRa mesh network, it is necessary to consider the link quality estimation (LQE) between neighbor nodes to choose reliable routes. In this paper, we propose a LQE method to classify the connection level between two nodes. The LQE method is developed based on the kernel support vector machine (kSVM), which is one of the machine learning techniques used in classification problems. Series of experiments were performed to collect a dataset consisting of received signal strength indicator (RSSI), signal-to-noise ratio (SNR) of received packets, and packet reception rate (PRR). The trained model shows a high prediction accuracy (mean = 95%) while using 10% of the dataset for training.

Keywords: IoT · Link quality estimation · Wireless sensor network · LoRa · RSSI · SNR · Support Vector Machine (SVM)

1 Introduction

Wireless sensor network (WSN) [7] has attracted much attention since its appearance by its applications based on many sensor nodes, which are capable of sensing and gathering information from the environment, then processing and transmitting those sensed data to the remote base station. The wireless sensor nodes can be deployed for different purposes [4] such as supervisory and security; environmental monitoring; smart living; smart agriculture; health care; military. Their main advantage is the ability to deploy in almost any kind of geography, including dangerous environments.

Currently, many wireless network technologies support WSN networks such as Wi-Fi, Bluetooth, Zigbee, LoRa, Sigfox, NB-IoT [2,12]. The LoRa technology

© Springer Nature Switzerland AG 2021
V. M. Vishnevskiy et al. (Eds.): DCCN 2021, LNCS 13144, pp. 92–102, 2021.
https://doi.org/10.1007/978-3-030-92507-9_9

proved consistent to create a Low-Power Wide-Area Network (LPWAN). In some cases, Sigfox offers longer-range communication compared to LoRa, but it has service subscription costs. Meanwhile, NB-IoT is a cellular-based technology and consumes much energy. Besides long-range capabilities (up to 15km), the LoRa technology also has advantages in battery life optimization, easy deployment, and robustness to interference. These features make LoRa a good choice for a vast number of WSN applications.

In the LoRa network in particular and the WSN network in general, the selection of optimal routing paths is always challenging due to the links' dynamic behavior [6,11]. The link quality must be suitable for critical industrial applications as they require sensor nodes to accurately measure the environment and communicate these data to other nodes without error. Many factors affect the spread of radio signals, such as properties of the transmission environment, which leads to multi-path propagation effects, noise, and interference by concurrent transmissions of other communication technologies or electromagnetic sources; the power of wireless signals. Typically, there are several methods to assess link quality: link estimation based on data-link layer parameter [8], link estimation based on physical layer parameter [13], and comprehensive link quality estimation [3].

Link quality estimation (LQE) is considered as a process to estimate the reliability level of the connection between nodes. The high quality links ensures the network quality of service (QoS) in decreasing the packet loss. The LQE can assist higher network protocols to mitigate and overcome the less reliable link. For instance, link quality estimation is an assistant for routing protocols to maintain routing tasks. Moreover, link quality estimation also can assist to maintain the network topology. With the high quality links, the mechanism is considered to control efficiently the network topology to maintain robust network connectivity [1].

The metrics used to evaluate the quality of the link mainly are RSSI, LQI, SNR, and PRR [15]. RSSI – received signal strength indicator represent the power of a received radio signal. The received packet is correctly decoded if the RSSI value is more than receiver sensitivity. Similar to RSSI, LQI – Link Quality Indicator is an integer value (from 0 to 255) used to estimate the link quality. The LQI value is defined by radio-chip manufacturer. Another metric is SNR – signal-to-noise ratio, which defines the level of received signal to the level of background noise. All three metrics, RSSI, LQI, and SNR are estimated based on hardware implementation. Moreover, the metric PRR – packet reception rate presents the ratio of received to sent packets over a defined window size, which can be implemented in the software. However, the LQI metric depends a lot on the hardware, and therefore, in this paper, we propose the LEQ method using RSSI, SNR, and PRR to evaluate the link quality. Furthermore, we will use a machine learning model trained by the input data collected by experiments in the laboratory environment [5] to evaluate and classify the links into different groups. From that result, we can choose the optimal link for routing tasks in each particular scenario.

2   Basic Estimation Metrics

2.1   Hardware-Based Estimation Methods

Based on hardware implementation, the following parameters such as RSSI, LQI, SNR can be obtained directly from the wireless transceiver without additional calculations. These parameters are saved in registers after receiving a packet.

Received Signal Strength Indicator. RSSI is an estimated measure of power strength that an RF device received from and another RF device. At long transmission distances, the signal gets weaker, leading to the probability of packet losses since the receiver can not decode the signal correctly. The signal is measured by the received signal strength indicator, which in most cases indicates how well a particular radio link can hear the remote wireless node. In the open space, the received signal power can be estimated using the following equation:

$$P_{rx} = P_{tx} + G_{tx} + G_{rx} + PL \tag{1}$$

where:

- P_{rx} is the expected received power or the received signal strength indicator,
- P_{tx} is devoted to the transmission power,
- G_{tx} and G_{rx} are the transmitting and receiving antenna gains,
- PL is represented as path loss.

The transmission power is reduced according to the increasing distance between receiver and transmitter. Therefore, if the obtained RSSI value is less than the receiver sensitivity, the signal could be incorrectly decoded. Moreover, the signal strength decreases due to the obstacles and environment. While implementing network planning, path loss models are taken into account to estimate the transmission range and reception power. Thus, there can be different values of RSSI with each time receiving packet. Hence, RSSI can be used with the other metrics to evaluate the link quality.

Signal to Noise Ratio. SNR is devoted to the differences in level between the received signal strength.

$$SNR(dB) = 10\log_{10}\frac{P_{signal}}{P_{noise}} = P_{signal}(dB) - P_{noise}(dB) \tag{2}$$

Signal to noise ratio defines the difference in level between the signal and the noise for a given signal level. The less noise is generated by the receiver, the better the signal-to-noise ratio.

Link Quality Indicator. LQI is a value used to quickly determine whether the link belongs to the reliable reception range. For example, this indicator is implemented in some transceivers (CC2420) used in multi-hop networks such as Zigbee to assess the link cost. LQI is required and described in the Zigbee and IEEE 802.15.4 standards. After receiving each packet, the LQI measurement results as an integer ranging from 0 to 255. The minimum and maximum LQI values correspond to the lowest and highest quality IEEE 802.15.4 signals detectable by the receiver.

However, these values can only be obtained from successfully received packets. In the case of packet loss, the link quality might be overestimated. Moreover, the measurement results vary unstable; hence, it is difficult to estimate the exact link quality if only one measurement is performed. Despite fast and cheap implementation in the hardware, these methods allow obtaining limited information about quality links for stable channels. Thus, using a combination of hardware and software metrics will improve the accuracy of link quality estimation.

2.2 Software-Based Estimation Methods

Software implementation allows to estimate some values such as packet reception rate (PRR) or packet delivery ratio (PDR), throughput, expected transmission number.

In a certain transmission period, PRR can be obtained by either direct calculation or approximation. PRR represents the ratio of the number of successfully received packets to the total number of packets transmitted. Therefore, a term of window size is considered to choose an interval of transmitted packets to calculate PRR. As a result, an accurate estimate can be obtained in a short time with high or low link quality. On the other hand, larger window size is required to obtain PRR more accurately.

Another method considers a number of expected transmissions as a metric for the link quality estimation. That means how many transmissions are required to transmit a packet successfully while considering the number of lost packets.

3   LoRa Link Quality Estimation

3.1 Experimental Measurement and Preprocessing

Series of experiments were performed at the "Internal Research, Development and Testing Center for new equipment, technologies and services" supported by "Rostelecom" and International Telecommunication Union. We used several LoRa Nodes (https://heltec.org/project/htcc-ab01/) in the experimental measurement, including sending nodes and a sink node. The sending nodes are located in the different ranges far from the sink node. The LoRa parameters are configured as following in Table 1.

Interval time between packet transmission was set randomly in range (500, 2000) ms. The sink node is fixed and connected to a computer to save the experimental data. On the other hand, the sending nodes move far away from the sink node at a low speed to measure signal power and noise changes related to the packet reception rate. After this experiment, we received 2500 packets for further preprocessing.

For each received packet, we obtained the RSSI and SNR values from the hardware. Moreover, the sequence number indexed to packet number also is

Table 1. Configuration parameter

Parameter	Value
Frequency	868 MHz
Bandwidth	250 kHz
Spreading factor	7
Coding rate	4/5
Transmission power	5 dBm

collected to calculate the PRR within a certain window size. As shown in Fig. 1 for an example of preprocessing within the window size = 10, we calculated the average RSSI and SNR values, and PRR according to the sequence number.

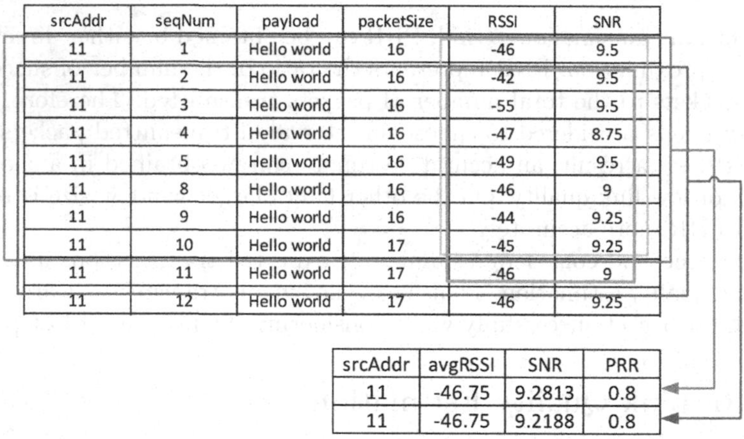

srcAddr	seqNum	payload	packetSize	RSSI	SNR
11	1	Hello world	16	-46	9.5
11	2	Hello world	16	-42	9.5
11	3	Hello world	16	-55	9.5
11	4	Hello world	16	-47	8.75
11	5	Hello world	16	-49	9.5
11	8	Hello world	16	-46	9
11	9	Hello world	16	-44	9.25
11	10	Hello world	17	-45	9.25
11	11	Hello world	17	-46	9
11	12	Hello world	17	-46	9.25

srcAddr	avgRSSI	SNR	PRR
11	-46.75	9.2813	0.8
11	-46.75	9.2188	0.8

Fig. 1. Data preprocessing with the window size = 10

Based on data analysis, we expect to use the average of RSSI and SNR values that could help us to estimate the link quality level. As shown in Fig. 2a, the RSSI values vary from −120 to −40 dBm and can be divided into four groups. On the other hand, the SNR values also change from −10 to 10 dB (Fig. 2b). Thus, approximately from this range, there are four levels that we expect to consider as link quality levels.

We consider four levels to assess the link quality, as shown in Table 2. The packet reception rate was used to label the link quality level [14].

In order to reduce the impact of range and model error, the avgRSSI and avgSNR are normalized so that the data are between 0 and 1. The normalization process corresponds to Eq. (3) as follows:

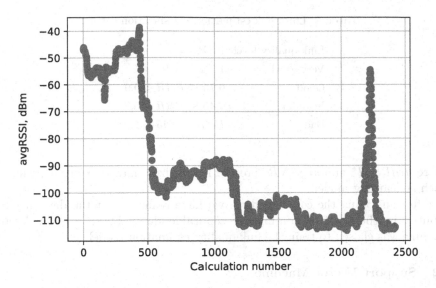

(a) The average RSSI

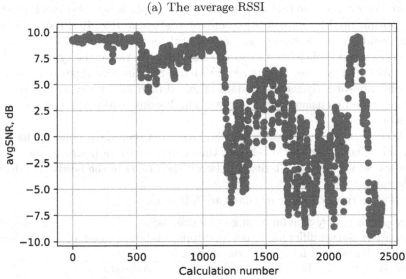

(b) The average SNR

Fig. 2. Preprocessed RSSI and SNR values

$$avgRSSI^* = \frac{avgRSSI - min(avgRSSI)}{max(avgRSSI) - min(avgRSSI)}$$

$$avgSNR^* = \frac{avgSNR - min(avgSNR)}{max(avgSNR) - min(avgSNR)}$$

(3)

Table 2. LoRa link estimation classification range

Link quality level	PRR
Very good	$PRR \geq 0.9$
Good	$0.75 \leq PRR < 0.9$
Medium	$0.45 \leq PRR < 0.75$
Bad	$PRR < 0.45$

where $avgRSSI^*$ and $avgSNR^*$ are the normalized data that are used in the machine learning model.

After analyzing the collected data, we have realized that the data are distributed in the form of non-linearly separable data. Thus, Support Vector Machine was chosen to train the link quality estimation model.

3.2 Support Vector Machine

Support Vector Machine (SVM) is a relatively simple supervised machine learning algorithm used for classification and/or regression [10]. It is more preferred for classification but is sometimes very useful for regression as well. Basically, SVM finds a hyper-plane that creates a boundary between the types of data.

In this paper, we use Kernelized SVM for non-linearly separable data. We have some non-linearly separable data in one dimension. We can transform this data into two-dimensions and the data will become linearly separable in two dimensions.

A kernel is nothing a measure of similarity between data points. The kernel function in a kernelized SVM tell you, that given two data points in the original feature space, what the similarity is between the points in the newly transformed feature space.

Pros of Kernelized SVM methods are following:

- they perform very well on a range of datasets;
- they are versatile: different kernel functions can be specified, or custom kernels can also be defined for specific datatypes;
- they work well for both high and low dimensional data.

SVM is a method to compute the best hyperplane that separates two distinct classes. SVM was originally designed for linear classification but can now be used for non-linear classification problems by applying the kernel. In this study, we implemented non-linear SVM with RBF (Radial Basis Function) as a kernel function since it can be generally used for all types of data.

The most popular RBF kernel is Gaussian Radial Basis function. Mathematically:

$$K\left(x, y\right) = \exp\left(-\gamma \parallel x - y \parallel^2\right), \gamma > 0 \tag{4}$$

where γ controls the influence of new features on the decision boundary. There is more influence of features on the decision boundary with higher value of γ.

3.3   LQE Model Based on SVM

Link quality estimation is a process of evaluating the link quality based on some metrics within a certain estimation window. The estimation window can be an interval time in seconds or a number of received packets. To have data for LQE, it is required to monitor links and measure (retrieve) data. Then, the link quality estimation model process the received data to give a link quality score. Hence, all steps of the LoRa link quality estimation can be illustrated in Fig. 3.

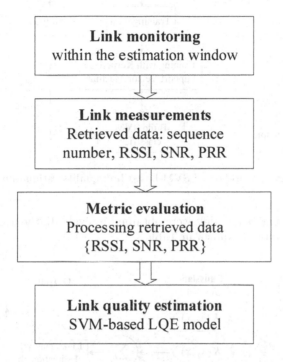

Fig. 3. Steps of LoRa LQE

The link quality estimation is converted to the multi-class classification problem, which can be solved using supervised learning models. Support vector machine (SVM) [10] is known as one of the efficient supervised learning models for solving multi-class classification. This paper proposes using SVM to train the collected data and predict the link quality level. The training and testing workflows are presented in Fig. 4. The data preprocessing and data normalization are described in the previous subsection. SVM training model is described in Fig. 5.

Figure 5 presents the SVM structure model using three parameters (avgRSSI, avgSNR, PRR) as the input data and Gaussian kernel for training. We used the programming language Python, and the library scikit-learn [9] to train and evaluate the proposed model. The dataset was divided into 10% for training and 90% for testing. We performed 50 training and testing times respectively. The

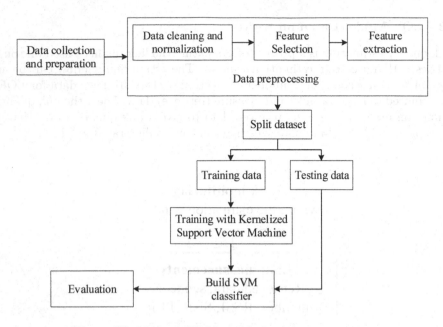

Fig. 4. Workflow of SVM-based link quality estimation

testing results of trained models were obtained to evaluate the average accuracy. The obtained results are shown in Table 3.

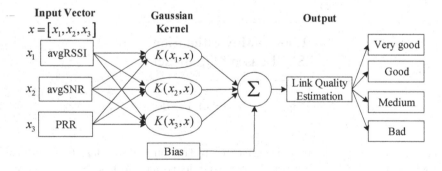

Fig. 5. SVM structure model

According to the evaluation results, the SVM model presents a high prediction accuracy with the mean = 95%. Thus, using three parameters such as {RSSI, SNR, PRR} the LoRa link quality can be estimated accurately using 10 probe packets, even though our dataset size is not large enough for training as usually can be seen in machine learning problems.

The results of link quality estimation might be mapped to a 5 point grading scale. Respectively, {2, 3, 4, 5} points correspond to the LQE with {Bad,

Table 3. The performance of link quality estimation model based on SVM

LQE	Precision	Recall	F1-Score
Very good	0.96	1.00	0.98
Good	0.93	0.93	0.93
Medium	0.98	0.92	0.95
Bad	1.00	0.87	0.93
Accuracy	95%		

Medium, Good, Very Good}. For instance of application in the mesh network, a reliable route is a route that has the sum of LQE grades is maximum. Moreover, it should be noted that we considered only one direction transmission (uplink). So the link quality level can be used to find the reliable uplink path in the LoRa mesh network.

4   Conclusion

In this paper, we considered a method for estimating the LoRa link quality. Link quality estimation is developed based on some metrics such as RSSI, SNR, PRR. The estimation problem was converted to a multi-class classification problem, which can be solved by using a kernelized support vector machine. An SVM-based LQE model was proposed and trained with collected data from the experiments. Despite using the small dataset for training and the number of probe packets, the proposed model has shown a high prediction accuracy (mean = 95%).

Acknowledgment. The publication has been prepared with the support of the grant from the President of the Russian Federation for state support of leading scientific schools of the Russian Federation according to the research project SS-2604.2020.9.

References

1. Baccour, N., et al.: Overview of Link Quality Estimation. Springer Briefs in Electrical and Computer Engineering, pp. 65–86. Springer, Heidelberg (2013). https://doi.org/10.1007/978-3-319-00774-8_3
2. Bertoldo, S., Carosso, L., Marchetta, E., Paredes, M., Allegretti, M.: Feasibility analysis of a LoRa-based WSN using public transport. Appl. Syst. Innov. **1**(4), 49 (2018)
3. Guo, Z.Q., Wang, Q., Li, M.H., He, J.: Fuzzy logic based multidimensional link quality estimation for multi-hop wireless sensor networks. IEEE Sens. J. **13**(10), 3605–3615 (2013)
4. Kandris, D., Nakas, C., Vomvas, D., Koulouras, G.: Applications of wireless sensor networks: an up-to-date survey. Appl. Syst. Innov. **3**(1), 14 (2020)

5. Kirichek, R., Koucheryavy, A.: Internet of Things laboratory test bed. In: Zeng, Q.-A. (ed.) Wireless Communications, Networking and Applications. LNEE, vol. 348, pp. 485–494. Springer, New Delhi (2016). https://doi.org/10.1007/978-81-322-2580-5_44

6. Kirichek, R., Vishnevsky, V., Pham, V.D., Koucheryavy, A.: Analytic model of a mesh topology based on LoRa technology. In: 2020 22nd International Conference on Advanced Communication Technology (ICACT), pp. 251–255. IEEE (2020)

7. Koucheryavy, A., Prokopiev, A., Koucheryavy, Y.: Self-organizing networks. SPb.: Lyubavich 312 (2011)

8. Luo, J., Yu, L., Zhang, D., Xia, Z., Chen, W.: A new link quality estimation mechanism based on LQI in WSN. Inf. Technol. J. **12**(8), 1626 (2013)

9. Parisi, L.: m-arcsinh: An efficient and reliable function for SVM and MLP in scikit-learn. arXiv preprint arXiv:2009.07530 (2020)

10. Patle, A., Chouhan, D.S.: SVM kernel functions for classification. In: 2013 International Conference on Advances in Technology and Engineering (ICATE). IEEE, January 2013. https://doi.org/10.1109/icadte.2013.6524743

11. Pham, V.D., Kisel, V., Kirichek, R., Koucheryavy, A., Shestakov, A.: Evaluation of a mesh network based on LoRa technology. In: 2021 23rd International Conference on Advanced Communication Technology (ICACT). IEEE, February 2021. https://doi.org/10.23919/icact51234.2021.9370792

12. Proskochylo, A., Vorobyov, A., Zriakhov, M., Kravchuk, A., Akulynichev, A., Lukin, V.: Overview of wireless technologies for organizing sensor networks. In: 2015 Second International Scientific-Practical Conference Problems of Infocommunications Science and Technology (PIC S&T), pp. 39–41. IEEE (2015)

13. Reijers, N., Halkes, G., Langendoen, K.: Link layer measurements in sensor networks. In: 2004 IEEE International Conference on Mobile Ad-Hoc and Sensor Systems (IEEE Cat. No. 04EX975), pp. 224–234. IEEE (2004)

14. Shu, J., Liu, S., Liu, L., Zhan, L., Hu, G.: Research on link quality estimation mechanism for wireless sensor networks based on support vector machine. Chin. J. Electron. **26**(2), 377–384 (2017). https://doi.org/10.1049/cje.2017.01.013

15. Sun, P., Zhao, H., Luo, D., Zhang, X.y., Zhu, J.: Study on measurement of link communication quality in wireless sensor networks. J.-China Inst. Commun. **28**(10), 14 (2007)

Identification Method for Endpoint Devices on Low-Power Wide-Area Networks Using Digital Object Architecture with Blockchain Technology Integration

Albina Pomogalova[1] , Dmitriy Sazonov[1]([⊠]) , Evgeny Donskov[2] ,
Alexey Borodin[3], and Ruslan Kirichek[1]

[1] The Bonch-Bruevich Saint-Petersburg State University of Telecommunications,
Saint-Petersburg, Russian Federation
kirichek@sut.ru

[2] St. Petersburg Institute for Informatics and Automation of the Russian Academy
of Sciences, Saint-Petersburg, Russian Federation

[3] PJSC Rostelecom, Moscow, Russian Federation
aleksey.borodin@rt.ru

Abstract. The paper explores the possibility of interaction between the
architecture of digital objects, as an identification system for endpoint
devices on Low-Power Wide-Area Networks, and blockchain technology,
as a particular case of using distributed ledger technologies. The work
covered the organization of the blockchain network, the architecture of
the model stand developed, as well as a number of studies reflecting the
results of the interaction of the identification system with the blockchain
network in terms of the time of creation of an entity in LHS and the
record thereof in the blockchain network, as well as the time of response
of the blockchain network in case of query of data from it. As part of this
work, the blockchain network acts as a tool to log all the changes of the
identification system in order to prevent counterfeiting and substitution
of the identification data of devices and sensors of the Internet of Things.

Keywords: Blockchain · DOA · IoT · Smart contract · Identification
system · Digital object architecture · Internet of Things

1 Introduction

The growth of various devices and sensors of the Internet of Things, the necessity of their organized interaction and accounting is a topical and long-term
problem since the beginning of the development of the Internet of Things as a

The publication has been prepared with the support of the grant from the President
of the Russian Federation for state support of leading scientific schools of the Russian
Federation according to the research project SS-2604.2020.9.

concept. Thus, until a few years ago, the number of such sensors and devices was in the thousands, whereas today it is in the millions and billions. Besides servicing devices of Internet of Things and debugging their interaction, one of the key problems remains identification of such devices. Earlier authors of the work carried out an extensive review and analysis of the existing systems of identification of narrowband wireless communication networks systems and devices of the Internet of Things, further confirming the need for the development of a single universal identifier for the Internet of Things.

The International Telecommunication Union (ITU-T) has formulated common requirements for an identification system that would address a number of existing problems. The authors' research [1,2] concluded that the DOA digital object architecture was suitable. The handle identifiers used in the DOA architecture are unique and consistent in the global namespace, guaranteeing the resolution of the digital object to the valid information when the client requests it.

But in order to meet all the requirements formulated by ITU-T, the identification system must be secure and ensure that there is no possibility of uncontrolled modification of the data of a digital entity. Part of this problem is solved by using a distributed administration system used in solutions based on the architecture of DOA digital objects. Administration is based on the use of asymmetric access keys.

However, in addition to distributed administration, it is required to add the ability to unambiguously establish a chain of changes in a digital entity at all stages of its life cycle.

For this reason, the authors of the work made an assumption about the possibility of using some of the properties and functions of distributed ledger technologies to prevent the substitution and forgery of identification data. The blockchain technology based on the Ethereum open source solution was used as a distributed ledger technology.

The main goal of this work is to study the possibility of interaction of the architecture of digital objects (DOA) with a blockchain network to control changes in the identification system. To achieve this goal, the authors set up a model stand and performed a number of experiments.

2   Related Works

The research carried out by the authors within the framework of the proposed concept showed that today there are practically no experiments related to the selected identification system. Existing works are devoted to research the possibility of combining the Internet of Things sphere and blockchain technology, including some questions of identification, but are more narrowly focused [5–7].

There is also a lot of research on the possibility of using blockchain technology to store confidential data [8–10]. Within the framework of this paper, the authors propose a more generalized solution, since the identification system used is applicable to many existing industries and acts as a trusted source of data storage, which is the first step in protecting identification data.

3   Features of Blockchain Network Setup

Today, there are a large number of different blockchain solutions in the world, which differ in the goals of use, performance, requirements for the parameters of validator devices and other devices that support the network, various consensus algorithms and features of the network architecture. As part of this work, the geth client was chosen as a blockchain network for the study, which allows setting up a private instance of the Ethereum blockchain network. This blockchain network operates on the Proof of Work consensus algorithm and is the first platform that made it possible to fully implement the idea of smart contracts. A smart contract within the selected blockchain network means a computer algorithm written in the Solidity programming language that allows performing various actions in the blockchain network, for example, writing and reading data. Data immutability is considered a key feature of blockchain networks, which is an important factor for any identification systems.

Data immutability means the fact that changing the instances of the block chain on all devices is practically impossible not only in terms of the number of devices containing the current instance of the block chain and constantly synchronizing with each other, but also because of the complexity of calculating each of the sequentially connected and cryptographically protected blocks for their complete replacement.

The Ethereum blockchain instance configured for research consists of three synchronized devices configured on the basis of three virtual machines. To configure the network, a genesis.json file was created, which contains all the key parameters of the future network, such as the network instance number, block generation complexity, forks occurring in the network and block numbers in which forks similar to the main Ethereum network will occur. Forks are required to use all existing functions of the blockchain network and smart contracts. For each virtual machine, miner accounts were also configured so that each of the three virtual machines performed not only the functions of a full node of the blockchain network, but also a miner that creates blocks and processes transactions. After configuring all three network nodes and synchronizing them, a smart contract was developed that performs the functions of recording and reading information about assigned identification numbers.

4   Model Stand Architecture

The model bench device interaction scheme is shown in Fig. 1. The internal interaction of the blockchain nodes (Geth Ethereum Node 1–3) is performed using the static-nodes.json file located on each of the virtual machines in the blockchain network containing the addresses of all three nodes on the network and the ports to which they should be addressed.

For external interaction with the blockchain network the web3.js library is used as well as Brownie - a Python-based framework for developing and testing smart contracts. Thus, an HTTP server was also configured for communication, which accepts Rest API requests and sends queries to the blockchain network to record new data or read existing data. On the Rest API client side queries are possible both to the blockchain network and to the LHS database.

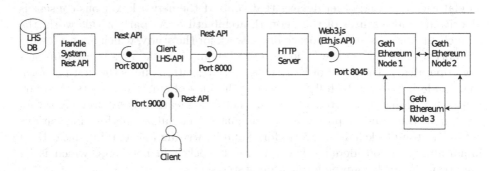

Fig. 1. Model stand interface diagram

The model stand developed makes it possible to evaluate the possibility and effectiveness of the application of blockchain network as a duplicate trusted distributed database, which acts as a reference repository.

5   Research Script

As part of the research, the authors evaluated the possibility of interaction between the blockchain network and the identification system of narrowband wireless communication devices and systems. The DOA digital object architecture has been chosen as the most efficient and effective system for identification of narrowband wireless communication networks systems and devices of the Internet of Things, according to previous authors' studies.

The Handle System was chosen as an implementation of the DOA concept. The key objective of the study was to assess the feasibility of validating all processes in the identification system using blockchain technology in particular to log all changes in the descriptor.

In the developed test system the main interaction of the client is via HTTP to Rest API service LHS-API, as shown in Fig. 1. This service acts as a "service-facade" providing end-to-end functionality for the client to interact with the Handle System infrastructure and the blockchain network.

The LHS-API service provides the client with the functionality to create, modify and read data of digital objects through interaction with the identifier handleID and Handle System. LHS-API also interacts via Rest Api with a blockchain platform to store data on changes in the meta-information of the digital object and the change chain.

6 Creation a New Handle System Descriptor to Save to Blockchain

Figure 2 presents a sequential query schema for the system, when adding a new descriptor and storing data in blockchain.

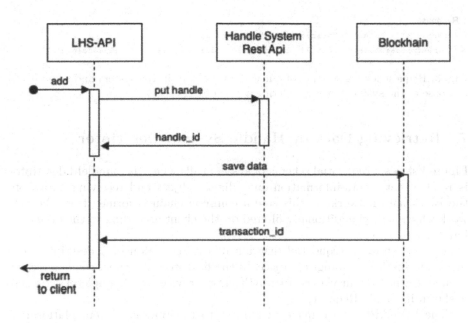

Fig. 2. Sequential diagram of the interaction of elements in the system when registering a new digital object

Figure 2 shows a sequential diagram of the interaction of elements in the system when registering a new descriptor of a digital object and adding it to the Blockchain.

At the first stage, the client calls the method to add a new descriptor (add) in the LHS-API service and transfers all the necessary data describing the meta-information of the digital object. An example of a request is shown below in Fig. 3 (Request).

At the next stage, the LHS-API service forms a request to the Handle System platform to register a new handle. In response, the Handle System platform returns a handleId - the identifier of the registered digital object. After receiving the identifier, the LHS-API service through the REST API contacts the Blockchain platform to register the new state of the digital object in the smart contract (save data). On successful access, the Blockchain system returns the ID of the successful transaction (transactionId). This identifier is returned to the client in the response of the LHS-API service (an example of the response is shown in Fig. 3 in the Response). In the reply it can be seen the saved descriptor and the blockchain transaction ID.

Request:
```
curl --location --request POST 'http://172.30.4.20:9000/handle?handleID=77.LOCAL/4'
--header 'Content-Type: application/json' \
--data-raw    '{"DESC":    "LoRa    Device    #1","DEVICE_ID":    "COD3F010","ACTION":
{"type":"rest","method":"POST","url":"https://test:8000/device/toggle?handleID=77.L
OCAL%2FCOD3F010"}}'
```

Response:
```
{"handle":"77.LOCAL/4","transaction":"<Transaction
'[0;m0x3c19def3bb4e185d85d9273355bd2ff82ba6b323388de53bc4ded444de1346c8[0;m'>"}
```

Fig. 3. Request for registration of a new digital object in the system and a successful response of the system when registering a new object

7   Retrieving Data by Handle System Descriptor

Figure 4 shows a sequential schema of system calls when the handleId descriptor is resolved into meta-information on a digital object and receiving data from the Blockchain network. In this case a common chain is formed from the network which is then additionally filtered on the client according to the necessary handleId.

Figure 4 shows a sequential diagram of the resolution of a descriptor into meta information on a digital object. In the first step, the client sends a resolve request to the system via the LHS-API entry service. An example of a request is shown in Fig. 5 (Request).

The LHS-API service makes a request to the Handle System platform to get the meta data stored in the LHS (get meta by id). If the Handle System platform responds successfully, the meta data of the digital object is returned in the response to the LHS-API (handle data).

Next, the LHS-API service makes a request to the Blockchain system via Rest Api to get the entire history of changes in the meta information of a digital object by its descriptor (get chain history). If the response is successful, the Blockchain platform returns a data array containing the entire history of changes stored in the smart contract. The LHS-API service filters and aggregates this data and returns the final response to the client (return to client). An example of a response is shown in Fig. 5 (Response). In the structure chainInfo all data is returned by modifying the descriptor in chronological order.

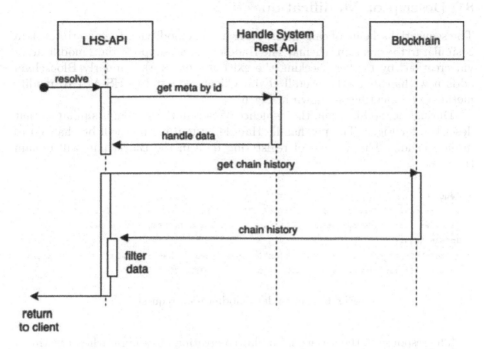

Fig. 4. Sequential diagram of the interaction of system elements when resolving a descriptor into meta information on a digital object

Request
curl --location --request GET 'http://172.30.4.20:9000/handle?handleID=77.LOCAL/4'

Response
{"handle":"77.LOCAL/4","values":{"handle":"77.LOCAL/4","responseCode":1,"values":[{
"data":{"format":"string","value":"LoRaDevice#1"},"index":1,"timestamp":"2021-07-07
T20:50:25Z","ttl":86400,"type":"DESC"},{"data":{"format":"string","value":"COD3F010
"},"index":2,"timestamp":"2021-07-07T20:50:25Z","ttl":86400,"type":"DEVICE_ID"},{"d
ata":{"format":"string","value":"{\"method\":\"POST\",\"type\":\"rest\",\"url\":\"h
ttps://test:8000/device/toggle?handleID=77.LOCAL%2FCOD3F010\"}"},"index":3,"timesta
mp":"2021-07-07T20:50:25Z","ttl":86400,"type":"ACTION"}]},"chain_info":[{"handle_id
":"77.LOCAL/4","data":"{\"handle\":\"77.LOCAL/4\",\"values\":[{\"index\":1,\"type\"
:\"DESC\",\"data\":{\"format\":\"string\",\"value\":\"LoRaDevice#1\"}},{\"index\":2
,\"type\":\"DEVICE_ID\",\"data\":{\"format\":\"string\",\"value\":\"COD3F010\"}},{\
"index\":3,\"type\":\"ACTION\",\"data\":{\"format\":\"string\",\"value\":\"{\\\"met
hod\\\":\\\"POST\\\",\\\"type\\\":\\\"rest\\\",\\\"url\\\":\\\"https://test:8000/de
vice/toggle?handleID=77.LOCAL%2FCOD3F010\\\"}\"}}]}","timestamp":1625691025}]}

Fig. 5. Input client request for data resolution by descriptor and successful result of data resolution by descriptor

8   Descriptor Modification

The sequential schema of calls in the system when modifying the descriptor data is similar to the creation schema. The Handle System supports data modification via creation (by further checking the existence of data). Similarly Blockchain adds new changes to the overall chain. Client request to LHS-API to modify metadata by identifier is shown in Fig. 6.

During the modification, the sequence of calls in the system is similar to that described for Fig. 2. The put handle Handle System method will be changed to update handle. The method of registering data in the Blockchain will remain the same.

```
Request
curl --location --request POST 'http://172.30.4.20:9000/handle?handleID=77.LOCAL/4'
--header 'Content-Type: application/json' --data-raw '{"DESC":"LoRa Device #1
Update handle
value","DEVICE_ID":"C0D3F010","ACTION":{"type":"rest","method":"POST","url":"https:
//test:8000/device/toggle?handleID=77.LOCAL%2FC0D3F010"}}'
```

Fig. 6. Handle data modification request

The response of the service is similar to creating a response when creating a response. ChainInfo can also see new data in the chain and shown on the Fig. 7.

```
38      },
39      "chain_info": [
40          {
41  .           "handle_id": "77.LOCAL/4",
42              "data": "{\"handle\":\"77.LOCAL/4\",\"values\":[{\"index\":1,\"type\":\"DESC\",\"data\":{\"format\":\"string\",\"value\":\"LoRa Device
                    #1\"}},{\"index\":2,\"type\":\"DEVICE_ID\",\"data\":{\"format\":\"string\",\"value\":\"C0D3F010\"}},{\"index\":3,\"type\":\"ACTION\",
                    \"data\":{\"format\":\"string\",\"value\":\"{\\\"method\\\":\\\"POST\\\",\\\"type\\\":\\\"rest\\\",\\\"url\\\":\\\"https://test:8000/
                    device/toggle?handleID=77.LOCAL%2FC0D3F010\\\"}\"}}]}",
43              "timestamp": 1625691825
44          },
45          {
46              "handle_id": "77.LOCAL/4",
47              "data": "{\"handle\":\"77.LOCAL/4\",\"values\":[{\"index\":1,\"type\":\"DESC\",\"data\":{\"format\":\"string\",\"value\":\"LoRa Device
                    #1 Update handle value\"}},{\"index\":2,\"type\":\"DEVICE_ID\",\"data\":{\"format\":\"string\",\"value\":\"C0D3F010\"}},{\"index\":3,
                    \"type\":\"ACTION\",\"data\":{\"format\":\"string\",\"value\":\"{\\\"method\\\":\\\"POST\\\",\\\"type\\\":\\\"rest\\\",
                    \\\"url\\\":\\\"https://test:8000/device/toggle?handleID=77.LOCAL%2FC0D3F010\\\"}\"}}]}",
48              "timestamp": 1625691840
49          }
50      ]
51  }
```

Fig. 7. Successful response of the system when modifying the metadata of a digital object

The following is an example of how the data was modified several times and the LHS-API output contained the entire history of data changes from the Blockchain on Fig. 8.

Eventually, the state of the descriptor will change again but the entire history of the changes in the chain can be seen on the Fig. 9.

Fifty tests were run to measure the response time of the LHS-API service. Figures 10, 11 shows the result of executing several requests to create a descriptor and their response time.

Request

```
curl --location --request POST 'http://172.30.4.20:9000/handle?handleID=77.LOCAL/4'
--header 'Content-Type: application/json' --data-raw '{"DESC":"LoRa Device Third
update","DEVICE_ID":"C0D3F010","ACTION":{"type":"rest","method":"POST","url":"https
://test:8000/device/toggle?handleID=77.LOCAL%2FC0D3F010"}}'
```

Fig. 8. Modified request data

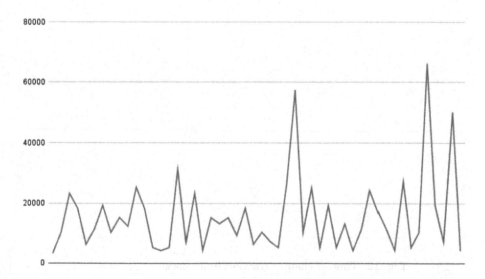

Fig. 9. The successful response of the system when re-modifying the descriptor contains the entire history of changes from the smart contract

Fig. 10. The LHS-API response time when creating a descriptor, as well as writing to a blockchain (in ms)

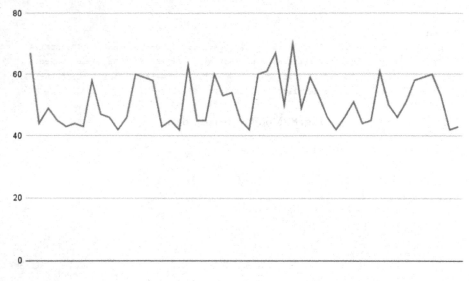

Fig. 11. The LHS-API response time when creating a descriptor, as well as writing to a blockchain (in ms)

Figure 12 shows the result of executing several requests to obtain metadata of a digital object and change history from a blockchain and their response time.

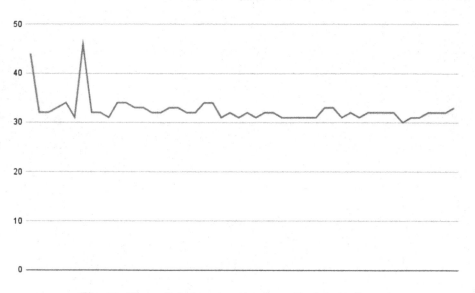

Fig. 12. Time of chain extraction from blockchain (in ms)

9   Conclusion

From the creation time schedule, it can be seen that the creation of a descriptor in LHS and then the subsequent synchronous recording of data in the chain can be as high as 60 s. Based on the results of testing the LHS system under similar conditions without the blockchain system [5] the service response time for creating a new descriptor is increased precisely because of synchronous writing in the blockchain. This is understandable, as the system needs to perform sufficiently resource- and time-consuming calculations to write a block. In order to speed up the response time, it is advisable to write to the system with an asynchronous query after successfully adding data to LHS.

The resolution time of the descriptor via the LHS-API with synchronous approach to the blockchain behind the given chain remains rather small, the read time of the blockchain is tens of milliseconds, which was sufficient performance of modern web services.

The time schedule of the blockchain response to read is presented separately. From the received time values it is evident that the greatest contribution to the total response time of the LHS-API service to the descriptor resolution is exactly the request to the blockchain.

The LHS service solves this problem by caching a descriptor resolution response to a digital object to reduce response time. At the same time, it is calculated in the system that the digital object is rarely modified and the data cached remains relevant for a sufficient period of time (up to 24 h). With the data recorded in the blockchain, it is not advisable to do so, as the chain can be supplemented continuously and it is necessary to see the current state of the system by blockchain casting.

References

1. Sazonov, D., Kirichek, R., Borodin, A.: Implementation of authentication and authorization system based on digital object architecture. In: 11TH International Congress on Ultra Modern Telecommunications and Control Systems and Workshops (ICUMT) (2019)
2. Sazonov, D., Kirichek, R.: Digital object architecture as an approach to identifying Internet of Things devices. In: Vishnevskiy, V.M., Samouylov, K.E., Kozyrev, D.V. (eds.) DCCN 2019. CCIS, vol. 1141, pp. 597–611. Springer, Cham (2019). https://doi.org/10.1007/978-3-030-36625-4_48
3. Sazonov, D., Kirichek, R.: Identification system model for energy-efficient long range mesh network based on digital object architecture. In: Vishnevskiy, V.M., Samouylov, K.E., Kozyrev, D.V. (eds.) DCCN 2020. CCIS, vol. 1337, pp. 497–509. Springer, Cham (2020). https://doi.org/10.1007/978-3-030-66242-4_39
4. Wang, H., Yu, X., Peng, J., Zhang, L., Xu, T.: Design of power material management system basing on Internet of Things identification and blockchain. In: 2020 IEEE Conference on Telecommunications, Optics and Computer Science (TOCS), pp. 160–163 (2020)

5. Kirks, T., Uhlott, T., Jost, J.: The use of blockchain technology for private data handling for mobile agents in human-technology interaction. In: 2019 IEEE International Conference on Cybernetics and Intelligent Systems (CIS) and IEEE Conference on Robotics, Automation and Mechatronics (RAM), pp. 445–450 (2019)
6. Mohammed, M.H.S.: A hybrid framework for securing data transmission in Internet of Things (IoTs) environment using blockchain approach. In: 2021 IEEE International IOT, Electronics and Mechatronics Conference (IEMTRONICS), pp. 1–10 (2021)
7. Roopak, T.M., Sumathi, R.: Electronic voting based on virtual ID of Aadhar using blockchain technology. In: 2020 2nd International Conference on Innovative Mechanisms for Industry Applications (ICIMIA), pp. 71–75 (2020)
8. Chalaemwongwan, N., Kurutach, W.: A practical national digital ID framework on blockchain (NIDBC). In: 15th International Conference on Electrical Engineering/Electronics, Computer, Telecommunications and Information Technology (ECTI-CON), pp. 497–500 (2018)
9. Choi, N., Kim, H.: Hybrid blockchain-based unification ID in smart environment. In: 22nd International Conference on Advanced Communication Technology (ICACT), pp. 166–170 (2020)
10. Kshetri, N., Voas, J.: Blockchain-enabled e-voting. IEEE Softw. **35**(4), 95–99 (2018)

Analytical Modeling of Distributed Systems

The Simulation of Finite-Source Retrial Queueing Systems with Two-Way Communication and Impatient Customers

János Sztrik⬤, Ádám Tóth⁽✉⁾⬤, Ákos Pintér⬤, and Zoltán Bács⬤

University of Debrecen, Debrecen 4032, Hungary
{sztrik.janos,toth.adam}@inf.unideb.hu,
apinter@science.unideb.hu, bacs.zoltan@econ.unideb.hu

Abstract. The aim of the paper is to analyze a M/M/1//N finite-source, two-way communication retrial queueing system with an unreliable server and impatient customers. In this model, every request in the source is eligible to generate customers when the server does not function but they are forwarded immediately to the orbit. Customers may depart from the system during its waiting in the orbit after a random time and they get back to the source. All random variables involved in the model construction are supposed to be independent of each other. The novelty of the investigation is to carry out a sensitivity analysis comparing various distributions of failure time on the performance measures such as the mean number of customers in the orbit, the mean waiting time of an arbitrary customer, probability of abandonment, etc. With the help of self-developed simulation program, results are illustrated graphically.

Keywords: Simulation · Blocking · Two-way communication · Sensitivity analysis · Finite-source queueing system · Unreliable server · Impatient customers

1 Introduction

Nowadays, network traffic increases in such a way that the design and optimisation of communication systems are required. This phenomenon can be followed both in the industrial sector like in the companies and in our homes due to the quick technological development and the great number of devices capable of IP communication. Therefore, researchers dedicate enough time to create new suitable models of telecommunication systems or adjust the current ones.

Retrial queues play quite an important role to depict real-life problems emerging from main telecommunication systems like telephone switching systems, call centers, computer networks, and computer systems. Investigating the available literature in the Internet many papers address topics related to retrial-queuing

The research was supported by the Thematic Excellence Programme (TKP2020-IKA-04) of the Ministry for Innovation and Technology in Hungary.

ⓒ Springer Nature Switzerland AG 2021
V. M. Vishnevskiy et al. (Eds.): DCCN 2021, LNCS 13144, pp. 117–127, 2021.
https://doi.org/10.1007/978-3-030-92507-9_11

systems with repeated calls. In [5,6,11,13] you can see some examples of it. In many areas of science analyzing these models can improve the efficiency of systems or bring about new advantageous features for example in the case of local-area networks with random access protocols and with multiple access protocols [1,10].

Speaking of two-way communication, it possesses favorable impacts on most of the systems. Because similarities can be observed with the operation of certain real-life systems ergo it is no wonder that models based on a two-way communication scheme are introduced in many papers. This is particularly suitable in the case of call centers where the service unit (or agent) performs other actions pertaining to selling, promoting, and advertising products apart from satisfying incoming calls. In our model, the server may perform that action (calling customers residing in the orbit) after some random time when it is functional and no request is under service. Examining such scenarios has a great influence on the utilization of the service unit (or workload of agents) that is an important aspect and extensively examined by several papers like [3,12].

Studying the related articles I found the assumption of having a service unit available all the time which is quite impractical regarding events in real-life applications of systems for example power outages, human error, or other failures. Although companies, providers want to ensure having fault-tolerant devices and services (the intention is to have high-availability scenarios), problems can occur at any time. Not to mention wireless communication where other factors could affect the transmission rate of the wireless channel and the forwarded information prone to undergo failure interruptions throughout transferring the packets. That is why random server breakdowns and repairs are centric topics so the inspection of these features alters undoubtedly the operation of systems, the system characteristics, and the performance measures. Finite-source retrial queues with server breakdowns have been studied in several papers like [4,9,11,19,20].

The main aim of this work is to investigate the operation of such a system containing a non-reliable service unit and customers which may leave the system without obtaining their service needs. The novelty of this investigation is to carry out a sensitivity analysis using different distributions of failure time on performance measures like the mean waiting time of an arbitrary customer, a customer leaving the system through the orbit, or the total utilization of the server. A simulation program is developed to accomplish our goal namely checking the effects of the distributions. Our program is based on SimPack toolkit [7] which is a collection of C and C++ libraries. Several approaches and algorithms are supported providing a set of utilities to build a working simulation from a model description. Simpack contains very basic building blocks, during the coding of the model several functions, random number generator, and features were integrated. With the help of this program, results are illustrated graphically. This paper is the natural continuation of [17].

2   Model Description and Notations

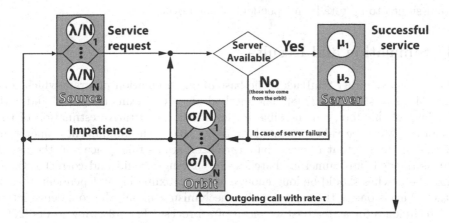

Fig. 1. System model

Figure 1 demonstrates the considered retrial queueing system of type $M/M/1//N$ which contains two-way communication feature and impatient customers. N customers are located in the finite-source where each of them can generate calls towards the server according to an exponential distribution with rate λ/N. In this model, every customer is characterized by an impatience feature that determines the maximum spent time of a customer in the orbit before leaving the system without completing its service requirement. This random variable also follows an exponential distribution with parameter τ. The model does not contain waiting queues therefore if the service unit is idle the service of an incoming customer starts immediately which is exponentially distributed with parameter μ. Upon its completion, request goes back to the source. Otherwise, the incoming customer is delivered to the orbit remaining in the system and after an exponentially distributed time with parameter σ/N they launch another attempt to reach the service facility. Our assumption is that every now and then the server breaks down according to gamma, hypo-exponential, hyper-exponential, Pareto, and lognormal distribution with different parameters but with the same mean value.

Throughout this period customers may proceed to produce their requests but they are transferred to the orbit immediately. The repair process is initiated instantaneously upon the failure of the server, which is also an exponentially distributed random variable with parameter γ_2. When the server breaks down during the service of a customer the execution will be cancelled and the customer returns to the orbit instantly. The feature of two-way communication is when the server becomes idle it may accomplish an outgoing call (secondary customers) after an exponentially distributed random time with rate ν that results in calling a customer in the orbit earlier. The service of these customers follows an exponential distribution with rate μ_2. Rates λ/N and σ/N are used because in [15,16] very similar systems are evaluated by an asymptotic method where N tends to infinity, and was proved that the number of customers in the system follows a normal distribution. All the random variables in the model creation are assumed to be totally independent of each other.

3   Simulation and Results

As mentioned earlier SimPack is the base of our simulation program which consists of a statistic package [8]. The method of batch means is applied and with the help of this tool, it is possible to perform a quantitative estimation of the mean and variance values of the desired variables. The fundamental operation of this method is that in every batch n observations take place and the useful run is divided into numerous batches. For having a valid and correct estimation the batches should be long enough and approximately independent of each other. This is one of the most popular mechanisms among the confidence interval techniques for a steady-state mean of a process. The following works [2,14] comprise very precise description and algorithm about batch means. The simulations are performed with a confidence level of 99.9%. The relative half-width of the confidence interval required to stop the simulation run is 0.00001.

3.1   Scenario 1

Four different distributions of failure time are used to investigate their effects on the main performance measures. To have a valid comparison we selected the parameters in such a way that the mean value and variance would be equal. Before that, a fitting process is necessary to be done to obtain the correct values of parameters. [18] describes in more detail the characteristics of the utilized distributions. In the first scenario, the squared coefficient of variation is greater than one therefore we utilized hyper-exponential, gamma, Pareto, and lognormal distributions and compared them with each other. Table 1 and Table 4 shows every values of the random variables including all the used input parameters of the various distributions of failure time as well (Table 2).

Table 1. Used numerical values of model parameters

N	λ/N	γ_2	σ/N	μ	μ_2	ν
100	0.01	1	0.01	1	1.2	0.02

Table 2. Parameters of failure time

Distribution	Gamma	Hyper-exponential	Pareto	Lognormal
Parameters	$\alpha = 0.6$	$p = 0.25$	$\alpha = 2.2649$	$m = -0.3081$
	$\beta = 0.5$	$\lambda_1 = 0.41667$	$k = 0.67018$	$\sigma = 0.99037$
		$\lambda_2 = 1.25$		
Mean	1.2			
Variance	2.4			
Squared coefficient of variation	1.6666666667			

Figure 2 shows the mean waiting time of an arbitrary customer in the function of arrival intensity. The disparity is quite obvious taking a closer look at the figure that represents the impact on the metrics using various distributions having the same first two moments. Customers spend by far more time in the orbit at Pareto distribution and the least at gamma distribution. Also the interesting maximum property characteristic of a finite-source retrial queueing system occurs despite the increasing arrival intensity.

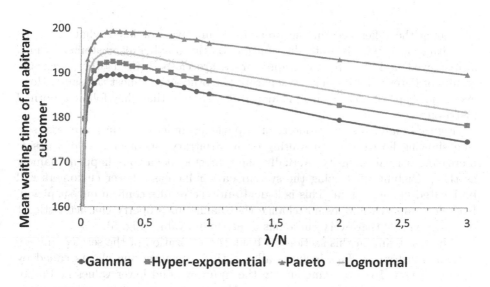

Fig. 2. Mean waiting time of an arbitrary customers

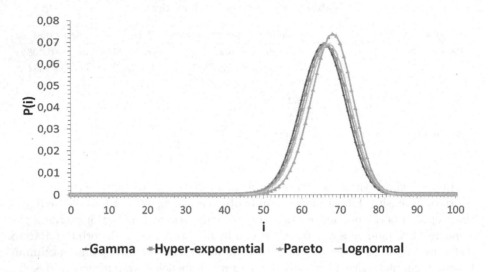

Fig. 3. Distribution of the number of customers in the system, $\lambda/N = 0.01$

Seeing the differences in the previous figure, we wondered what could be the situation of steady-state distribution of the number of customers which is displayed in Fig. 3. In the graph curves are close to each other, basically results of Pareto distribution are separate the others are almost the same. However, the shape of each of them is identical assuming that they follow a normal distribution.

Figure 4 highlights the property of impatience under different parameter setting showing how the mean waiting of an arbitrary customer develops beside increasing arrival intensity. Actually the expected behaviour happens namely as the probability of leaving the system earlier increases fewer customers will be located in the system. This is logical and the results confirm our suspicion. However, impatience does not change the maximum property characteristic, it is clearly visible that every curve has a maximum value (Fig. 5).

The last figure in this section is about the utilization of the service unit by outgoing customers. This includes all the time spent serving clients called by the idle server. By examining closely the figure we find lower values at Pareto distribution meaning that fewer number of outgoing customers are under service and regarding the others the received values are near to each other. With the increment of arrival intensity the utilization of the service unit by outgoing customers increases as well but after 0.01 it slowly decreases which is true for every investigated cases.

3.2   Scenario 2

After analysing the results of Scenario 1, we wondered if the same phenomena that occurred in the previous section would also show up with other parameter settings. Here the parameters of each distribution, which can be seen in Table 4, have been chosen so that the squared coefficient of variation would be less than one. Instead of hyper-exponential distribution we utilize hypo-exponential distribution to carry out a sensitivity analysis. The other parameters remain untouched (see Table 3), only the parameters of failure time differ between the two scenarios.

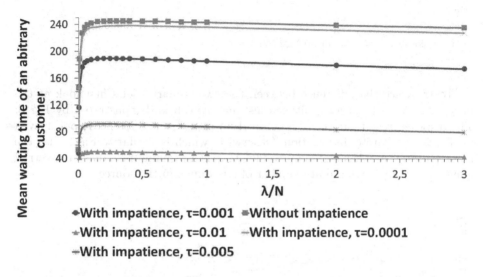

Fig. 4. The effect of impatience on the mean waiting time

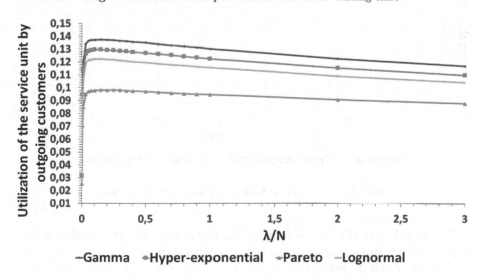

Fig. 5. Utilization of the service unit by outgoing customers

Table 3. Used numerical values of model parameters

N	λ/N	γ_2	σ/N	μ	μ_2	ν
100	0.01	1	0.01	1	1.2	0.02

Table 4. Parameters of failure time

Distribution	Gamma	Hypo-exponential	Pareto	Lognormal
Parameters	$\alpha = 1.3846$	$\mu_1 = 1$	$\alpha = 2.5442$	$m = -0.08948$
	$\beta = 1.1538$	$\mu_2 = 5$	$k = 0.7283$	$\sigma = 0.7373$
Mean	1.2			
Variance	1.04			
Squared coefficient of variation	0.72222222			

To truly see the difference between the two scenarios let's first look at the mean waiting time of an arbitrary customer which is demonstrated by Fig. 6. The achieved results are nearly identical no significant differences appear even in the case of Pareto distribution. Otherwise, which is similar to Fig. 2 that the mean waiting time has maximum value. This is a common phenomena of retrial queuing systems having finite number of customers in the source.

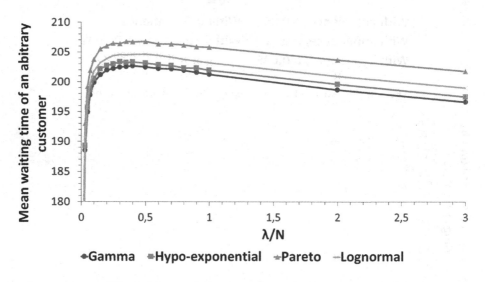

Fig. 6. Mean waiting time of an arbitrary customer

The next Figure (Fig. 7), as in the previous scenario, exhibits the distribution of the number of customers in the system with $\lambda/N = 0.01$ so what is the probability $(P(i))$ having exactly i customer in the system. Compared to Fig. 3, the curves are almost totally identical even in the case of Pareto distribution

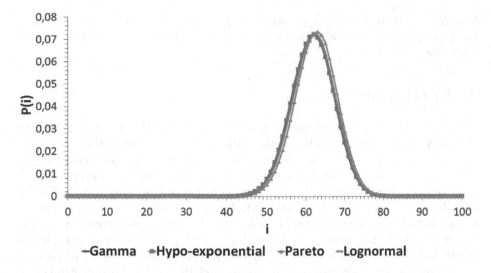

Fig. 7. Distribution of the number of customers in the system, $\lambda/N = 0.01$

and it can be also said that from the look of them they might follow normal distribution under this parameter setting as well.

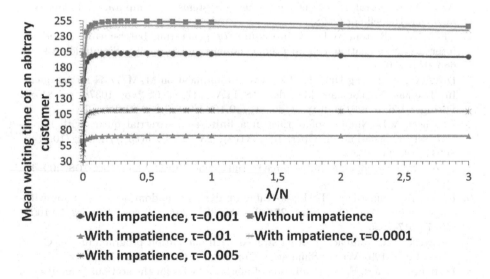

Fig. 8. The effect of impatience on the mean waiting time

To emphasize the effect of impatience Fig. 8 illustrates the mean waiting of an arbitrary customer using different values of impatience intensities. The obtained results are not surprising as this intensity grows the customers tend to spend less time in the system averagely which is quite logical. In comparison to Fig. 4, the tendency of difference seems to be the same, but the values in this scenario are higher.

4   Conclusion

We presented a finite-source retrial queueing system with an unreliable server that may call in requests residing in the orbit (two-way communication property) and impatient customers. The obtained results demonstrate the effect of impatience on the visualized performance measures indicating that customers with less patience depart much earlier without reaching the service facility. Results also display the influence of various distributions of failure time on the performance measures when the squared coefficient of variation was greater than one despite the fact that mean and variance are equal. In the case of less than one significant differences do not appear, curves almost totally overlap each other. In the future we plan to complete that system including other features like trying out other distributions or introducing disaster failure, or including more capacity of service.

References

1. Artalejo, J., Corral, A.G.: Retrial Queueing Systems: A Computational Approach. Springer, Heidelberg (2008)
2. Chen, E.J., Kelton, W.D.: A procedure for generating batch-means confidence intervals for simulation: checking independence and normality. Simulation **83**(10), 683–694 (2007)
3. Dragieva, V., Phung-Duc, T.: Two-way communication M/M/1//N retrial queue. In: Thomas, N., Forshaw, M. (eds.) ASMTA 2017. LNCS, vol. 10378, pp. 81–94. Springer, Cham (2017). https://doi.org/10.1007/978-3-319-61428-1_6
4. Dragieva, V.I.: Number of retrials in a finite source retrial queue with unreliable server. Asia-Pac. J. Oper. Res. **31**(2), 23 (2014). https://doi.org/10.1142/S0217595914400053
5. Falin, G., Artalejo, J.: A finite source retrial queue. Eur. J. Oper. Res. **108**, 409–424 (1998)
6. Fiems, D., Phung-Duc, T.: Light-traffic analysis of random access systems without collisions. Ann. Oper. Res. **277**(2), 311–327 (2017). https://doi.org/10.1007/s10479-017-2636-7
7. Fishwick, P.A.: Simpack: getting started with simulation programming in C and C++. In: In 1992 Winter Simulation Conference, pp. 154–162 (1992)
8. Francini, A., Neri, F.: A comparison of methodologies for the stationary analysis of data gathered in the simulation of telecommunication networks. In: Proceedings of MASCOTS '96 - 4th International Workshop on Modeling, Analysis and Simulation of Computer and Telecommunication Systems, pp. 116–122 (Feb 1996)
9. Gharbi, N., Nemmouchi, B., Mokdad, L., Ben-Othman, J.: The impact of breakdowns disciplines and repeated attempts on performances of small cell networks. J. Comput. Sci. **5**(4), 633–644 (2014)

10. Kim, J., Kim, B.: A survey of retrial queueing systems. Ann. Oper. Res. **247**(1), 3–36 (2015). https://doi.org/10.1007/s10479-015-2038-7
11. Krishnamoorthy, A., Pramod, P.K., Chakravarthy, S.R.: Queues with interruptions: a survey. TOP **22**(1), 290–320 (2012). https://doi.org/10.1007/s11750-012-0256-6
12. Kuki, A., Sztrik, J., Tóth, Á., Bérczes, T.: A contribution to modeling two-way communication with retrial queueing systems. In: Dudin, A., Nazarov, A., Moiseev, A. (eds.) ITMM/WRQ -2018. CCIS, vol. 912, pp. 236–247. Springer, Cham (2018). https://doi.org/10.1007/978-3-319-97595-5_19
13. Kumar, B.K., Vijayalakshmi, G., Krishnamoorthy, A., Basha, S.S.: A single server feedback retrial queue with collisions. Comput. Oper. Res. **37**(7), 1247–1255 (2010)
14. Law, A.M., Kelton, W.D.: Simulation Modeling and Analysis. McGraw-Hill Education (1991)
15. Nazarov, A., Sztrik, J., Kvach, A.: A survey of recent results in finite-source retrial queues with collisions. In: Dudin, A., Nazarov, A., Moiseev, A. (eds.) ITMM/WRQ -2018. CCIS, vol. 912, pp. 1–15. Springer, Cham (2018). https://doi.org/10.1007/978-3-319-97595-5_1
16. Nazarov, A., Sztrik, J., Kvach, A., Bérczes, T.: Asymptotic analysis of finite-source M/M/1 retrial queueing system with collisions and server subject to breakdowns and repairs. Ann. Oper. Res. **277**(2), 213–229 (2018). https://doi.org/10.1007/s10479-018-2894-z
17. Sztrik, J., Tóth, Á., Pintér, Á., Bács, Z.: The simulation of finite-source retrial queueing systems with two-way communications to the orbit and blocking. In: Vishnevskiy, V.M., Samouylov, K.E., Kozyrev, D.V. (eds.) DCCN 2020. CCIS, vol. 1337, pp. 171–182. Springer, Cham (2020). https://doi.org/10.1007/978-3-030-66242-4_14
18. Toth, A., Sztrik, J., Kuki, A., Berczes, T., Efrosinin, D.: Reliability analysis of finite-source retrial queues with outgoing calls using simulation. In: 2019 International Conference on Information and Digital Technologies (IDT), pp. 504–511, June 2019
19. Tóth, A.., Bérczes, T.., Sztrik, J.., Kuki, A.., Schreiner, W..: The simulation of finite-source retrial queueing systems with collisions and blocking. J. Math. Sci. **246**(4), 548–559 (2020). https://doi.org/10.1007/s10958-020-04759-4
20. Zhang, F., Wang, J.: Performance analysis of the retrial queues with finite number of sources and service interruptions. J. Korean Stat. Soc. **42**(1), 117–131 (2012). https://doi.org/10.1016/j.jkss.2012.06.002

Asymptotic Waiting Time Analysis of a M/GI/1 RQ System

Anatoly Nazarov[ID] and M. V. Samorodova[(✉)][ID]

Institute of Applied Mathematics and Computer Science, National Research Tomsk
State University, 36 Lenina Avenue, 634050 Tomsk, Russian Federation

Abstract. Main objective of this article is waiting time in M/GI/1 RQ
system. We investigate this process by the use of asymptotic analysis
method under heavy load condition. As the main result of our research
we obtain asymptotic characteristic function of the waiting time distribution. Within the framework of the problem posed, the asymptotic
probability distribution of the number of returns of the request to the
orbit in the considered RQ system was obtained, and also an analytical expression for the asymptotic conditional characteristic function of
the limiting number of returns of the request to the orbit was obtained,
which has the form of the characteristic function of exponential distribution with a random parameter, the distribution of which is determined
by the value of the limiting number of requests in the orbit that has a
gamma distribution.

Keywords: Waiting time distribution · Retrial queue · Number of
returns

1 Introduction

In RQ systems it is hard to deal with the waiting time distribution because of
random order service of customers. Despite the objective difficulties, attempts
to investigate the waiting time in $M/GI/1$ RQ - systems were undertaken by
Artalejo, Gomez-Corral, Falin, Fricker, Neuts, Nobel, Lee, Kim, Kim. In their
studies, they mainly used either numerical methods or simulation modeling.

In our research we use analytical methods to obtain asymptotic results under
heavy load condition for both waiting time distribution and number of returns
of the request to the orbit. In section two the basics of considered mathematical
model are presented. Kolmogorov equations for investigated processes are determined in section three. Sections four, five and six are dedicated to asymptotic
analysis of the distribution of the number of request in the orbit, the distribution
of the number of returns of the request and the distribution of the waiting time.

2 Mathematical Model

We consider a $M/GI/1$ retrial queuing (RQ) system. Requests arrival process is
a Poisson process with intensity $\tilde{\nu} = \rho\lambda$. Arrived request occupies the server and

© Springer Nature Switzerland AG 2021
V. M. Vishnevskiy et al. (Eds.): DCCN 2021, LNCS 13144, pp. 128–139, 2021.
https://doi.org/10.1007/978-3-030-92507-9_12

the service starts instantly if the server was empty at the arrival moment. The service time of the request follows a common probability law with arbitrary distribution function $B(x)$. In case the server was not empty at the arrival moment the request go to the orbit.

Each request from the orbit after a random delay retries to occupy the server. This random delay has exponential distribution with rate σ. If the server occur busy again at the retrial moment the request immediately returns to the orbit in order to make one more retrial attempt.

Let's denote waiting time by W and define it as the length of the interval from the moment the request arrives in the system till the start of the service. As counterparts of the waiting time process we consider: the number of transitions of the request to the orbit $\tilde{\nu}$ and the number of returns of the request to the orbit from the moment t until the start of the service $\nu(t)$. If r is the probability that the server is busy at the request arrival moment then

$$\tilde{\nu} = \begin{cases} 0, & \text{with probability } (1-r), \\ 1 + \nu(t), & \text{with probability } r. \end{cases}$$

Using above notations, the characteristic function of the waiting time W can be presented as follows:

$$G(u) = E\left\{e^{juW}\right\} = (1-r) + r\sum_{n=0}^{\infty} E\left\{e^{juW}/\tilde{\nu} = 1 + n\right\} P\left\{\nu(t) = n\right\}$$

$$= (1-r) + r\sum_{n=0}^{\infty} \left(\frac{\sigma}{\sigma - ju}\right)^{1+n} P\left\{\nu(t) = n\right\}. \tag{1}$$

3   Kolmogorov's Equations

Let's denote the number of requests in the orbit at time t by $i(t)$ and the state of the server at time t by $k(t)$, then:

$$k(t) = \begin{cases} 0, & \text{if the server is idle,} \\ 1, & \text{if the server is busy.} \end{cases}$$

We introduce a process $y(t)$ - the elapsed service time at the moment t for a request standing on the server, and the conditional rate $\mu(x) = \frac{B'(x)}{1-B(x)}$ of service of a request standing on the server in case that the elapsed service time is equal to x.

We do not define process $y(t)$ when the server is free. Thus, we investigate a random process with a variable number of components $\{k(t), i(t), y(t)\}$, which forms a continuous time Markov process. Let's assume that the stationary probability distribution of the states of this process is exist.

Let's denote:

$$P_0(i,t) = P\{k(t) = 0, i(t) = i\},$$

$$P_1(i,y,t) = \frac{\partial P\{k(t) = 1, i(t) = i, y(t) < y\}}{\partial y}.$$

Let's derive a system of Kolmogorov differential equations for $P_0(i,t)$ and $P_1(i,y,t)$:

$$P_0(i, t+\Delta t) = (1 - \tilde{\lambda}\Delta t)(1 - i\sigma\Delta t)P_0(i,t) + \Delta t \int_0^\infty P_1(i,y,t)\mu(y)dy + o(\Delta t),$$

$$P_1(i, y+\Delta t, t+\Delta t) = (1 - \tilde{\lambda}\Delta t)(1 - \mu(y)\Delta t)P_1(i,y,t) + \tilde{\lambda}\Delta t P_1(i-1,y,t) + o(\Delta t),$$

$$\int_0^{\Delta t} P_1(i,x,t)dx = \tilde{\lambda}\Delta t P_0(i,t) + (i+1)\sigma\Delta t P_0(i+1,t) + o(\Delta t),$$

$$\lim_{\Delta t \to 0} \frac{P_0(i, t+\Delta t) - P_0(i,t)}{\Delta t} = -(\tilde{\lambda} + i\sigma)P_0(i,t) + \int_0^\infty P_1(i,y,t)\mu(y)dy,$$

$$\lim_{\Delta t \to 0} \frac{P_1(i, y+\Delta t, t+\Delta t) - P_1(i,y,t)}{\Delta t} = -(\tilde{\lambda} + \mu(y))P_1(i,y,t) + \tilde{\lambda}P_1(i-1,y,t),$$

$$P_1(i, 0, t) = \tilde{\lambda}P_0(i,t) + (i+1)\sigma P_0(i+1,t).$$

As a result we get the following system for $P_0(i,t)$ and $P_1(i,y,t)$:

$$\frac{\partial P_0(i,t)}{\partial t} = -(\tilde{\lambda} + i\sigma)P_0(i,t) + \int_0^\infty P_1(i,y,t)\mu(y)dy,$$

$$\frac{\partial P_1(i,y,t)}{\partial t} + \frac{\partial P_1(i,y,t)}{\partial y} = -(\tilde{\lambda} + \mu(y))P_1(i,y,t) + \tilde{\lambda}P_1(i-1,y,t),$$

$$P_1(i,0,t) = \tilde{\lambda}P_0(i,t) + (i+1)\sigma P_0(i+1,t).$$

In stationary regime for $P_0(i,t)$ and $P_1(i,y,t)$ we get:

$$-(\tilde{\lambda} + i\sigma)P_0(i) + \int_0^\infty P_1(i,y)\mu(y)dy = 0,$$

$$\frac{\partial P_1(i,y)}{\partial y} = -(\tilde{\lambda} + \mu(y))P_1(i,y) + \tilde{\lambda}P_1(i-1,y), \tag{2}$$

$$P_1(i,0) = \tilde{\lambda}P_0(i) + (i+1)\sigma P_0(i+1).$$

Let's introduce steady-state partial characteristic functions:

$$H_0(u) = \sum_{i=0}^\infty e^{jui}P_0(i), \quad H_1(u,y) = \sum_{i=0}^\infty e^{jui}P_1(i,y)$$

and represent system (2) as follows:

$$-\tilde{\lambda}H_0(u) + j\sigma\frac{\partial H_0(u)}{\partial u} + \int\limits_0^\infty H_1(u,y)\mu(y)dy = 0,$$

$$\frac{\partial H_1(u,y)}{\partial y} = ((e^{ju}-1)\tilde{\lambda} - \mu(y))H_1(u,y), \tag{3}$$

$$H_1(u,0) = \tilde{\lambda}H_0(u) - j\sigma e^{-ju}\frac{\partial H_0(u)}{\partial u}.$$

Then let's integrate the second equation of system (3) and add it with the first equation, finally we obtain the following equality:

$$\tilde{\lambda}H_1(u) + e^{-ju}j\sigma H'_0(u) = 0.$$

Thus, a system of equations has been composed for $H_0(u)$ and $H_1(u,y)$:

$$-\tilde{\lambda}H_0(u) + j\sigma\frac{\partial H_0(u)}{\partial u} + \int\limits_0^\infty H_1(u,y)\mu(y)dy = 0,$$

$$\frac{\partial H_1(u,y)}{\partial y} = ((e^{ju}-1)\tilde{\lambda} - \mu(y))H_1(u,y), \tag{4}$$

$$H_1(u,0) = \tilde{\lambda}H_0(u) - j\sigma e^{-ju}\frac{\partial H_0(u)}{\partial u},$$

$$\tilde{\lambda}H_1(u) + e^{-ju}j\sigma H'_0(u) = 0.$$

Characteristic function for $\nu(t)$ in stationary mode can be represented as follows:

$$G(u) = E\left\{e^{ju\nu(t)}\right\} = \sum_{i=0}^\infty\left[G_0(i,u)P_0(i) + \int\limits_0^\infty G_1(i,u,y)P_1(i,y)dy\right],$$

where $G_0(i,u)$ and $G_1(i,u,y)$ are conditional characteristic functions:

$$G_0(i,u,t) = E\left\{e^{ju\nu(t)}/k(t) = 0, i(t) = i\right\},$$

$$G_1(i,u,y,t) = E\left\{e^{ju\nu(t)}/k(t) = 1, i(t) = i, y(t) = y\right\}.$$

Let us compose a system of inverse Kolmogorov equations for the conditional characteristic functions $G_0(i,u,t)$ and $G_1(i,u,y,t)$:

$$G_0(i,u,t-\Delta t) = (1-\tilde{\lambda}\Delta t)(1-i\sigma\Delta t)G_0(i,u,t) + \tilde{\lambda}\Delta tG_1(i,u,0,t)$$
$$+(i-1)\sigma\Delta tG_1(i-1,u,0,t) + \sigma\Delta t + o(\Delta t),$$

$$G_1(i,u,y-\Delta t,t-\Delta t) = (1-\tilde{\lambda}\Delta t)(1-\sigma\Delta t)(1-\mu(y)\Delta t)G_1(i,u,y,t)$$
$$+\tilde{\lambda}\Delta tG_1(i+1,u,y,t) + e^{ju}\sigma\Delta tG_1(i,u,y,t) + \mu(y)\Delta tG_0(i,u,t) + o(\Delta t),$$

$$\lim_{\Delta t \to 0} \frac{G_0(i, u, t - \Delta t) - G_0(i, u, t)}{\Delta t} = -(\tilde{\lambda} + i\sigma)G_0(i, u, t) + \tilde{\lambda}G_1(i, u, 0, t)$$

$$+(i-1)\sigma G_1(i-1, u, 0, t) + \sigma,$$

$$\lim_{\Delta t \to 0} \frac{G_1(i, u, y - \Delta t, t - \Delta t) - G_1(i, u, y, t)}{\Delta t} = -(\tilde{\lambda} + \sigma + \mu(y))G_1(i, u, y, t)$$

$$+\tilde{\lambda}G_1(i+1, u, y, t) + e^{ju}\sigma G_1(i, u, y, t) + \mu(y)G_0(i, u, t),$$

$$-\frac{dG_0(i, u, t)}{dt} = -(\tilde{\lambda} + i\sigma)G_0(i, u, t) + \tilde{\lambda}G_1(i, u, 0, t)$$

$$+(i-1)\sigma G_1(i-1, u, 0, t) + \sigma,$$

$$-\left(\frac{dG_1(i, u, y, t)}{dy} + \frac{dG_1(i, u, y, t)}{dt}\right) = -(\tilde{\lambda} + \sigma(1 - e^{ju}) + \mu(y))G_1(i, u, y, t)$$

$$+\tilde{\lambda}G_1(i+1, u, y, t) + \mu(y)G_0(i, u, t).$$

In stationary regime for $G_0(i, u)$ and $G_k(i, u, y)$ we get:

$$-(\tilde{\lambda} + i\sigma)G_0(i, u) + \tilde{\lambda}G_1(i, u, 0) + (i-1)\sigma G_1(i-1, u, 0) + \sigma = 0, \quad (5)$$

$$\frac{dG_1(i, u, y)}{dy} - (\tilde{\lambda} + \sigma(1 - e^{ju}) + \mu(y))G_1(i, u, y)$$
$$+\tilde{\lambda}G_1(i+1, u, y) + \mu(y)G_0(i, u) = 0. \quad (6)$$

4   Asymptotic Analysis of the Distribution of the Number of Requests in the Orbit

Taking into account that $\tilde{\lambda} = \rho\lambda$, and , as will be shown below, that, $\lambda b_1 = 1$, let's make in (4) the following substitutions $\rho = 1-\varepsilon$, $u = \varepsilon w$, $H_0(u) = \varepsilon F_0(w, \varepsilon)$, $H_1(u, y) = F_1(w, y, \varepsilon)$ and get:

$$-\lambda(1 - \varepsilon)\varepsilon F_0(w, \varepsilon) + j\sigma\frac{\partial F_0(w, \varepsilon)}{\partial w} + \int_0^\infty F_1(w, y, \varepsilon)\mu(y)dy = 0,$$

$$\frac{\partial F_1(w, y, \varepsilon)}{\partial y} = -\left[\lambda(1 - \varepsilon)(1 - e^{j\varepsilon w}) + \mu(y)\right]F_1(w, y, \varepsilon), \quad (7)$$

$$F_1(w, 0, \varepsilon) = \lambda(1 - \varepsilon)\varepsilon F_0(w, \varepsilon) - j\sigma e^{-j\varepsilon w}\frac{\partial F_0(w, \varepsilon)}{\partial w},$$

$$(1 - \varepsilon)\lambda F_1(w, \varepsilon) + e^{-j\varepsilon w}j\sigma\frac{\partial F_0(w, \varepsilon)}{\partial w} = 0.$$

Let $\varepsilon \to 0$, denote

$$\lim_{\varepsilon \to 0} F_0(w, \varepsilon) = F_0(w),$$

$$\lim_{\varepsilon \to 0} F_1(w, y, \varepsilon) = F_1(w, y).$$

Passing to the limit in (7), we get a system with respect to $F_0(w)$ and $F_1(w, y)$:

$$\int_0^\infty F_1(w, y)\mu(y)dy = -j\sigma\frac{\partial F_0(w)}{\partial w},$$

$$\frac{\partial F_1(w, y)}{\partial y} = -\mu(y)F_1(w, y), \quad F_1(w, 0) = -j\sigma\frac{\partial F_0(w)}{\partial w}, \tag{8}$$

$$\lambda F_1(w) + j\sigma\frac{\partial F_0(w)}{\partial w} = 0.$$

Solving the Cauchy problem for the second equation of system (8), we get:

$$F_1(w, y) = -j\sigma F'_0(w)\left(1 - B(y)\right),$$
$$F_1(w) = -j\sigma F'_0(w)b_1.$$

Substituting in the last equation of system (8) we obtain:

$$(1 - \lambda b_1)\, j\sigma F'_0(w) = 0.$$

Accordingly, $\lambda b_1 = 1$.

Let's assume that $F_0(w, \varepsilon)$ and $F_1(w, y, \varepsilon)$ may be presented in the form of a decomposition:

$$F_0(w, \varepsilon) = F_0(w) + \varepsilon f_0(w) + o(\varepsilon^2),$$
$$F_1(w, y, \varepsilon) = F_1(w, y) + \varepsilon f_1(w, y) + o(\varepsilon^2).$$

and substitute this decompositions in (7):

$$-\lambda(1 - \varepsilon)\varepsilon\left[F_0(w) + \varepsilon f_0(w)\right] + j\sigma\frac{\partial\left[F_0(w) + \varepsilon f_0(w)\right]}{\partial w}$$

$$+ \int_0^\infty \left[F_1(w, y) + \varepsilon f_1(w, y)\right]\mu(y)dy = o(\varepsilon^2),$$

$$\frac{\partial\left[F_1(w, y) + \varepsilon f_1(w, y)\right]}{\partial y} = -\left[\lambda(1 - \varepsilon)(1 - e^{j\varepsilon w})\right.$$ \tag{9}

$$\left. +\mu(y)\right]\left[F_1(w, y) + \varepsilon f_1(w, y)\right] + o(\varepsilon^2),$$

$$F_1(w, 0) + \varepsilon f_1(w, 0) = \lambda(1 - \varepsilon)\varepsilon\left[F_0(w) + \varepsilon f_0(w)\right]$$

$$-j\sigma e^{-j\varepsilon w}\frac{\partial\left[F_0(w) + \varepsilon f_0(w)\right]}{\partial w} + o(\varepsilon^2),$$

$$(1 - \varepsilon)\lambda\left[F_1(w) + \varepsilon f_1(w)\right] + e^{-j\varepsilon w}j\sigma\frac{\partial\left[F_0(w) + \varepsilon f_0(w)\right]}{\partial w} = o(\varepsilon^2).$$

First equation of system (9) can be transformed as follows:

$$-\lambda F_0(w) + j\sigma\frac{\partial f_0(w)}{\partial w} + \int_0^\infty f_1(w, y)\mu(y)dy = 0.$$

Second equation of system (9) can be transformed as follows:

$$\frac{\partial \left[F_1(w,y) + \varepsilon f_1(w,y) \right]}{\partial y} = - \left[\lambda(1-\varepsilon)(1-e^{j\varepsilon w}) \right.$$

$$+ \mu(y) \left] \left[F_1(w,y) + \varepsilon f_1(w,y) \right] + o(\varepsilon^2), \right.$$

$$-\mu(y)F_1(w,y) + \varepsilon\frac{\partial f_1(w,y)}{\partial y} = -\lambda(1-\varepsilon)(1-e^{j\varepsilon w})F_1(w,y)$$

$$-\mu(y)F_1(w,y) - \left[\lambda(1-\varepsilon)(1-e^{j\varepsilon w}) + \mu(y) \right] \varepsilon f_1(w,y) + o(\varepsilon^2).$$

After performing some actions on the equation we get:

$$\frac{\partial f_1(w,y)}{\partial y} = \lambda(1-\varepsilon)(jw)F_1(w,y) - \left[\lambda(1-\varepsilon)(1-e^{j\varepsilon w}) + \mu(y) \right] f_1(w,y) + o(\varepsilon^2).$$

Passing to the limit we obtain:

$$\frac{\partial f_1(w,y)}{\partial y} = \lambda jw F_1(w,y) - \mu(y)f_1(w,y).$$

Third equation of system (9) can be transformed as follows:

$$F_1(w,0) + \varepsilon f_1(w,0) = \lambda(1-\varepsilon)\varepsilon \left[F_0(w) + \varepsilon f_0(w) \right]$$

$$-j\sigma e^{-j\varepsilon w}\frac{\partial \left[F_0(w) + \varepsilon f_0(w) \right]}{\partial w} + o(\varepsilon^2),$$

$$-j\sigma\frac{\partial F_0(w)}{\partial w} + \varepsilon f_1(w,0) = \lambda(1-\varepsilon)\varepsilon \left[F_0(w) + \varepsilon f_0(w) \right]$$

$$-j\sigma(1-j\varepsilon w)\frac{\partial \left[F_0(w) + \varepsilon f_0(w) \right]}{\partial w} + o(\varepsilon^2),$$

$$-j\sigma\frac{\partial F_0(w)}{\partial w} + \varepsilon f_1(w,0) = \lambda(1-\varepsilon)\varepsilon \left[F_0(w) + \varepsilon f_0(w) \right]$$

$$-j\sigma\frac{\partial \left[F_0(w) + \varepsilon f_0(w) \right]}{\partial w} + j\sigma j\varepsilon w\frac{\partial \left[F_0(w) + \varepsilon f_0(w) \right]}{\partial w} + o(\varepsilon^2).$$

After performing some actions on the equation we get:

$$f_1(w,0) = \lambda(1-\varepsilon)F_0(w) - j\sigma\frac{\partial f_0(w)}{\partial w} + j\sigma jw\frac{\partial F_0(w)}{\partial w} + o(\varepsilon^2).$$

Passing to the limit we obtain:

$$f_1(w,0) = \lambda F_0(w) - j\sigma\frac{\partial f_0(w)}{\partial w} + j\sigma jw\frac{\partial F_0(w)}{\partial w}.$$

Forth equation of system (9) can be transformed as follows:

$$(1 - \varepsilon)\lambda \left[F_1(w) + \varepsilon f_1(w)\right] + e^{-jew} j\sigma \frac{\partial\left[F_0(w) + \varepsilon f_0(w)\right]}{\partial w} = o(\varepsilon^2),$$

$$(1 - \varepsilon)\lambda \left[F_1(w) + \varepsilon f_1(w)\right] + j\sigma \frac{\partial\left[F_0(w) + \varepsilon f_0(w)\right]}{\partial w}$$
$$- jewj\sigma \frac{\partial\left[F_0(w) + \varepsilon f_0(w)\right]}{\partial w} = o(\varepsilon^2),$$

$$\lambda\varepsilon f_1(w) - \varepsilon\lambda \left[F_1(w) + \varepsilon f_1(w)\right] + j\sigma \frac{\partial\varepsilon f_0(w)}{\partial w}$$
$$- jewj\sigma \frac{\partial\left[F_0(w) + \varepsilon f_0(w)\right]}{\partial w} = o(\varepsilon^2),$$

passing to the limit we obtain:

$$f_1(w) - F_1(w) + j\sigma b_1 \frac{\partial f_0(w)}{\partial w} - jwj\sigma b_1 \frac{\partial F_0(w)}{\partial w} = 0,$$

$$j\sigma b_1 \frac{\partial f_0(w)}{\partial w} + (jw - 1)F_1(w) + f_1(w) = 0.$$

After all equations transformation we obtain the following system:

$$j\sigma f_0'(w) + \int_0^\infty f_1(w, y)\mu(y)dy = \lambda F_0(w),$$

$$\frac{\partial f_1(w, y)}{\partial y} + \mu(y)f_1(w, y) = \lambda jw F_1(w, y), \tag{10}$$

$$f_1(w, 0) = \lambda F_0(w) + j\sigma jw F_0'(w) - j\sigma f_0'(w),$$

$$j\sigma b_1 f_0'(w) + (jw - 1)F_1(w) + f_1(w) = 0.$$

Let us solve the Cauchy problem with respect to $f_1(w, y)$ and after simple transformations for $f_1(w) = \int_0^\infty f_1(w, y)dy$ we get:

$$f_1(w) = j\sigma jw F_0'(w) \left(b_1 - \frac{b_2}{2b_1}\right) + F_0(w) - j\sigma b_1 f_0'(w).$$

Substituting the resulting expression into the last equation of system (10) we get:

$$\left(1 - jw \frac{b_2}{2b_1{}^2}\right) F_0'(w) + \frac{1}{j\sigma b_1} F_0(w) = 0.$$

Solving this equation we obtain:

$$F_0(w) = \left(1 - \frac{b_2}{2b_1{}^2} jw\right)^{-\frac{2b_1}{\sigma b_2}}.$$

Let's denote $\Phi(w) = \left(1 - \frac{b_2}{2b_1{}^2}jw\right)^{-\left(\frac{2b_1}{\sigma b_2}+1\right)}$, then:

$$F_1(w, y) = \frac{1}{b_1}\Phi(w)\left(1 - B(y)\right), \qquad (11)$$

here $F_1(w, y)$ is conditional asymptotic characteristic function under the condition that $y(t) = y$. Equality (11) will be used in further analysis of the number of returns of the request to the orbit, which will be considered in detail in the next section. Integrating (11) over y, we find asymptotic characteristic function under the heavy load condition:

$$F_1(w) = \frac{1}{b_1}\Phi(w)b_1 = \Phi(w) = \left(1 - \frac{b_2}{2b_1{}^2}jw\right)^{-\left(\frac{2b_1}{\sigma b_2}+1\right)}.$$

$F_1(w)$ has the form of gamma distribution with density:

$$f_{\alpha,\beta}(x) = \frac{\alpha^\beta}{\Gamma(\beta)}x^{\beta-1}e^{-\alpha x}, x \geq 0, \qquad (12)$$

where $\alpha = \frac{2b_1{}^2}{b_2}$, $\beta = \frac{2b_1}{\sigma b_2} + 1$.

Note that the expression for $F_1(w)$ match with the result obtained in [9], in which in order to find the asymptotic characteristic function under heavy load condition was used the method of residual service time. Accordingly, [9] does not contain the result (11) for the conditional asymptotic characteristic function $F_1(w, y)$, under the condition that the elapsed service time $y(t) = y$.

5   Asymptotic Analysis of the Distribution of the Number of Returns of the Request to the Orbit

Taking into account that $\tilde{\lambda} = \rho\lambda$, making substitutions $\rho = 1 - \varepsilon$, $u = \varepsilon w$, $i\varepsilon = x$, $G_0(i, u) = g_0(x, w, \varepsilon)$, $G_1(i, u, y) = g_1(x, w, y, \varepsilon)$ and multiplying (5) by ε we get:

$$-(\varepsilon(1-\varepsilon)\lambda + x\sigma)g_0(x, w, \varepsilon) + \varepsilon(1-\varepsilon)\lambda g_1(x, w, 0, \varepsilon)$$
$$+(x - \varepsilon)\sigma g_1(x - \varepsilon, w, 0, \varepsilon) + \varepsilon\sigma = 0,$$
$$\frac{dg_1(x, w, y, \varepsilon)}{dy} - \left[(1-\varepsilon)\lambda + \mu(y) + \sigma(1 - e^{j\varepsilon w})\right]g_1(x, w, y, \varepsilon) \qquad (13)$$
$$+\mu(y)g_0(x, w, \varepsilon) + (1-\varepsilon)\lambda g_1(x + \varepsilon, w, y, \varepsilon) = 0.$$

Let $\varepsilon \to 0$, denote

$$\lim_{\varepsilon \to 0} g_0(x, w, \varepsilon) = g_0(x, w),$$
$$\lim_{\varepsilon \to 0} g_1(x, w, y, \varepsilon) = g_1(x, w, y).$$

For the functions $g_0(x, w)$ and $g_1(x, w, y)$ we can get the following expression:

$$-x\sigma g_0(x, w) + x\sigma g_1(x, w, 0) = 0,$$

$$\frac{dg_1(x, w, y)}{dy} = \mu(y) \left[g_1(x, w, y) - g_0(x, w) \right],$$

$$\frac{d\left[g_1(x, w, y) - g_0(x, w) \right]}{dy} = \mu(y) \left[g_1(x, w, y) - g_0(x, w) \right].$$

From the obtained equations follows that $g_0(x, w) = g_1(x, w, 0) = g_1(x, w, y)$. This functions are equal, so let's denote them by $g(x, w)$.

Let us write system (13) in the following form:

$$-x g_0(x, w, \varepsilon) + x g_1(x, w, 0, \varepsilon) = \varepsilon \frac{\partial \left[x g(x, w) \right]}{\partial x} - \varepsilon + O(\varepsilon^2),$$

$$\frac{dg_1(x, w, y, \varepsilon)}{dy} - \mu(y) g_1(x, w, y, \varepsilon) + \mu(y) g_0(x, w, \varepsilon) = -\sigma j \varepsilon w g(x, w) \quad (14)$$

$$-\varepsilon \lambda \frac{\partial g(x, w)}{\partial x} + O(\varepsilon^2).$$

Let $w = 0$ in (11), then $F_1(0, y) = \frac{1}{b_1} (1 - B(y))$. Multiplying the second equation of system (14) by $F_1(0, y)$ and integrating we obtain:

$$- g_1(x, w, 0, \varepsilon) + g_0(x, w, \varepsilon) = -\sigma b_1 j \varepsilon w g(x, w) - \varepsilon \lambda b_1 \frac{\partial g(x, w)}{\partial x} + O(\varepsilon^2). \quad (15)$$

Then let's divide the first equation of system (14) by x, add it with Eq. (15) and get the equation with respect to $g(x, w)$:

$$g(x, w) - \sigma x b_1 j w g(x, w) - 1 = 0.$$

Finally, the limiting conditional characteristic function will get the following form:

$$g(x, w) = \frac{1/\sigma x b_1}{1/\sigma x b_1 - jw}.$$

Let us pass from the variable w to the variable u, applying the equality $w = \frac{u}{\varepsilon} = \frac{u}{1-\rho}$ and obtain:

$$h(x, u) = \frac{1/\sigma x b_1}{1/\sigma x b_1 - j\frac{u}{1-\rho}} = \frac{(1-\rho)/\sigma x b_1}{(1-\rho)/\sigma x b_1 - ju}.$$

The conditional characteristic function $h(x, u)$ of the limit value $\nu(t)$ defines an exponential distribution with a random parameter $\alpha = \frac{1-\rho}{\sigma x b_1}$, where x is the value of a random variable having a gamma distribution with density (12). Passing from the conditional characteristic function to the unconditional one we get:

$$\tilde{G}(u) = \int_0^\infty h(x, u) f_{\alpha, \beta}(x) dx = \int_0^\infty \frac{(1-\rho)/\sigma x b_1}{(1-\rho)/\sigma x b_1 - ju} f_{\alpha, \beta}(x) dx.$$

The density of such distribution will have the following form:

$$\tilde{P}(z) = \int\limits_0^\infty \frac{(1-\rho)}{\sigma x b_1} e^{-\frac{(1-\tilde{\rho})}{\sigma x b_1} z} f_{\alpha,\beta}(x) dx. \tag{16}$$

6   Asymptotic Probability Distribution of the Waiting Time of the Customer in the Orbit

Using the found distribution density (16), we compose a discrete approximation:

$$P(n) = \tilde{P}(n) \cdot \left(\sum_{m=0}^\infty \tilde{P}(m) \right)^{-1},$$

where $P(n)$ - discrete approximation of asymptotic probability distribution of $\nu(t)$. Let's substitute the resulting distribution into (1):

$$G(u) = (1-r) + r \sum_{n=0}^\infty \left(\frac{\sigma}{\sigma - ju} \right)^{1+n} P(n).$$

Thus we have found asymptotic characteristic function $G(u)$ of the waiting time W for the case of RQ system $M/GI/1$ under the heavy load condition.

7   Conclusion

We managed to obtain an analytical form of the asymptotic characteristic function of the waiting time distribution for the RQ system $M/GI/1$ under the heavy load condition. Also, by the use of the asymptotic method we obtained the distributions of the number of returns of the request to the orbit and the number of requests in the orbit.

References

1. Artalejo, J.R., Gómez-Corral, A.: Waiting time analysis of the M/G/1 queue with finite retrial group. Naval Res. Logistics (NRL) **54**(5), 524–529 (2007)
2. Falin, G., Fricker, C.: On the virtual waiting time in an M/G/1 retrial queue. J. Appl. Probab. **28**(2), 446–460 (1991)
3. Falin, G.I., Templeton, J.G.C.: Retrial Queues. Chapman and Hall, London (1997)
4. Gomez-Corral, A., Ramalhoto, M.: On the waiting time distribution and the busy period of a retrial queue with constant retrial rate. Stochast. Model. Appl. **3**, 37–47 (2000)
5. Kim, J., Kim, B.: A survey of retrial queueing systems. Ann. Oper. Res. **247**(1), 3–36 (2015). https://doi.org/10.1007/s10479-015-2038-7
6. Lee, S.W., Kim, B., Kim, J.: Analysis of the waiting time distribution in M/G/1 retrial queues with two way communication. Ann. Oper. Res. (1), 1–14 (2020). https://doi.org/10.1007/s10479-020-03717-2

7. Moiseeva, E., Nazarov, A.: Asymptotic Analysis of RQ-Systems M/Gi/1 on Heavy Load Condition. In: Proceedings of the IV International Conference Problems of Cybernetics and Informatics (PCI 2012), pp. 164–166. IEEE (2012)
8. Neuts, M.: The joint distribution of the virtual waiting time and the residual busy period for the M/G/1 queue. J. Appl. Probab. **5**, 224–229 (1968)
9. Nobel, R., Tijms, H.: Waiting-time probabilities in the M/G/1 retrial queue. Stat. Neerl. **60**(3), 73–78 (2006)
10. Phung-Duc, T.: Retrial queueing models: a survey on theory and applications. arXiv preprint arXiv:1906.09560, (2019)

Computational Algorithm for an Analysis of a Single-Line Queueing System with Arrived Alternating Poisson Flow

Alexander M. Andronov[1] , Iakov M. Dalinger[2] ,
and Nadezda Spiridovska[1(✉)]

[1] Transport and Telecommunication Institute, Lomonosov Street, 1,
Riga 1019, Latvia
spiridovska.n@tsi.lv
[2] Saint-Petersburg State University of Civil Aviation, Pilotov Street, 38,
Saint-Petersburg 196210, Russia
http://www.tsi.lv

Abstract. The $M/G/1$ queue is considered for a case when an alternating Poisson flow takes place on the input. The analysis is based on an embedded Markov chain, built on the instants of service ending. Various algorithms are elaborated for the calculation of the distribution of the system' states and various numerical indices.

Keywords: Poisson flow · Random environment · Pollaczek-Khinchine formula

1 Introduction

The Markov-modulated $M/G/1$ queueing system is the object of the consideration a long time [1–8]. The gotten results have usually an analytical character and their numerical realization encounters big difficulties. In the paper [6] only "The results come out as close matrix parallels of the Pollaczek-Khinchine formula without using transforms or complex variables." But the waiting time distribution is studied here only. It should be noted that the stationary case is considered usually.

We consider a simplified variant of the queueing system, when the alternating Poisson flow [9, 10], takes place on the input. It allows us to suggest the numerical algorithms for the calculation not only the mean indices (of the queue and the waiting time) but the corresponding stationary and non-stationary distributions too.

Also, there exists a random environment, in which the flow operates. The environment is described as continuous time alternating Markov chain (MC) $J(t) \in \{0,1\}$. The sojourn times in the alternating two states are independent random variables having the exponential distributions with parameters μ_0 and μ_1 correspondingly. If the ith state of the random environment takes place, that the Poisson flow with intensity λ_i arrives.

© Springer Nature Switzerland AG 2021
V. M. Vishnevskiy et al. (Eds.): DCCN 2021, LNCS 13144, pp. 140–152, 2021.
https://doi.org/10.1007/978-3-030-92507-9_13

The Poisson flow of claims enters on one-line queueing system with infinite queue. The service time has continuous distribution with density $b(t), t \geq 0$, and the finite mean b and the distribution function $B(t)$. Let $X(t)$ be the number of claims in the line at time t.

Our aim is the investigation of the distribution of the two-dimension process $Y(t) = (J(t), X(t)), t \geq 0$. We use the semiregenerative approach for that [11]. The embedded Markov chain, which is built on the time moments of services ending, is considered. The state of this chain is determined as the pair (i, n), where $i \in \{0, 1\}$ is the state of the environment, n is the number of customers, whose are remained in the system. The corresponding stationary distribution is calculated. Further some state (i, n) is fixed and there are calculated the expectations of sojourn time of the process $Y(t)$ in all own states, until the state (i, n) of embedded chain takes place. That allows calculating the distribution of the process $Y(t)$.

The paper is organized as follows. Preliminary knowledge's are gotten in section No. 2. Section No. 3 contains the analysis of the idle period of the queueing system. Embedded Markov chain, built on the instants of service ending, is considered in section No. 4. The stationary distribution of the process $Y(t) = (J(t), X(t))$ is presented in section No. 5. Section No. 6 contains a numerical example. Section No. 7 concludes the paper with final remarks.

2   Preliminary

We will use some results from the paper [10]. Let i and j be the initial and final states of MC $J(t)$ on interval $(0, t)$. The density $f_{i,j}(\tau, t)$ of the sojourn time $T_i(t)$ in the initial state i during time t jointly with probability that the final state equals j is calculated as follows for $0 < \tau < t$:

$$f_{i,j}(\tau, t) = \sum_{\eta=0}^{\infty} \mu_i \frac{1}{\eta! \eta!} \left(\mu_i \mu_j \tau(t-\tau) \right)^{\eta} \exp(-t\mu_j) \exp\left(-\tau(\mu_i - \mu_j) \right), j \neq i; \quad (1)$$

$$f_{i,i}(\tau, t) = \sum_{\eta=0}^{\infty} \frac{1}{(\eta+1)!} (\tau\mu_i)^{\eta+1} \exp(-\tau\mu_i) \mu_j \frac{1}{\eta!} \left(\mu_j(t-\tau) \right)^{\eta} \exp\left(-(t-\tau)\mu_j \right).$$

$$(2)$$

The last distribution has a singular component at the point t:

$$P\{T_i(t) = t | J(0) = i\} = \exp(-\mu_i t), t \geq 0. \quad (3)$$

It is the probability, that the initial state i takes place during the whole interval $(0, t)$.

Now we are able to calculate the expectation of the sojourn time in the initial state i:

$$E(T_i(t) | J(0) = i) = \int_0^t \tau \left(f_{i,not(i)}(\tau, t) + f_{i,i}(\tau, t) \right) d\tau + t \exp(-\mu_i t). \quad (4)$$

The stationary probabilities of the states $\pi_i = \lim_{t \to \infty} P\{J(t) = i\}, i = 0, 1$, are calculated as follows:

$$\pi_0 = \frac{\mu_1}{\mu_0 + \mu_1}, \ \pi_1 = \frac{\mu_0}{\mu_0 + \mu_1}. \tag{5}$$

Arrived Poisson flow $N(t)$ has intensity λ_i if the external random environment has state $i = 0, 1$. Therefore the stationary intensity of customers arrivals, conditioned by the ith state of the random environment, is $\Lambda_i = \pi_i \lambda_i$, $i = 0, 1$. Let $\Lambda = \Lambda_0 + \Lambda_1$ be the total stationary intensity of the arrivals.

The load coefficient is calculated as follows:

$$\rho = \Lambda b = (\pi_0 \lambda_0 + \pi_1 \lambda_1) b = (\mu_1 \lambda_0 + \mu_0 \lambda_1) \frac{b}{\mu_0 + \mu_1}. \tag{6}$$

We suppose that $\rho < 1$, so the stationary distribution exists.

Also $N(t)$ means a number of arrivals on the interval $(0, t)$. Formulas (1) - (3) allow to calculate the probability $P_{i,j}(n, t)$ of n arrivals during interval $(0, t)$ and to have state j at the instant t, if initially state i takes place. Actually, for $n = 0, 1, \ldots, t \geq 0$:

$$P_{i,j}(n, t) = \int_0^t f_{i,j}(\tau, t) \frac{1}{n!} \left(\lambda_i \tau + \lambda_j (t - \tau) \right)^n \exp(-(\lambda_i \tau + \lambda_j (t - \tau))) d\tau$$

$$+ \delta(i, j) \exp(-t\mu_i) \frac{1}{n!} (\lambda_i t)^n \exp(-\lambda_i t), \tag{7}$$

where

$$\delta(i, j) = \begin{bmatrix} 0, & \text{if} \ \ i \neq j, \\ 1, & \text{if} \ \ i = j. \end{bmatrix} \tag{8}$$

Finally, the probability $P_i(n, t)$ of n arrivals during interval $(0, t)$ is calculated as

$$P_i(n, t) = P_{i,i}(n, t) + P_{i,not(i)}(n, t), t \geq 0, n = 0, 1, \ldots, \tag{9}$$

where $not(i) = 1 - i, i = 0, 1$.

Now we can calculate various numerical indices. For example, the expectation jointly with probability that the last state equals j:

$$E_{i,j}(N(t)) = \mu_i \exp(-t\mu_j) \sum_{\eta=0}^{\infty} \frac{1}{\eta! \eta!} (\mu_i \mu_j)^\eta$$

$$\times \int_0^t (\tau(t - \tau))^\eta (\lambda_i \tau + \lambda_j (t - \tau)) \exp\left(-\tau(\mu_i - \mu_j) \right) d\tau, j \neq i, \tag{10}$$

$$E_{i,i}(N(t)) = \exp(-t\mu_{not(i)}) \sum_{\eta=0}^{\infty} \frac{1}{(\eta+1)!\eta!} (\mu_i\mu_{not(i)})^{\eta+1}$$

$$\times \int_0^t (t-\tau)^\eta \tau^{\eta+1}(\lambda_i\tau + \lambda_{not(i)}(t-\tau)) \exp\left(-\tau(\mu_i - \mu_{not(i)})\right)d\tau \tag{11}$$

$$+\lambda_i t \exp(-t\mu_i).$$

The full expectation is as follows:

$$E_i(N(t)) = E_{i,0}(N(t)) + E_{i,1}(N(t)). \tag{12}$$

3 Analysis of the Idle Period

Our main process $Y(t)$ is described by two components: the state $J(t)$ of the random environment at the time t and the number $X(t)$ of claims in the line: $Y(t) = (J(t), X(t))$. This process has two stages: the idle stage, when the customers absent ($X(t) = 0$) and the busy stage when some customer is served ($X(t) > 0$).

Let us calculate the density $h_{i,j}(t)$ of the idle time T jointly with the probability that the jth state of random environment takes place at the instant of the idle period' ending, if initially state i takes place. It follows from the formulas (1) and (2):

$$h_{i,j}(t) = \lambda_j \exp(-\lambda_j t) \int_0^t f_{i,j}(\tau,t) \exp\left(-\tau(\lambda_i - \lambda_j)\right)d\tau, i \neq j, \tag{13}$$

$$h_{i,i}(t) = \lambda_i \exp(-\lambda_j t) \int_0^t f_{i,i}(\tau, t) \exp\left(-\tau(\lambda_i - \lambda_j)\right)d\tau. \tag{14}$$

Let

$$h_i(t) = h_{i,i}(t) + h_{i,not(i)}(t). \tag{15}$$

The last formula allows to calculate the mean time of the idle period, if the environment has the ith state initially:

$$E(T|J(0) = i) = \int_0^{\infty} th_i(t)dt. \tag{16}$$

Additionally to formula (16), we will use the mean sojourn time TS in the initial state i during the idle period. We denote $E(TS|J(0) = i)$ this mean time. Obviously:

$$E(TS|J(0) = i) = \int_0^{\infty} P_{i,i}(0,t)dt. \tag{17}$$

The mean sojourn time in the opposite state during the idle period equals $E(T|J(0) = i) - E(TS|J(0) = i)$.

4   Embedded Markov Chain

We consider the discrete time Markov chain, which is built on the instants of service ending. We call "steps" these instants. Let $Pr_{n,m}(i,j)$ be the probability that the random environment has state j and the number of claims equals m if on previous step the state i takes place and the number of the claims equals n. We have for $n > 0$:

$$Pr_{n,m}(i,j) = \int_0^\infty P_{i,j}(m - n + 1, t)b(t)dt, n > 0, m \geq n - 1. \tag{18}$$

Now we consider the case $n = 0$. Let $Ptr(i,j)$ be the probability, that the idle period is ended in the state j, if its initial state was i:

$$Ptr(i,j) = \int_0^\infty h_{i,j}(t)dt, i, j = 0, 1. \tag{19}$$

Note that $Ptr(i,0) + Ptr(i,1) = 1$.
Further

$$Pr_{0,m}(i,j) = Ptr(i,0) \int_0^\infty P_{0,j}(m,t)b(t)dt + Ptr(i,1) \int_0^\infty P_{1,j}(m,t)b(t)dt, m \geq 0. \tag{20}$$

Now we give the matrix representation for the last formula. The odd rows and columns corresponds to the value 1 of i and j, the even rows and columns - to value 0 of i and j. Namely, for $n, m = 0, 1, \ldots$:

$$\begin{aligned} MPr_{2n+1,2m+1} &= Pr_{n,m}(1,1) \\ MPr_{2n+1,2m} &= Pr_{n,m}(1,0) \\ MPr_{2n,2m+1} &= Pr_{n,m}(0,1) \\ MPr_{2n,2m} &= Pr_{n,m}(0,0) \end{aligned} \tag{21}$$

The stationary distribution of the states' probabilities is presented by means of the eigenvector of matrix MPr^T which corresponds to the unit eigenvalue. We denote the corresponding eigenvector by $\psi = (\psi_0, \psi_1, \ldots)$. Normalizing this vector, we have the vector of the stationary probabilities:

$$q = (q_0, q_1, \ldots) = \left(\sum_j \psi_j \right)^{-1} \psi, \tag{22}$$

where q_{2m+j} is the stationary probability that state j and m claims take place when a service is completed.

5 Stationary Distribution of the Process $Y(t) = (J(t), X(t))$

Now our aim is to calculate the stationary distribution of the state probabilities for the continuous time. Let $p_{j,m}$ be the stationary probability of the state for the process $Y(t) = (J(t), X(t))$:

$$p_{j,m} = \lim_{t \to \infty} P\{J(t) = j, X(t) = m\}.$$

We begin with states $(0, 0)$ and $(1, 0)$ of the idle period. The mean sojourn time $E(TS|J(0) = j)$ in the initial state j during the idle period is calculated by formula (17). The mean sojourn time $TI(j)$ in the state $(j, 0)$ during the idle period is calculated as follows:

$$E(TI(j)) = \frac{q_j}{q_j + q_{1-j}} E(TS|J(0) = j) + \frac{q_{1-j}}{q_j + q_{1-j}} (E(T|J(0) = 1 - j) - E(TS|J(0) = 1 - j)).$$

Therefore, the stationary probability of the state $(j, 0)$ is the following:

$$p_{j,0} = \frac{E(TI(j))}{E(TI(j)) + E(TI(1 - j))} (1 - \rho), j = 0, 1. \tag{23}$$

The stationary mean time of the idle period is calculated as follows:

$$E(T) = E(TI(0)) + E(TI(1)). \tag{24}$$

Now we can calculate the mean time $E(W)$ of the busy period. Because

$$\rho = \frac{E(W)}{E(T) + E(W)},$$

then

$$E(W) = \frac{\rho}{1 - \rho} E(T),$$

where the load coefficient is calculated by the formula (6).

The mean time of the whole cycle is calculated as follows:

$$E(H) = E(T) + E(W) = \frac{1}{1 - \rho} E(T). \tag{25}$$

Further we consider the general case. Let state $J(0) = i, X(0) = n, n \geq 0$, takes place in the instant, when a service is completed. We denote $\gamma_{i,n}(j, m), m \geq n, j = 0, 1$, the mean time, when the line contents m claims and the environment has the jth state during a new service. We have for $n > 0$:

$$\gamma_{i,n}(j, m) = \int_0^\infty P_{i,j}(m - n, t)(1 - B(t)) dt, j = 0, 1, m = n, n + 1, \ldots. \tag{26}$$

Now we consider the case $n = 0$. If $m = 0$, then

$$\gamma_{i,0}(j,0) = \left[\begin{array}{ll} E(TS|J(0) = i, & \text{if } i = j, \\ E(T|J(0) = i) - E(TS|J(0) = i), & \text{if } i \neq j. \end{array} \right. \tag{27}$$

If $m > 0$, then

$$\gamma_{i,0}(j,m) = Ptr(i,0)\gamma_{0,1}(j,m) + Ptr(i,1)\gamma_{1,1}(j,m), i,j = 0,1. \tag{28}$$

We have for states $(j,0)$ of the idle period:

$$p_{j,0} = \frac{1}{E(H)}(q_j\gamma_{j,0}(j,0) + q_{1-j}\gamma_{1-j,0}(j,0)), j = 0,1.$$

The total formula is the following:

$$p_{j,m} = \frac{1}{E(H)}\sum_{n=0}^{m}\sum_{i=0}^{1} q_{2n+i}\gamma_{i,n}(j,m), j = 0,1, m = 0,1,\ldots. \tag{29}$$

The last formula allows calculating the stationary distribution of the number of the claims in the system:

$$Pr_m = \lim_{t\to\infty} P\{X(t) = m\} = p_{0,m} + p_{1,m}, m = 0,1,\ldots. \tag{30}$$

A checking of the calculation's correctness is the currying-out of the equalities

$$\sum_{m=0}^{\infty} p_{j,m} = \pi_j, j = 0,1, \tag{31}$$

$$\sum_{m=1}^{\infty} Pr_m = \rho. \tag{32}$$

6   Example

Practical calculations imply a substitute of the infinite number of addends in sums and the infinite upper limit in integrals by finite values. This number is denoted as $nmax$ and $tmax$ correspondingly. It is set in such a way that its increase doesn't change a gotten result. We set for considered example $nmax = 13$ and $tmax = 12$.

The following initial data is fixed for the input flow: $\lambda = \left(\begin{smallmatrix}1\\2\end{smallmatrix}\right)$, $\mu = \left(\begin{smallmatrix}0.2\\0.3\end{smallmatrix}\right)$ The service time has symmetric triangular distribution [12] with the mean $b = 0.55$:

$$b(t) = \left[\begin{array}{ll} b^{-2}t, & 0 \leq t \leq b, \\ b^{-2}(2b - t), & b \leq t \leq 2b, \\ 0, & t \notin (0, 2b), \end{array} \right.$$

$$
B(t) = \left[
\begin{array}{ll}
0, & t < 0, \\[2mm]
\dfrac{1}{2} b^{-2} t^2, & 0 \le t \le b, \\[3mm]
b^{-2} \left(2bt - \dfrac{1}{2} t^2 \right) - 1, & b \le t \le 2b, \\[3mm]
1, & t \ge 2b.
\end{array}
\right.
$$

The stationary probabilities of the states of the random environment (5) have the following values:

$$
\pi_0 = \frac{\mu_1}{\mu_0 + \mu_1} = 0.6, \quad \pi_1 = \frac{\mu_0}{\mu_0 + \mu_1} = 0.4.
$$

The mean intensity of the arrived flow

$$
\Lambda = \pi_0 \lambda_0 + \pi_1 \lambda_1 = \frac{1}{\mu_0 + \mu_1} (\mu_1 \lambda_0 + \mu_0 \lambda_1) = 1.4.
$$

The load coefficient is calculated by formula (6):

$$
\rho = (\mu_1 \lambda_0 + \mu_0 \lambda_1) \frac{b}{\mu_0 + \mu_1} = 0.77.
$$

We see that $\rho < 1$, so the stationary distribution exists.

The densities $f_{i,j}(\tau, t)$ of the sojourn time $T_i(t)$ in the initial state $i = 1$ during time $t = 40$ jointly with the probability of the final state j are presented on Fig. 1.

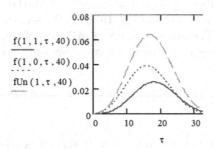

Fig. 1. Graphs of the densities $f_{1,j}(\tau, 40)$

The graphs of the probabilities $P_i(n, t)$ (see formula (9)) are presented on Fig. 2. Here $Pr(i, n, t, nmax) = P_i(n, t), i = 0$ or $1, t = 3$ or $t = 10$, and $nmax = 10$.

The densities $h(t, i, j) = h_{i,j}(t)$ of idle time for initial state $i = 0$ jointly with probability of final state j are presented on Fig. 3.

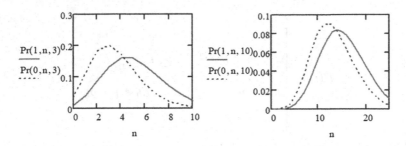

Fig. 2. The graphs of the probabilities

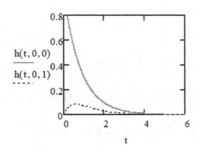

Fig. 3. The densities $h(t, 0, j)$

Formula (16) gives the following values for the mean time of the idle period, if the environment has the ith state initially: $E(T|J(0) = 0) = 0.926, E(T|J(0) = 1) = 0.556$.

Let us present matrices $Pr(i, j) = (Pr_{n,m}(i, j)), i, j = 0, 1$, these describe the transition probabilities (20) of the embedded Markov chain. The five rows of matrix $Pr(0, 0)$ are presented in the Table 1. We see, that the zero row differs from other rows only. The other rows are translation of the first row to the right. Therefore these matrices can present by two rows only, see Tables 2, 3 and 4.

Table 1. The five firstly rows of the matrix $Pr_{n,m}(0,0)$

n\m	0	1	2	3	4	5	6
0	0.468	0.233	0.071	0.016	0.003	5.816e−4	9.597e−5
1	0.540	0.266	0.079	0.018	0.003	5.716e−4	8.273e−5
2	0.000	0.540	0.266	0.079	0.018	0.003	5.716e−4
3	0.000	0.000	0.540	0.266	0.079	0.018	0.003
4	0.000	0.000	0.000	0.540	0.266	0.079	0.018

Table 2. The two firstly rows of the matrix $Pr_{n,m}(0,1)$

n\m	0	1	2	3	4	5	6
0	0.081	0.069	0.036	0.014	4.644e−3	1.363e−3	3.426e−4
1	0.039	0.032	0.015	0.006	1.646e−3	4.837e−4	1.147e−4

Table 3. The two firstly rows of the matrix $Pr_{n,m}(1,0)$

n\m	0	1	2	3	4	5	6
0	0.112	0.072	0.029	0.001	2.552e−3	6.318e−4	1.622e−4
1	0.059	0.048	0.023	0.008	2.468e−3	6.393e−4	1.721e−4

Table 4. The two firstly rows of the matrix $Pr_{n,m}(1,1)$

n\m	0	1	2	3	4	5	6
0	0.293	0.256	0.139	0.058	0.020	5.762e−3	1.482e−3
1	0.325	0.284	0.155	0.064	0.022	6.422e−3	1.653e−3

Now formula (21) allows calculating matrix MPr. Formula (22) gives stationary distribution of the states probabilities for the embedded Markov chain. It is presented on the Table 5.

Table 5. The Stationary probabilities of the states of the embedded Markov chain

State (j,m)	$(0,0)$	$(1,0)$	$(0,1)$	$(1,1)$	$(0,2)$	$(1,2)$
Probability q_{j+2m}	0.146	0.059	0.123	0.078	0.082	0.071
State (j,m)	$(0,3)$	$(1,3)$	$(0,4)$	$(1,4)$	$(0,5)$	$(1,5)$
Probability q_{j+2m}	0.055	0.058	0.039	0.045	0.029	0.035
State (j,m)	$(0,6)$	$(1,6)$	$(0,7)$	$(1,7)$	$(0,8)$	$(1,8)$
Probability q_{j+2m}	0.022	0.027	0.017	0.020	0.013	0.016
State (j,m)	$(0,9)$	$(1,9)$	$(0,10)$	$(1,10)$	$(0,11)$	$(1,11)$
Probability q_{j+2m}	9.603e−3	0.012	7.136e−3	9.79e−−3	5.045e−3	7.899e−3

Now we can calculate different indices. The mean time of the idle period is calculated by formula (24):

$$E(T) = \frac{q_0}{q_0 + q_1} E(T|J(0) = 0) + \frac{q_1}{q_0 + q_1} E(T|J(0) = 1)$$

$$= \frac{0.146}{0.146 + 0.059} \times 0.926 + \frac{0.059}{0.146 + 0.059} \times 0.556 = 0.819.$$

The mean time $E(W)$ of the busy period is the following:

$$E(W) = \frac{\rho}{1 - \rho} E(T) = \frac{0.77}{1 - 0.77} 0.819 = 2.741.$$

Therefore, these sum $E(H) = E(T) + E(W) = 3.56$.

The stationary probabilities of the states for the process $Y(t) = (J(t), X(t))$ are presented in the Table 6.

Table 6. The Stationary probabilities of the states $\{p_{j,m}\}$

State (j, m)	$(0,0)$	$(1,0)$	$(0,1)$	$(1,1)$	$(0,2)$	$(1,2)$
Probability	0.179	0.051	0.144	0.069	0.087	0.064
State (j, m)	$(0,3)$	$(1,3)$	$(0,4)$	$(1,4)$	$(0,5)$	$(1,5)$
Probability	0.054	0.052	0.037	0.041	0.027	0.031
State (j, m)	$(0,6)$	$(1,6)$	$(0,7)$	$(1,7)$	$(0,8)$	$(1,8)$
Probability	0.020	0.024	0.015	0.018	0.012	0.014
State (j, m)	$(0,9)$	$(1,9)$	$(0,10)$	$(1,10)$	$(0,11)$	$(1,11)$
Probability	8.754e−3	0.011	6.583e−3	8.628e−3	4.8e−3	6.844e−3

We see that the stationary probability of the idle system gives the right value: $0.179 + 0.051 = 1 - \rho = 0.23$. The tests (31) and (32) are fulfilled.

Mean number of the claims in the system equals 2.77. It is interesting to compare it with the case, when the random environment absents and the arrival Poisson flow has the constant intensity 1.4, how it takes place in our example. Let us use Pollaczek-Khintchine formula [13] for that:

$$E(X) = \lambda^2 b_2 \frac{1}{2(1 - \rho)} + \rho,$$

where b_2 is the second moment of the service time.

For our initial data

$$b_2 = \int_0^\infty t^2 b(t) dt = 0.353$$

and we have the expected result: $E(X) = 2.274$.

Therefore, the factor of the random intensity increases the mean number of the customers on

$$\frac{2.77 - 2.274}{2.274} \times 100\% = 21.8\%.$$

Simultaneously, we can control the precision of our calculations. Our calculations for $\lambda_0 = \lambda_1 = 1.4$ give that mean number of the customers in the system equals 2.239. The mistake equals

$$\frac{2.274 - 2.239}{2.274} \times 100\% = 1.5\%.$$

It verifies the authenticity of the presented results.

7   Conclusions

In this work, we have presented a new numerical algorithms for calculating the main characteristics of the queueing system $M/G/1$, which has an alternating Poisson flow on the input. The elaborated algorithms can be used in modern telecommunication and data processing systems, performing a processing of calls or data packages in the random external environment. The authors intent to continue current investigation and to transfer the used approach on the $G/G/1$ queue.

References

1. van Hoorn, M.H., Seelen, L.P.: The SPP/G/1 queue: a single server queue with a switched Poisson process as input process. Oper. Res. Spektrum **5**(4), 207–218 (1983). https://doi.org/10.1007/BF01719844
2. Regterschot, G.J.K., De Smit, J.H.A.: The queue M/G/1 with Markov modulated arrivals and services. Math. Oper. Res. **11**(3), 465–483 (1986)
3. Neuts, M.F.: Structured stochastic matrices of MG-1 type and their applications. Dekker (1989)
4. Rossiter, M.H.: The switched Poisson process and the SPP/G/1 queue. In: Bonatti, M. (ed.) ITC-12, pp. 1406–1412. North-Holland (1989)
5. Takine, T., Takahashi, Y.: On the relationship between queue lengths at a random instant and at a departure in the stationary queue with BMAP arrivals. Communications in statistics. Stochast. Models **14**(3), 601–610 (1998)
6. Asmussen, S.: Ladder heights and the Markov-modulated M/G/1 queue. Stochast. Processes Appl. **37**(2), 313–326 (1991)
7. Takine, T.: A new recursion for the queue length distribution in the stationary BMAP/G/1 queue. Stoch. Model. **16**(2), 335–341 (2000)
8. Du, Q.: A monotonicity result for a single-server queue subject to a Markov-modulated Poisson process. J. Appl. Probability, 1103–1111 (1995)
9. Fischer, W., Meier-Hellstern, K.: The Markov-modulated Poisson process (MMPP) cookbook. Perform. Eval. **18**(2), 149–171 (1993)
10. Andronov, A.M., Dalinger, I.M.: Poisson flows with alternating intensity and their application. Autom. Control. Comput. Sci. **54**(5), 403–411 (2020). https://doi.org/10.3103/S0146411620050028

11. Tang, L.C., Prabhu, N.U., Pacheco, A.: Markov-modulated processes and semiregenerative phenomena. World Scientific (2008)
12. Sleeper, A.: Six sigma distribution modeling. McGraw Hill Professional (2007)
13. Gnedenko, B.V., Kovalenko, I.N.: Introduction to queueing theory. Birkhäuser (1989)

Analysis of a Batch Service Queueing System Associated with Inventory Transport

Achyutha Krishnamoorthy[1]($\boxtimes$) and Anu Nuthan Joshua[2,3]

[1] Centre for Research in Mathematics, CMS College, Kottayam 686001, India
[2] Department of Mathematics, Union Christian College, Aluva 683102, India
[3] Department of Mathematics, Cochin University of Science and Technology, Cochin 682022, Kerala, India

Abstract. This paper considers a single server batch service queueing system in which service is provided in at most k stages. A batch of customers is admitted for service under GBS - rule (with minimum batch size a and maximum batch size b). The customers have the choice to leave the system after service completion from any stage. The service time of a customer in a stage depends on the stage as well as the number of customers being served in that stage. The stability of such a system is investigated and key performance measures computed based on system state distribution. The time required to provide service to a batch of specified size is analyzed. An optimization problem is considered and a numerical example provided.

Keywords: Batch service · GBS -rule · Inventory transport · k-stage queueing system

1 Introduction

In this paper, a queueing problem which arises in the transport of inventory, encountered by most e-commerce firms is considered. Due to restrictions imposed by cost and time all items to be shipped from a fixed source in a particular route will be shipped together, in batches of a certain minimum batch size (which forms the lower threshold a). Also, there will be a certain capacity restriction for the transport vessel(server), so the maximum batch size for service (which forms the upper threshold b) should be specified. So if the source is denoted by node 0, the terminal node(the last station in this route) by $k + 1$, the inventory will be delivered at nodes $1, 2 ..., k$, in addition to the terminal node. It is of interest to know the time required to provide service to a batch of a particular size. For simplicity, we have modeled the problem as a single server queueing system, though in real life it is a multiserver queueing system. The demands for inventory at the source are arrivals to this queueing system. The service is assumed to be provided in at most k stages as the service ends when the entire

© Springer Nature Switzerland AG 2021
V. M. Vishnevskiy et al. (Eds.): DCCN 2021, LNCS 13144, pp. 153–166, 2021.
https://doi.org/10.1007/978-3-030-92507-9_14

inventory is delivered. Inventory will be delivered at nodes provided at the end of a stage. Though the literature on batch service queueing systems is extensive (an outline of these can be found in [1,2]), the batch service model discussed in this paper is not analyzed yet. The problem considered here can be regarded as an inverse to the problem considered in [3].

2   Model Description and Formulation

Consider a single server system in which the arrivals are governed by a Poisson process with rate λ. Service is provided in batches according to GBS - rule - i.e., if at least a customers are waiting in queue (but not more than the maximum service batch size b), all of them are taken for batch service. If more than b customers are waiting, b are taken for service and the rest of the customers (if any) have to wait in the queue. The service in stage i, if j customers are in the batch at the beginning of the stage i, is exponential with rate μ_i^j. There can be at most k stages for service. The above queueing system can be pictorially represented as follows (Fig. 1):

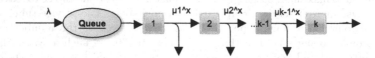

Fig. 1. Queueing system with k stages of service ('x' is used to denote the batch size as it is a variable)

The system described above can be studied as a Markov chain $(N_q(t), N_s(t), P(t))$, where $N_q(t), N_s(t), P(t)$ respectively denote the number of customers in queue, number of customers undergoing a batch service, and the stage in which the service process is in, at time t.

The state-space of such a system is $\Delta = \{(n_q), 0 \leq n_q \leq a-1\}^1 \cup \{(n_q, n_s, p); n_q \geq 0, 1 \leq n_s \leq a-1, 2 \leq p \leq k\}^2 \cup \{(n_q, n_s, p); n_q \geq 0, a \leq n_s \leq b, 1 \leq p \leq k\}^3$. The first set of states indicate that the server is idle while the other two sets indicate that the service process is on, with the difference that states in the second set are reached by the departure of customers from any of the intermediate stages.

The transitions and corresponding transition rates are given as follows:

1. **Transitions due to arrival**

$n_q \rightarrow n_q + 1; 0 \leq n_q \leq a-2$, with rate λ
$n_q \rightarrow (0, a, 1); n_q = a-1$, with rate λ
$(n_q, n_s, p) \rightarrow (n_q + 1, n_s, p); 1 \leq n_s \leq b, 1 \leq p \leq k$, with rate λ

2. **Transitions due to completion of service in stage p; $p \neq k$ with customers not leaving the system thereafter**

$$(n_q, n_s, p) \rightarrow (n_q, n_s, p+1); 1 \leq n_s \leq b, 1 \leq p \leq k, \text{ with rate } \mu_p^{n_s}$$

3. **Transitions due to completion of service in stage p with j customers leaving the system thereafter**

$(n_q, n_s, p) \rightarrow (n_q, n_s - j, p+1); 1 \leq n_s \leq b, 1 \leq p \leq k, \text{ with rate } \mu_p^{n_s}$ if $j \neq n_s; p \neq k$

$(n_q, n_s, p) \rightarrow (n_q - b, b, 1); 1 \leq n_s \leq b, 1 \leq p \leq k, \text{ with rate } \mu_p^{n_s}$ if $j = n_s$ and $n_q \geq b$

$(n_q, n_s, p) \rightarrow (0, n_q, 1); 1 \leq n_s \leq b, 1 \leq p \leq k, \text{ with rate } \mu_p^{n_s}$ if $j = n_s$ and $n_q \geq a$

$(n_q, n_s, p) \rightarrow (n_q); 1 \leq n_s \leq b, 1 \leq p \leq k, \text{ with rate } \mu_p^{n_s}$ if $j = n_s$ and $n_q < a$

The infinitesimal generator matrix of this Markov chain when the level is redefined using a (i.e., $l(\underline{0}) = \{l(0), l(1), l(2)...l(a-1)\}$, $l(\underline{1}) = \{l(a), l(a+1), l(a+2)...l(2a-1)\}$, $l(\underline{2}) = \{l(2a), l(2a+1), l(a+2)...l(3a-1)\}$ and so on) is given by

$$Q = \begin{bmatrix} A_{00} & A_{01} & & & & \\ A_{10} & A_1 & A_0 & & & \\ A_{20} & A_2 & A_1 & A_0 & & \\ A_{30} & A_3 & A_2 & A_1 & A_0 & \\ A_{40} & A_4 & A_3 & A_2 & A_1 & A_0 \\ & & \ddots & & \ddots & \ddots \end{bmatrix}$$

The generator matrix is of $GI/M/1$ type queue.

Define Let $s' = \frac{a+b}{a}$, if it is an integer and $s' = \lceil \frac{a+b}{a} \rceil + 1$, otherwise.

Then,

$$A_{i0} = \mathbf{0}; i \geq s', A_i = \mathbf{0}; i > s'$$

$$A_0 = \begin{bmatrix} 0 & 0 & & \ldots & 0 \\ 0 & 0 & & \ldots & 0 \\ \vdots & & \ddots & \ddots & \vdots \\ B_0 & 0 & \ldots & & 0 \end{bmatrix}$$

$$B_0 = \lambda I$$

$$A_1 = \begin{bmatrix} B_1 & B_0 & & \ldots & 0 \\ 0 & B_1 & B_0 & \ldots & 0 \\ \vdots & & \ddots & \ddots & \\ & & & B_1 & B_0 & 0 \\ 0 & \ldots & & & B_1 & B_0 \end{bmatrix}$$

$A_2, A_3, ..., A_{s'}$ are matrices such that $A_2 + A_3 + ... + A_{s'}$ have a block B_2 in each row, while all other entries are $\mathbf{0}$ and $A_0 + A_1 + A_2 + A_3 + ... + A_{s'}$ is a block circulant matrix.

$$B_2 = \begin{bmatrix} 0 & \cdots & e'_k(1) \otimes D_1 \\ 0 & \cdots & e''_k(1) \otimes D_2 \\ \vdots & & \vdots \\ 0 & & e'_k(1) \otimes D_{k-1} \\ 0 \cdots & & e'_k(1) \otimes D_k \end{bmatrix}$$

$$D_j = [\mu_2^j, \mu_3^j, ... \mu_k^j]'; j < a$$

and

$$D_j = [\mu_1^j, \mu_2^j, \mu_3^j, ... \mu_k^j]'$$

$$\mathcal{B}_1 = \begin{bmatrix} C_1^1 & & & & & & \\ C_3^2 & C_1^2 & & & & & \\ C_3^3 & C_3^3 & C_1^2 & & & & \\ \vdots & \vdots & \vdots & \ddots & & & \\ C_3^{a-1} & C_3^{a-1} & C_3^{a-1} & \cdots & C_1^{a-1} & & \\ C_4^a & C_4^a & C_4^a & \cdots & C_4^a & C_2^a & \\ C_4^{a+1} & C_4^{a+1} & C_4^{a+1} & \cdots & C_4^a & C_5^{a+1} & C_5^{a+1} \\ \vdots & \vdots & \vdots & & \vdots & & \ddots \\ C_4^b & C_4^b & C_4^b & \cdots & C_4^a & C_5^b & \cdots C_2^b \end{bmatrix} - \lambda I$$

$$\mathcal{C}_1^j = \begin{bmatrix} -(j+1)\mu_2^j & (j+1)\mu_2^j & & \\ & -(j+1)\mu_3^j & (j+1)\mu_3^j & \\ & & \ddots & \ddots & \\ & & & & -(j+1)\mu_k^1 \end{bmatrix}$$

$$\mathcal{C}_2^j = \begin{bmatrix} -(j+1)\mu_1^j & \mu_1^j & & \\ & -(j+1)\mu_2^j & \mu_2^j & \\ & & \ddots & \ddots \\ & & & -(j+1)\mu_k^j \end{bmatrix}$$

$$\mathcal{C}_3^j = \begin{bmatrix} \mu_1^j & & & \\ & \mu_2^j & & \\ & & \ddots & \\ & & & \mu_{k-1}^j \end{bmatrix}$$

$$\mathcal{C}_4^j = \begin{bmatrix} C_3^j \\ \mathbf{0} \end{bmatrix}$$

$$\mathcal{C}_5^j = \begin{bmatrix} 0 & C_4^j \end{bmatrix}$$

For instance, in the case when $a = 2, b = 4, k = 4$, the state space is

$$\{0, 1\} \cup \{(n_q, 1, p); n_q \geq 0, 2 \leq p \leq 4\}^2 \cup \{(n_q, n_s, p); n_q \geq 0, 2 \leq n_s \leq 4, 1 \leq p \leq 4\}^3$$

The generator matrix is of the form,

$$\mathcal{Q}_1 = \begin{bmatrix}
A_{00} & A_{01} & & & & \\
A_{10} & A_1 & A_0 & & & \\
A_{20} & 0 & A_1 & A_0 & & \\
0 & A_3 & 0 & A_1 & A_0 & \\
0 & 0 & A_3 & 0 & A_1 & A_0 \\
& & & \ddots & & \ddots & \ddots
\end{bmatrix}$$

$$A_{00} = \begin{bmatrix} B_{10} & \lambda I \\ 0 & B_{10} \end{bmatrix}, A_{10} = \begin{bmatrix} B_{20} & 0 \\ B_{30} & 0 \end{bmatrix}, A_{20} = \begin{bmatrix} B_{40} & 0 \\ 0 & B_{40} \end{bmatrix}$$

$$A_0 = \begin{bmatrix} 0 & 0 \\ B_0 & 0 \end{bmatrix}, A_1 = \begin{bmatrix} B_1 & B_0 \\ 0 & B_1 \end{bmatrix}, A_2 = 0, A_3 = \begin{bmatrix} B_2 & 0 \\ 0 & B_2 \end{bmatrix}.$$

The individual matrices are of the form,

$$B_0 = \lambda I$$

$$\mathcal{B}_1 = \begin{bmatrix}
C_1^1 - \lambda I & & & \\
C_4^2 & C_2^2 - \lambda I & & \\
C_4^3 & C_5^3 & C_5^3 - \lambda I & \\
C_4^4 & C_4^4 & C_5^4 & C_2^4 - \lambda I
\end{bmatrix}$$

where,

$$C_1^1 = \begin{bmatrix}
-2\mu_2^1 & \mu_2^1 & \\
& -2\mu_3^1 & \mu_3^1 \\
& & -2\mu_4^1
\end{bmatrix}$$

$$C_2^j = \begin{bmatrix}
-(j+1)\mu_1^j & \mu_1^j & & \\
& -(j+1)\mu_2^j & \mu_2^j & \\
& & -(j+1)\mu_3^j & \mu_3^j \\
& & & -(j+1)\mu_4^j
\end{bmatrix}$$

$$C_4^j = \begin{bmatrix}
\mu_1^j & & \\
& \mu_2^j & \\
& & \mu_3^j \\
0 & 0 & 0
\end{bmatrix}$$

$$C_5^j = [0 | C_4^j]$$

$$\mathcal{B}_2 = \begin{bmatrix} 0\ 0\ 0\ e_4'(1) \otimes D_1 \\ 0\ 0\ 0\ e_4'(1) \otimes D_2 \\ 0\ 0\ 0\ e_4'(1) \otimes D_3 \\ 0\ 0\ 0\ e_4'(1) \otimes D_4 \end{bmatrix}$$

,

$$D_1 = [\mu_2^1, \mu_3^1, \mu_4^1]', D_j = [\mu_1^j, \mu_2^j, \mu_3^j, \mu_4^j]'$$

for $j = 2, 3, 4$

2.1   Stability Condition

Theorem 1. *The queueing system under consideration is stable iff*

$$\lambda < b.[\sum_{j=a}^{b}\sum_{i=1}^{k} \pi_{\tilde{i},j}\mu_i^j + \sum_{j=1}^{a-1}\sum_{i=2}^{k} \pi_{\tilde{i},j}\mu_i^j]$$

Proof. Let $\pi = (\pi_1, \pi_2, \ldots, \pi_a)$ denote the steady-state probability vector of the generator matrix, $A = \sum_{i=0}^{s'} A_i$. Then, π satisfies

$$\pi A = \mathbf{0} \text{ and } \pi e = 1.$$

A is a block circulant matrix and hence

$$\pi = \frac{1}{a}(\mathbf{e}'(a) \otimes \tilde{\pi}),$$

where $\tilde{\pi}$ is a row vector that satisfies

$$\tilde{\pi}(\sum_{i=0}^{2} B_i) = \mathbf{0}.$$

The queueing system is stable (see Neuts [4]) if and only if

$$\pi A_0 e < \sum_{i=2}^{i=\infty}(i-1)\pi A_i e$$

or

$$\frac{1}{a}\tilde{\pi}B_0 e < \frac{b}{a}\tilde{\pi}B_2 e$$

,

$$\tilde{\pi}B_0 e = \tilde{\pi}\lambda I e = \lambda$$

$$\tilde{\pi}B_2 e = \sum_{j=a}^{b}\sum_{i=1}^{k} \pi_{\tilde{i},j}\mu_i^j + \sum_{j=1}^{a-1}\sum_{i=2}^{k} \pi_{\tilde{i},j}\mu_i^j$$

2.2   Steady-State Probability Vector

Theorem 2. *The steady-state probability vector x can be computed as follows:*

$$\mathbf{x_i} = \mathbf{x_1} R^{i-1}, i \geq 2,$$

where R is the minimal non-negative solution to the equation

$$\sum_{i=0}^{s'} R^i A_i = \mathbf{0}.$$

and the vectors x_0 and x_1 satisfy

$$\sum_{i=0}^{s'-1} x_i A_{i0} = \mathbf{0},$$

$$\mathbf{x_0} A_{01} + \sum_{i=1}^{s'} x_i A_i = \mathbf{0},$$

subject to $xe = 1$, gives

$$\mathbf{x_0} e + \mathbf{x_1} [I - R]^{-1} e = 1.$$

Proof. Assuming that the stability condition is satisfied we determine the steady state probability vector for the Markov chain, $(N_q(t), N_s(t), P(t))$. Let $\mathbf{x}(n_q, n_s, p) = \lim_{t \to \infty} (N_q(t) = n_q, N_s(t) = n_s, P(t) = p)$. Then $\mathbf{x}$ satisfies

$$\mathbf{x}Q = \mathbf{0}, \mathbf{x}e = 1.$$

Partitioning $\mathbf{x}$ to $(\mathbf{x_0}, \mathbf{x_1}, \mathbf{x_2}, \mathbf{x_3}...)$, we see that $\mathbf{x}$ is such that

$$\mathbf{x_i} = \mathbf{x_1} R^{i-1}, i \geq 2,$$

where R is the minimal non-negative solution to the equation

$$\sum_{i=0}^{s'} R^i A_i = \mathbf{0}.$$

The boundary equations are given as

$$\sum_{i=0}^{s'-1} \mathbf{x}_i A_{i0} = \mathbf{0},$$

$$\mathbf{x_0} A_{01} + \sum_{i=1}^{s'} \mathbf{x}_i A_i = \mathbf{0}.$$

The normalizing condition, $\mathbf{x}e = 1$, gives

$$\mathbf{x_0} e + \mathbf{x_1} [I - R]^{-1} e = 1.$$

3   Performance Measures

- Expected queue length, $E_Q = \sum_{n_q=0}^{\infty} n_q(x_{n_q}.e)$.
- Probability that the server is idle, $P_I = \sum_{n_q=0}^{a-1} x_{n_q}$.
- Probability that the server serves a batch of size $'m', P_m = \sum_{n_q=0}^{a-1} \sum_{p=2}^{k} x_{(n_q,m,p)} + \sum_{n_q=a}^{\infty} \sum_{p=1}^{k} x_{(n_q,m,p)}$.
- Expected number of customers served in a batch, $ES = \sum_m m.P_m$.
- Probability that service at phase 1 starts immediately on completion of current batch departure at phase j, $P = \sum_{n_q \geq a} \sum_{n_s} x_{(n_q,n_s,j)}$.

4   Related Distributions

4.1   Analysis of Service Times

In this section, we compute the Laplace Stieltjes transforms of service times. As the service rates in a particular stage depends on both the number undergoing service as well as the stage, the service times differ.

Let X be the service time of a batch of n_s customers.

$$X = X_1 + X_2 + ... + X_k$$

where, $X_1, X_2, ...X_k$ are service times in stages $1, 2, 3...k$ respectively.

- Case 1
  If the entire batch leave the system from Stage 1, then $X = X_1$ the LST of service times is
  $$\frac{\mu_1^{n_s}}{s - \mu_1^{n_s}}.$$

- Case 2
  If the entire batch leave the system from Stage 2, then $X = X_1 + X_2$ the LST of service times is
  $$\frac{\mu_1^{n_s}}{s - \mu_1^{n_s}} \sum_{i_1=1}^{n_s} \frac{\mu_2^i}{s - \mu_2^i}.$$

- Case 3
  If the entire batch leave the system from Stage j, then $X = X_1 + X_2 + ..X_j$ the LST of service times is

  $$\frac{\mu_1^{n_s}}{s - \mu_1^{n_s}} \sum_{i_1=1}^{n_s} \frac{\mu_2^{i_1}}{s - \mu_2^{i_1}} \sum_{i_2=1}^{i_1} \frac{\mu_3^{i_2}}{s - \mu_3^{i_2}} ... \sum_{i_{k-1}=1}^{i_{k-2}} \frac{\mu_k^{i_{k-1}}}{s - \mu_k^{i_{k-1}}}.$$

4.2 Analysis of Waiting Times

In this section, we compute the expected waiting time of a tagged customer (TC) in the queue. If with the arrival of TC to an idle server, the service could be initiated his waiting time is 0. Otherwise, we need to consider the following cases:

- **Case 1**

  Suppose on arrival a tagged customer (TC) finds himself in the state (h), where $h \geq 1$. The new service is initiated only when a customers accumulate in the system. Since the arrival of customers follows Poisson process, the interarrival times are exponentially distributed with parameter λ. The waiting time of the customer is gamma distributed with parameters $(\lambda, a - h)$.

- **Case 2**

  Suppose on arrival a tagged customer(TC) finds himself in the state (h, n_s, p), where $h \geq 1, a \leq n_s \leq b, 1 \leq p \leq k$, i.e., the position of the TC is h in queue, while n_s customers are undergoing service in p^{th} stage. The waiting time of the TC is the time to absorption of Markov process with state space $\{(n_q, n_s, p); 1 \leq n_q \leq h, 1 \leq n_s \leq a - 1, 2 \leq p \leq k\} \cup \{(n_q, n_s, p); 1 \leq n_q \leq h, a \leq n_s \leq b, 1 \leq p \leq k\}$ onto $*_1$ (indicating the initiation of service to TC) or onto $*_2$ (indicating the TC getting into an idle state) plus the time for initiation of next batch service as mentioned in the previous case.

  The generator matrix for this Markov process is

  $$\mathcal{Q}_2 = \begin{bmatrix} W & W_1 & W_2 \\ \mathbf{0} & \mathbf{0} & \mathbf{0} \end{bmatrix}$$

  $$\mathcal{W} = I_h \otimes \begin{bmatrix} C_1^1 & & & & & & & \\ C_3^2 & C_1^2 & & & & & & \\ C_3^3 & C_3^3 & C_1^2 & & & & & \\ \vdots & \vdots & \vdots & \ddots & & & & \\ C_3^{a-1} & C_3^{a-1} & C_3^{a-1} & \cdots & C_1^{a-1} & & & \\ C_4^a & C_4^a & C_4^a & \cdots & C_4^a & C_2^a & & \\ C_4^{a+1} & C_4^{a+1} & C_4^{a+1} & \cdots & C_4^a & C_5^{a+1} & C_2^{a+1} & \\ \vdots & \vdots & \vdots & & \vdots & \vdots & & \ddots \\ C_4^b & C_4^b & C_4^b & \cdots & C_4^a & C_5^b & & \cdots C_2^b \end{bmatrix}$$

  $$\mathcal{G}_1 = \begin{bmatrix} D_1 \\ D_2 \\ \vdots \\ D_k \end{bmatrix}$$

  $$\mathcal{W}_1 = \begin{bmatrix} \mathbf{0} \\ e_{h-(a-1)} \otimes \mathcal{G}_1 \end{bmatrix}$$

  $$\mathcal{W}_2 = \begin{bmatrix} e_{a-1} \otimes \mathcal{G}_1 \\ \mathbf{0} \end{bmatrix}$$

Theorem 3. *The expected waiting time of a TC who arrives when the system in an idle state $(h); h \geq 1$ is $\frac{a-h}{\lambda}$. The expected waiting time of a TC who arrives when the system is busy is either: a. $\beta_1(-W)^{-2}W_1$. (If there are enough customers to initiate the service of batch including the TC) b. $\beta_1(-W)^{-2}W_2 + \frac{a-h'}{\lambda}$, (otherwise, provided h' is the position of the customer in queue, once the Markov process enters the idle state). Here, β_1 is the initial probability vector with 1 in the position of state (h, n_s, p) with all other entries 0.*

4.3   Conditional Distribution of Server Return Times

In this subsection, we analyze the distribution of time taken by the server to complete the service to $n_s; a \leq n_s \leq b$ customers and then return to stage 1. The service might be completed from first stage, or second stage or from any of the subsequent stages (depending on when the last customer currently in service leaves the system). Assuming that the server completes service from j^{th} stage, we find the distribution of server return times.

Consider the Markov chain, $(N_s(t), P(t); t \geq 0)$ on state space $\{(n_s, 1), (n_s, 2), (n_{s-1}, 2), \ldots (1, 2), (n_s, 3), (n_{s-1}, 3), \ldots (1, 3) \ldots (n_s, j), (n_{s-1}, j), \ldots (1, j)\} \cup \{\Delta\}$, where Δ denote the absorbing state indicating the server returning to stage 1 after service completion. The infinitesimal generator matrix of this Markov chain is,

$$Q_3 = \begin{bmatrix} T_j & T_j^0 \\ \mathbf{0} & 0 \end{bmatrix}$$

$$T_j = \begin{bmatrix} T_1^1 & T_1^2 & & & & \\ & T_2^1 & T_2^2 & & & \\ & & T_3^1 & T_3^2 & & \\ & & & \ddots & \ddots & \\ & & & & T_{j-1}^1 & T_{j-1}^2 \\ & & & & & T_j^1 \end{bmatrix}$$

$T_1^1 = -n_s\mu_1^{n_s}, T_1^2 = \mu_1^{n_s}.e', \quad T_i^j = diag[-n_s\mu_i^{n_s}, -(n_s - 1)\mu_i^{n_s}, \ldots - 3\mu_i^{n_s}, -2\mu_i^{n_s}, -1\mu_i^{n_s}],$

$$T_i^2 = \begin{bmatrix} \mu_i^{n_s} & \mu_i^{n_s} & \cdots & \mu_i^{n_s} \\ & \mu_i^{n_s-1} & \cdots & \mu_i^{n_s-1} \\ & & \ddots & \vdots \\ & & & \mu_i^2 & \mu_i^2 \\ & & & & \mu_i^1 \end{bmatrix}$$

for $i = 2, 3, ..j$

$$T_j^0 = \begin{bmatrix} 0 \\ 0 \\ 0 \\ \vdots \\ 0 \\ T_0 \end{bmatrix}$$

$$T_0 = \begin{bmatrix} \mu_j^{n_s} \\ \mu_j^{n_s-1} \\ \vdots \\ \mu_j^2 \\ \mu_j^1 \end{bmatrix}$$

i.e., the conditional distribution of server return times, follow phase type distribution $PH(\alpha_j, T_j)$, where $\alpha_j = (1, 0, ...0)$, is a row vector of order $(j-1).n_s + 1$.

Theorem 4. *The expected server return time is* $\sum_{j=1}^{k} \alpha_j (-T_j)^{-1} e.P_j$, *where* P_j *denotes the probability that all the* n_s *customers who joins service at stage 1, leave the system from stage* j.

5   Numerical Illustration

5.1   Effect of λ on Performance Measures

As an illustration, we consider a 4 stage queueing system, with minimum service batch size $a = 2$ and maximum service batch size $b = 4$. First, we analyze **the effect of dependence of service rates on batch sizes.** We consider 3 situations, by assuming that the service rates are directly proportional to the stage in which the process is in:

A. $\mu_i^j = j * i$, Service rate directly propotional to batch size
B. $\mu_i^j = i$, Service rate is independent of batch size
C. $\mu_i^j = i/j$, Service rate inversely propotional to batch size (Table 1).

Table 1. Service Rates, μ_i^j for various size, j and stage, i in Scenario I

	A				B				C			
	m = 1	m = 2	m = 3	m = 4	m = 1	m = 2	m = 3	m = 4	m = 1	m = 2	m = 3	m = 4
1	0	0.6689	1.0033	1.3378	0	1.0101	1.0101	1.0101	0	1.2821	0.8547	0.6410
2	0.6689	1.3378	2.0067	2.6756	2.0202	2.0202	2.0202	2.0202	5.1882	2.5641	1.7064	1.2821
3	1.0033	2.0067	3.0100	4.0134	3.0303	3.0303	3.0303	3.0303	7.6983	3.8462	2.5641	1.9231
4	1.3378	2.6756	4.0134	5.3512	4.0404	4.0404	4.0404	4.0404	10.2564	5.1282	3.4188	2.5641

For the purpose of comparison, we have normalised the weighted average service rate to be $\mu^* = 10$. The results so obtained are presented graphically (Fig. 2).

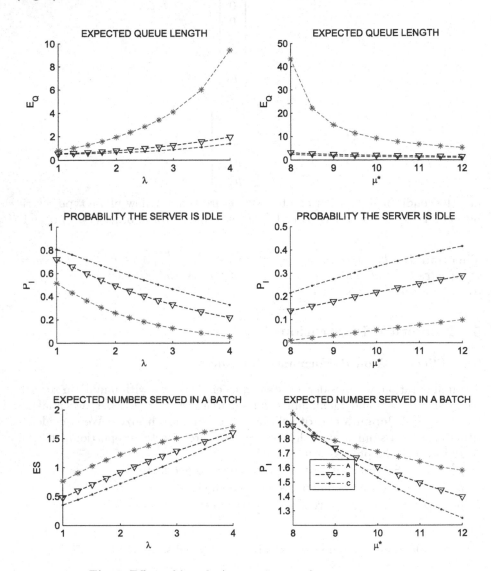

Fig. 2. Effect of λ and μ^* on various performance measures

– The increase in expected queue length and expected number of customers served in a batch, is relatively higher when the service rate is directly propotional to batch size. It is least when service rate is inversely propotional to batch size. This could be explained as follows: Customers leave after service completion from intermediate stages; so, as stage advances, the number in a batch is likely to decrease. The individual service rates are higher when the batch size is smaller, so service times are shorter and the expected queue length decreases.

- The expected queue length and expected number of customers served in a batch, increases with λ and decreases with μ^*. On the other hand, as λ increases, the idleness percentage decreases and as μ^* increases the idleness percentage increases.

5.2   An Optimization Problem

Based on performance measures, we construct a revenue/profit function which is of importance for computing the optimal batch size, for a particular value of λ and μ^*. We define a revenue function, $F(n_s) = R.n_s - ESR_{n_s}.C_{n_s} - E_Q C_Q - P_I C_I$ where,

R, is the average revenue for providing service to a single customer per unit time
ESR_{n_s} is the expected server return time, if the transport vehicle starts with n_s customers
C_{n_s} is the cost for providing service to a batch of size n_s, per unit time
C_Q is the holding cost of a customer in queue, per unit time
C_I is the cost associated with idleness of server per unit time
We fix the revenue/costs as follows:
$R = \$15, C_2 = \$4, C_3 = \$6, C_4 = \$8, C_Q = \$2, C_I = \$1$. The arrival rate is $\lambda = 1$ and weighted average service rate is $\mu^* = 10$. We compute the value of $F(n_s)$ in 3 different scenarios (as mentioned in previous subsection). The results are tabulated as follows (Table 2):

Table 2. Effect of n_s on Profit Function, when service rates depend on batch size.

Situation	$n_s = 2$	$n_s = 3$	$n_s = 4$
A	26.3882	**41.9406**	26.1592
B	27.6362	**42.6581**	37.0188
C	27.9877	**42.8966**	36.7983

The maximum profit is indicated in bold font. In all the three scenarios, the maximum profit is obtained when $n_s = 3$. The profits are more or less same for same value of n_s, except in Scenario A.

6   Conclusions

In this paper, we have analyzed a batch service queueing system in which server provides service in batches, and customers can leave after service completion from intermediate stages. Attributes of the queueing model under consideration like expected waiting times and expected server return times are analyzed. The dependence of service rates on stages as well as batch sizes is observable in many real life situations. For e.g., if the size of the batch is higher, the server increases its service rate to ensure that the inventory is delivered without much delay and the costs associated with it are reduced. The model could be extended to multiserver queueing systems and to systems in which server moves amongst the various stages collecting and distributing inventory.

References

1. Banerjee, A., Gupta, U.C., Chakravarthy, S.R.: Analysis of a finite-buffer bulk-service queue under Markovian arrival process with batch-size dependent service. Comput. Oper. Res. **60**, 138–149 (2015)
2. D'Arienzo, M.P., Dudin, A.N., Dudin, S.A., Manzo, R.: Analysis of a retrial queue with group service of impatient customers. J. Ambient. Intell. Humaniz. Comput. **11**(6), 2591–2599 (2019). https://doi.org/10.1007/s12652-019-01318-x
3. Krishnamoorthy, A., Joshua, A.N., Vishnevsky, V.: Analysis of a k-stage bulk service queuing system with accessible batches for service. Mathematics **9**(5), 559 (2021). https://doi.org/10.3390/math9050559
4. Neuts, M.F.: A general class of bulk queues with poisson input. Ann. Math. Stat. **38**, 759–770 (1967)

The Analytical Method of Transient Behavior of the $M|M|1|n$ Queuing System for Piece-Wise Constant Information Flows

Konstantin Vytovtov$^{(\boxtimes)}$ ⓘ, Elizaveta Barabanova ⓘ,
and Vladimir Vishnevsky ⓘ

V.A. Trapeznikov Institute of Control Sciences RAS,
Profsoyuznaya 65 Street, Moscow, Russia

Abstract. The analytical method for studying the behavior of queuing system with abruptly changing information flows is presented in this paper for the first time. The proposed method is based on the conception of fundamental matrix of the Kolmogorov system and allows studying the transient and stationary modes of functioning different types of queuing system. The transient behavior of the $M|M|1|n$ system with constant flows and piecewise constant arrival and service rates of flows are considered. It is obtained the analytical expressions for the state probabilities of the queuing system as in transient and in stationary mode. The conception of transient time for the queuing system has been introduced and the formulas for calculation of such transient performance metrics as throughput, a number of received and processed packets have been obtained for the first time.

Keywords: Queuing system · Kolmogorov equations · Transient mode

1 Introduction

The queuing theory is widely-used for modeling of processes in telecommunication networks [1–5]. Now various classes of queuing systems including different types of single- and multi-channel systems as well as single- and multi-phase systems have been investigated in sufficient detail [1–5]. As a rule calculations of basic characteristics of queuing systems, such as blocking probability, the size of buffer and others are carried out in a steady-state mode, when $t \to \infty$. This mode occurs some time after the system starts functioning. And the mode from the beginning of system functioning before transition it in stationary mode when its behavior depends on time is called as the transient mode [6–16]. In addition, the system turns out to be in a transitional mode also when its normal functioning is disrupted. Note that, although the stationary mode is used to analyze most

The reported study was funded by RFBR, project number 19-29-06043.

queuing systems, in some cases it is not describing behaviour of one's adequately. Such situations often arise in the network when arrival rate of packets rises or the service rate decreases sharply. For example, "broadcast storm" when traffic intensity increases sharply is a fairly common situation in computer networks. Analogously subscriber load in telephone networks increases when emergence situation occurs. Also service rate decreases sharply in a moment when a system starts to reboot. For the above-described cases analyzing transient behavior of queuing network allows us not only to predict operation of telecommunication equipment but also to identify critical situations that can completely disable the network. Transient modes have been considered quite deeply in radio engineering, electrodynamics, optics, etc. The transient behavior of the queuing system $M|M|1$ has been considered by Prabhu [7] in 1965 and Cohen [6] in 1982 years for the first time. Harrison was the first who pays attention to the study of the transient behavior of a computer network [8]. The author has considered the non-steady-state mode of a closed homogeneous queuing network and he has developed a convergent iterative method for the solution of the Kolmogorov forward equations for Gordon-Newell type queuing networks. The transient mode of queuing networks has been considered in a number of other works [9–16]. For example in [10] the author investigated the transient behavior of $M|M|n$ system that is used to describe the call centre. The author has proposed the approximate analytical method for solving the system of homogeneous Kolmogorov equations. In [11] the authors have presented the method for analyzing the transient state of the queuing system $M|M|1$ and the approximate calculation of the state probabilities. Their approach is based on the representation of the system in the form of an inhomogeneous Markov chain with continuous time and the number of the states much less than the ones of original system. In this paper the authors used approximate analytical method for solving Kolmogorov equation system describing the considered queuing system. The transient mode of a single-server queuing system with a finite buffer and correlated input flows has been considered in [12]. But the authors present only approximate numerical solution of corresponding different equations by the Runge-Kutta method. Despite the transient mode of queuing systems has been considered in a number of the scientific works [6–17], its analysis has been carried out by using numerical methods of solving Kolmogorov equations only. These are, for example, the Runge-Kutta method [12,17], boundary value methods [17] and others. All of them have a number of disadvantages. First of all, they are approximate, the accuracy of calculations caring out by these methods is not always controlled, numerical methods are very problematic to use when solving so-called inverse problems also. Besides the dependence of the system states on time for real values of arrival and service rates and the special cases when the arrival rate exceeds the service rate practically have not been considered early. Moreover, the case of time-dependent incoming flows has hardly been studied. In particular, an abrupt change in arrival rate is analytically investigated in the presented work for the first time. In this paper, the accurate analytical method for the analysis of transient behavior of queuing systems is proposed for the first time.

The method is based on concept of the fundamental matrix of the Kolmogorov equation system. The transient mode when the system is turned on is considered (Sect. 3). Here analytical expressions for the probabilities of system states as functions of time are presented and expressions that determine the time of the transient mode are received. Throughput of the system, the number of packets received and processed during the transient mode are calculated also. The mathematical model of queuing system for the case of a jump-like change in arrival and service rates at an arbitrary time instant by an arbitrary value is presented in Sect. 4. The fundamental matrix of Kolmogorov equations for this case is written in analytical form in elementary functions. A rigorous proof that this matrix correctly describes the behaviour of the system is given. Numerical calculations in accordance with the constructed models are presented in Sect. 5. The calculation results for the limiting cases fully correspond to the previously obtained analytical and numerical results.

2   The Statement of the problem

In this work, the transient behavior of a single-channel queuing system with finite buffers $M|M|1|n$ is considered. Identical requests arrive according to a time-homogeneous Poisson process at rate λ, and the service time of a request is exponentially distributed at rate μ. A request from the incoming stream waits for the start of service in a buffer if the channel is busy and the buffer is free. If all n waiting places are occupied, then the next request lost. The state transition diagram of such a process is shown in Fig. 1. Here S_0 is the state when the service channel and the buffer are empty, S_1 is the state when the service channel is busy and the buffer is empty, S_2 is the service channel is busy and there is one request (one packet) in the buffer, S_{n+1} is the service channel is busy and there are n requests in the buffer (n packages). As you can see, the process is the birth-and-death process, where $\lambda_i = \lambda$, $0 \leq i \leq n$, $\mu_i = \mu$, $1 \leq i \leq n+1$. The number of requests in the system and the number of states of the process describing its operation can take values on the set $\{0, 1, ..., n+1\}$, $n \in Z$.

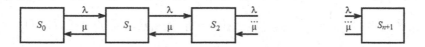

Fig. 1. The state transition diagram.

The main aim of this work is studying both the transient and steady-state modes of the system. First of all, the transient time, throughput and the probability of packet loss in the transient mode are investigated. In addition, the influence of jumps of arrival and service rates on the characteristics of the system is investigated in the paper also. In this case, the jumps of $\lambda(t)$ and $\mu(t)$ can occur at some fixed times, and the arrival and service rates are unchanged between the jumps. Thus in the case under consideration, $\lambda(t)$ and $\mu(t)$ are described by the expressions

$$\lambda(t) = \begin{cases} \lambda_1, t_0 \leq t \leq t_1 \\ \lambda_2, t_1 \leq t \leq t_2 \\ \cdots\cdots\cdots \\ \lambda_K, t_{K-1} \leq t \leq t_K \end{cases} \quad ; \mu(t) = \begin{cases} \mu_1, t_0 \leq t \leq t_1 \\ \mu_2, t_1 \leq t \leq t_2 \\ \cdots\cdots\cdots \\ \mu_K, t_{K-1} \leq t \leq t_K \end{cases} \tag{1}$$

where on intervals with constant λ_i and μ_i, incoming Poisson flows and an exponential distribution of service time are assumed. Here to study such systems, the method of the fundamental matrix of Kolmogorov equations is proposed for both the case of constant and jump-like rates of arrival and service of claims.

The system of differential equations for a single-channel queuing system with finite buffers obtained by using Δt−method can be written in the form

$$\begin{cases} \dfrac{dP_0}{dt} = -\lambda P_0(t) + \mu P_1(t) \\[2em] \dfrac{dP_i}{dt} = \lambda P_{i-1}(t) - (\lambda + \mu) P_i(t) + \mu P_{i+1}(t) \; 1 \leq i \leq n \\[2em] \dfrac{dP_{n+1}}{dt} = \lambda P_n(t) - \mu P_{n+1}(t) \end{cases} \tag{2}$$

Here $P_i(t)$ is the probability of the i-th state at time t (the probability of i packets are in the buffer). The analytical solution [16] of the system (1) is following

$$P_i(t) = \sum_{j=1}^{n+1} P_{ij}(t) = \sum_{j=1}^{n+1} \xi_{ij} A_i \exp(\gamma_j t) \tag{3}$$

for an arbitrary finite number n of the equations in (1). Here γ_j are the roots of the characteristic equation of system (1), A_i are the integration constants determined by the initial conditions, ξ_{ij} are the coefficients expressing the probabilities $P_i(i \neq 0)$ in terms of P_0. The probabilities P_{ij} in (3) can be expressed in terms of the probability P_{0j} by using the relations

$$P_{ij}(t) = -\frac{1}{b_{i-1,i}^{(j)}} \left(b_{i-1,i-2}^{(j)} P_{i-2,j}(t) + b_{i-1,i-1}^{(j)} P_{i-1,j}(t) \right) = \xi_{ij} P_{oj}(t) \tag{4}$$

where $b_{kl}^{(j)}$ are elements of the matrix

$$\mathbf{B}^{(j)} = \mathbf{A} - \mathbf{I}\gamma_j \tag{5}$$

where $\mathbf{A}$ is the coefficient matrix of system (1), $\mathbf{I}$ is the unit matrix. The index j in (2) and (3) corresponds to the index j of the eigenvalue of (4).

3 Transient Behavior of the System $M|M|1|n$ with Constant Flows

A fundamental matrix $\mathbf{M}$ of a linear differential equation system of the order $n+1$ (1) is a matrix of the $n+1$-th rank and it relates the probabilities of the system states at an arbitrary time t to these probabilities at the initial time moment t_0:

$$\mathbf{U}(t) = \mathbf{M}_{n+2\times n+2}(t)\mathbf{U}(t_0) \tag{6}$$

Here $\mathbf{U}(t) = (P_i)^T$ is the $1 \times n+2$-vector of the system states, where $i = \overline{0, n+1}$, T is the transpose operator.

In accordance with the definition of a fundamental matrix [18–20], the elements of the $\mathbf{M}_{n+2\times n+2}(t)$ for a queuing system $M|M|1|n$ are

$$m_{kl} = \sum_{q=0}^{n+1} (-1)^{k+q} \xi_{lq} \frac{\Delta_{kq}}{\Delta} \exp(\gamma_q t) \tag{7}$$

where

$$\Delta = \begin{pmatrix} 1 & 1 & 1 & 1 & 1 & 1 \\ \xi_{21} & \xi_{22} & \xi_{23} & \xi_{24} & \cdots & \xi_{2,n+2} \\ \xi_{31} & \xi_{32} & \xi_{33} & \xi_{34} & \cdots & \xi_{3,n+2} \\ \xi_{41} & \xi_{42} & \xi_{43} & \xi_{44} & \cdots & \xi_{4,n+2} \\ \cdots & \cdots & \cdots & \cdots & \cdots & \cdots \\ \xi_{n+2,1} & \xi_{n+2,2} & \xi_{n+2,3} & \xi_{n+2,4} & \cdots & \xi_{n+2,n+2} \end{pmatrix} \tag{8}$$

Δ_{kq} is the minor element of matrix Δ with indices kq.

Thus, here, for the first time, an exact analytical expression for the fundamental matrix of system (1) with constant flux intensities was found, which makes it possible to fully describe the behavior of the system, both in transient and stationary modes.

In the limiting case, as $t \to \infty$, the obtained expressions (5)–(7) are reduced to the well-known formulas [1–5]. Indeed, one of the roots γ_0 of the characteristic equation of system (1) is always zero, and the rest γ_i, $i = \overline{1, n+1}$, are negative. Then, for all roots except γ_0, as $t \to \infty$, we have $exp(\gamma_q t) \to 0$.

$$P_i(t) = \sum_{j=0}^{n+1} m_{ij} P_i(t_0) = \sum_{j=0}^{n+1} (-1)^i \xi_{j0} \frac{\Delta_{io}}{\Delta} P_i(t_0) \tag{9}$$

Obviously, at the moment of turning on the system the buffer is free. Therefore, the initial conditions are $P_0(t_0) = 1, P_i(t_0) = 0$, $i = \overline{1, n+1}$. Then, after algebraic transformations (8) it is reduced to the form

$$P_i(t \to \infty) = \frac{\lambda^i}{(n+1)\mu^i} \tag{10}$$

Another important problem of transient mode investigation is determination of transient time. The proposed method allows to calculate the throughput of the system and the probability of packet loss in this mode also. It follows from (2) and (6) that the state probability is described by the sum of n decreasing exponential functions $(\exp(\gamma_q t), \gamma_q < 0, i = \overline{1, n+1})$ and a constant term, since one of the roots of the characteristic equation γ_0 is equal to zero. As a result, over time, the probabilities of the states tend to the values determined by (9). To describe the rate of this process, we introduce the concept of the time constant of the system's transient mode as the maximum time for which the probability of a state changes by e times. So the probability changes from $\exp(\gamma_i t_0) = 1$ if $t_0 = 0$ to $\exp[\gamma_i(t_0 + \tau_i)] = e^{-1}$ if $\tau_i = 1/ \mid \gamma_i \mid, i = \overline{1, n+1}$.

$$\tau = max\{\tau_i, i = \overline{1, n+1}\} = \max\{\frac{1}{\mid \gamma_i \mid}, i = \overline{1, n+1}\} \tag{11}$$

In next numerical calculations, it will be shown that the transient mode is almost completely ends and the system goes over to the steady-state mode if

$$\tau_{tr} = (3 \div 5)\tau = (3 \div 5) \max(\tau_i, i = \overline{1, n+1}) \tag{12}$$

The transient instantaneous throughput is calculated as

$$A(t) = (1 - P_{n+1}(t))\lambda \tag{13}$$

where $P_{n+1}(t)$ is the instantaneous probability of loss which is calculated using (6) and (7). The number of arrive N_a and served N_s packets during the transient mode are

$$N_a = \lambda \tau_{tr} \tag{14}$$

$$N_s = \mu \tau_{tr} \tag{15}$$

Thus, in this section, a mathematical model of the $M|M|1|n$ system with constant arrival and service rates is proposed. This model describes the behaviour of the system in both transient and steady-state modes.

4   Transient Behavior of the $M|M|1|n$ Queue with Piecewise Constant Arrival and Service Rates of the Flows

First of all, let us show that the resulting matrix of a system with piecewise constant parameters is equal to the multiplication of the matrices of intervals with constant parameters. Indeed, let the process be described by a fundamental matrix on the first interval with constant transition probabilities. Then, on this interval, the states of the system at an arbitrary moment of time are found as

$$\mathbf{U}(t) = \mathbf{M}_1(t - t_0)\mathbf{U}(t_0) \tag{16}$$

And for $t = t_1$ we have

$$\mathbf{U}(t_1) = \mathbf{M}_1(\Delta t_1)\mathbf{U}(t_0) \tag{17}$$

Analogously, for the second interval, we can write

$$\mathbf{U}(t) = \mathbf{M}_2(t - t_1)\mathbf{U}(t_1) \tag{18}$$

where $\mathbf{M}_2 = (t - t_1)$ is the fundamental matrix of the system on the second interval with constant rates of flows. Then, substituting (15) into (16), we obtain

$$\mathbf{U}(t) = \mathbf{M}_2(t - t_1)\mathbf{M}_1(\Delta t_1)\mathbf{U}(t_0) \tag{19}$$

And for $t = t_2$ we have

$$\mathbf{U}(t_2) = \mathbf{M}_2(\Delta t_2)\mathbf{M}_1(\Delta t_1)\mathbf{U}(t_0) \tag{20}$$

Continuing this procedure for the third interval, we write

$$\mathbf{U}(t) = \mathbf{M}_3(t - t_2)\mathbf{U}(t_2)\mathbf{U}(t) = \mathbf{M}_3(t - t_2)\mathbf{M}_2(\Delta t_2)\mathbf{M}_1(\Delta t_1)\mathbf{U}(t_0) \tag{21}$$

In the general case, for K intervals, we obtain

$$\mathbf{U}(t) = \mathbf{M}_K(t - t_{K-1}) \prod_{i=K-1}^{1} \mathbf{M}_i(\Delta t_i)\mathbf{U}(t_0) \tag{22}$$

Then the fundamental matrix describing the behavior of the system on the K-th interval has the form

$$\mathbf{M}_K = \mathbf{M}_K(t - t_{K-1}) \prod_{i=K-1}^{1} \mathbf{M}_i(\Delta t_i) \tag{23}$$

Thus, in this section, we have found the fundamental matrix of the system (1) with arbitrary piecewise constant flow rates in an analytical form. Note that the method also allows us to describe the dependence of the state probabilities on time for arbitrary laws of change $\lambda(t)$ and $\mu(t)$.

5   Numerical Calculations

5.1   Transient Analysis for a Time-Independent Queuing System

First of all, let us show that the presented method allows to carry out calculations for a large number of system state probabilities in the transient mode. For example, Fig. 2 shows the dependence of the probabilities of two hundred states on time for $\lambda = 8.3 \times 10^3$ packet/s, $\mu = 5 \times 10^3$ packet/s. Further, this section presents the results of numerical calculations and their analysis for the $M|M|1|n$ queue, where $n = \overline{1,3}$ for $i = 5$ states, where $i = \overline{0,4}$.

Transient modes are considered for the cases of constant ρ and abruptly varying ρ. In addition, the analysis of dependence of the state probabilities including the probability of failures on time for the cases $\rho > 1$, $\rho = 1$, $\rho < 1$ is carried

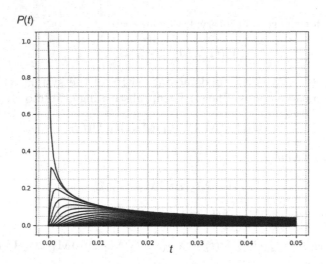

Fig. 2. Time dependence of the probability of two hundred states of the $M|M|1|n$ queue.

out. Also the throughput of the system and the time of the transient mode are calculated.

Figure 3 shows the calculation results of the state probabilities dependences on time at $\lambda = 8.3 \times 10^5$ packet/s, $\mu = 5 \times 10^5$ packet/s [5] and initial conditions $(P_0(0), P_1(0), P_2(0), P_3(0), P_4(0))^T = (1, 0, 0, 0, 0)^T$, meaning that the system buffer is completely free at the moment of time $t = 0$. The indicated λ corresponds to the maximum arrival rate of packets in the 10 Gbit/s communication channel when the size of packet is 1500 bytes [5]. The service rate of a particular switch μ depends on its performance and in this case corresponds to the average performance of a local area network switch. Thus, in this case $\rho = 1.66 > 1$, which is typical for example, for a "broadcast storm". Note that such a mode of operation for real systems is critical and can lead to fatal errors in information processing and require a complete reboot of the system [21]. In Fig. 3 $P_0(t)$ is the probability that the service channel is idle and the buffer is empty, $P_1(t)$ is the probability that the service channel is busy but the buffer is empty $P_1(t)$, $P_2(t)$, $P_3(t)$ are the probabilities that there are one and two packets in the buffer correspondingly and $P_4(t)$ is the probability that there are three packets in the buffer and the next arrived packet will be lost.

In the transient mode, first of all, the bursts of state probabilities and the time of the transient mode are of interest. Studying these probability bursts in the transient mode, we can say that (Fig. 3) the maximum probability that the one packet is processing by the switch and the buffer is empty in this moment of time $P_{1max}(t = 0.125 \times 10^{-5}) = 0.33$. And it is three times higher the steady-state probability $P_1(t \to \infty) = 0.1$. Similarly, the maximum probability that one packet is in the buffer in the transient mode $P_{2max}(t = 0.25 \times 10^{-5}) = 0.22$. And it is almost one and a half times higher the steady-state probability $P_2(t \to$

∞) = 0.16. An increase in the arrival rate λ leads to an increase in the amplitude of probability bursts: $P_{1max} = 0.35$ and $P_{2max} = 0.24$ if $\lambda = 15 \times 10^5$ packet/s. This situation can lead to buffer overflow and packet loss if taking into account the values of state probabilities in stationary mode only.

The second most important problem of transient analysis analogously as in radio electronics is to find the transient time of the system. In this case, in accordance with the theory described above, $\tau_{tr} = 1.74s$. Note that at the selected arrival rate λ and service rate μ during the transient time $N_a = \lambda \tau_{tr} \approx$ 15 packets are arrived in the switch and $N_s = \mu \tau_{tr} \approx 9$ packets are serviced by the switch. However, the probability of packet loss under the considered conditions is not large in comparison with the steady-state mode and increases according to the hyperexponential law (see Fig. 3 $p_4(t)$).

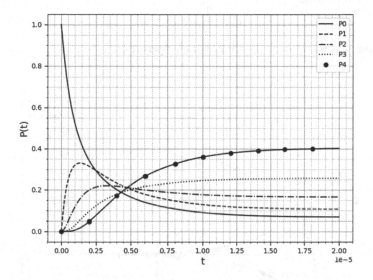

Fig. 3. Dependences of the states probabilities of the switch buffer

In the stationary mode probabilities of states are $P_0(t = 1, 74 \cdot 10^{-5}) = 0.07$, $P_1(t = 1, 74 \cdot 10^{-5}) = 0.1$, $P_2(t = 1, 74 \cdot 10^{-5}) = 0.16$, $P_3(t = 1, 74 \cdot 10^{-5}) = 0.25$, $P_4(t = 1, 74 \cdot 10^{-5}) = 0.4$. The throughput of the network in stationary state is $A(t = 1, 74 \cdot 10^{-5}) = (1 - 0.4) \cdot 8.3 \cdot 10^5 = 486 \, \text{Kb/s}$. These results correspond to the ones obtained in previous works by well-known methods [1].

Figure 4 shows the dependence of states for the case $\rho = 1$. Here $\lambda = \mu = 5 \cdot 10^5$ packet/s. During this time, $N_a = N_s = \lambda \tau_{tr} = \mu \tau \approx 13$ packets are arrived and serviced. In this case, at the beginning of the transient mode, there is also a probability burst up to $P_{1max}(t = 0.25 \cdot 10^{-5}) = 0.31$ which 1.55 times more steady-state probability $P_1(t \rightarrow \infty) = 0.2$. The transient time in comparison with the previous case increased and amounted to $t_{tr} = 2, 6 \cdot 10^{-5}$ s, which also corresponds to the results obtained known methods [1].

In addition, the steady-state probabilities take the same values $p_1(2,6 \cdot 10^{-5}) = p_2(2,6 \cdot 10^{-5}) = p_3(2,6 \cdot 10^{-5}) = p_4(2,6 \cdot 10^{-5}) = 0.2$, *the throughput is* $A(t = 2,6 \cdot 10^{-5}) = (1 - 0.2) \cdot 5 \cdot 10^5 = 391\,\text{Kb/s}$, which also corresponds to the results obtained by well-known methods. Indeed, in [1] the formula is presented

$$P_i = \rho^i \frac{1 - \rho}{1 - \rho^{n+1}}, 0 \le i \le n \tag{24}$$

where n is the number of states. For $\rho = 1$, it gives the uncertainty $0/0$ and the authors propose to use the L'Hospital's rule to find the probabilities of states. Here we use the binomial decomposition $a^m - b^m = (a - b)(a^{m-1} + a^{m-2}b + \ldots + ab^{m-2} + b^{m-1})$ and we obtain a simple expression for the case under consideration in a stationary mode.

$$P_i = \frac{1}{n}, 0 \le i \le n \tag{25}$$

For five states, we have $p_i = 0.2$, which corresponds to Fig. 4.

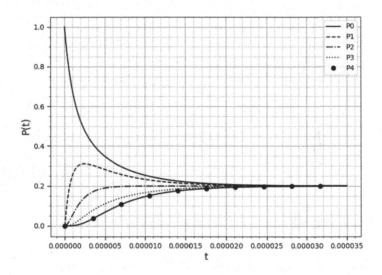

Fig. 4. Dependences of the state probabilities of the switch buffer.

Figure 5 shows the third case, when $\rho < 1$, where $\lambda = 2 \cdot 10^5$ packet/s and $\mu = 5 \cdot 10^5$ packet/s, for $\rho = 0.4$. For this, small transition probability bursts are also observed, however, over time they disappear. The transition time in comparison with the previous cases increased slightly and amounted to $2,7 \cdot 10^{-5}$ s. During this time, $N_a = \lambda\tau_{tr} \approx 6$ packets are arrived to the switch and $N_s = \mu\tau_{tr} \approx 14$ packets can be served.

The steady-state probabilities are $\pi_0 = P_0(t \to \infty) = 0.6$, $\pi_1 = P_1(t \to \infty) = 0.24$, $\pi_2 = P_2(t \to \infty) = 0.1$, $\pi_3 = P_3(t \to \infty) = 0.04$, $\pi_4 = P_4(t \to \infty) = 0.02$, the throughput is $A(t = 2,7 \cdot 10^{-5}) = (1 - 0.02) \cdot 2 \cdot 10^5 = 191$ Kb/s. These values correspond to calculation results carried out by well-known formulas [1–5].

The analysis of transient behavior of the queuing system shows that when λ decreases, the transient time increases, and, accordingly, the number of packets processed during this time increases too. In addition, with decreasing λ, the probability bursts decrease in the transient mode. It should be noted that the obtained numerical results of the steady-state probabilities of the system completely coincide with the numerical solutions obtained earlier [1–5].

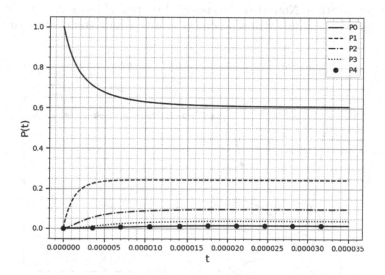

Fig. 5. Dependences of the states probabilities of the switch buffer.

5.2 Transient Analysis for a Time-Dependent Queuing System

In this section, we analyze the distribution of packets in the system when arrival rate of packets changes abruptly at some moments of time. So it will be consider how jumps of λ affect the transient mode for three cases $\rho > 1$, $\rho = 1$, $\rho < 1$ and initial conditions $((P_0(0), P_1(0), P_2(0), P_3(0), P_4(0))^T = (1, 0, 0, 0, 0)^T$ as in case with time-independent flows.

Let us consider the first case when $\rho > 1$, $\lambda_0 = 8.3 \cdot 10^5$ packet/s and $\mu_0 = 5 \cdot 10^5$ packet/s. This is the case when a network has been overloaded before the arrival rate jump can degrade network performance even more (Fig. 4). For example, this situation occurs in the network as a result of a "broadcast storm". Let at time $t_1 = 1 \cdot 10^{-5}$ s there is a jump from λ_0 to $\lambda_1 = 2.5 \cdot 10^6$ packet/s, $\mu_0 = \mu_1 = 5 \cdot 10^5$ packet/s.

From the distribution of state probabilities dependencies in the transient mode (Fig. 6) it is seen that the jump at $t = t_1$ affected the resulting transient time, which decreased to $1.5 \cdot 10^{-5}$ s. Indeed, the resulting transient mode includes two modes. The first one is determined by the process of system starting at the moment of time $t = t_0$, and the second one depends on the jump of λ at the moment of time $t = t_1$, when the previous transient mode has not ended yet. In this case, the time of the second transient mode is less than the time of the first one, since $\lambda_1 \gg \lambda_0$ (see the previous subsection).

Further, the steady-state probabilities take the following values: $\pi_0 = P_0(t \rightarrow \infty) = 0$, $\pi_1 = P_1(t \rightarrow \infty) = 0.01$, $\pi_2 = P_2(t \rightarrow \infty) = 0.03$, $\pi_3 = P_3(t \rightarrow \infty) = 0.16$, $\pi_4 = P_4(t \rightarrow \infty) = 0.8$, the throughput is $A(t = 1.5 \cdot 10^{-5}) = (1 - 0.8) \cdot 2.5 \cdot 10^6 = 488\,\text{Kb/s}$. In this case, jump of arrival packet rate in the mode when the switch has been overload leads to sharp packet losses in the network (up to 80%). Note that the steady-state probabilities depend on λ and μ after the jump and do not depend on the state of the system before the jump.

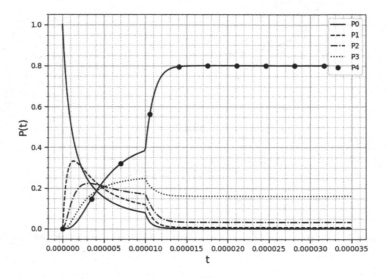

Fig. 6. Dependencies of the states probabilities of the buffer for $\rho > 1$ and jump of λ.

Figure 7 shows the calculation results of states probabilities for the case $\rho = 1$ before the jump. As in the previous case the jump from $\lambda_0 = 5 \cdot 10^5$ to $\lambda_1 = 2.5 \cdot 10^6$ packets/s occurs in the moment of time $t_1 = 1 \cdot 10^{-5}$ s, when the previous transient mode has not been over yet. In this case, the jump of arrival rate λ also leads to a decrease of transient period to $\tau_{tr} = 2 \cdot 10^{-5}$ s due to an increase of the λ. At the same time, it is obvious that the jump also influenced the redistribution of the probabilities of the system states in the stationary mode: $\pi_0 = P_0(t \rightarrow \infty) = 0$, $\pi_1 = P_1(t \rightarrow \infty) = 0.01$, $\pi_2 = P_2(t \rightarrow \infty) = 0.03$, $\pi_3 = P_3(t \rightarrow \infty) = 0.16$, $\pi_4 = P_4(t \rightarrow \infty) = 0.8$, the throughput is $A(t = 2 \cdot 10^{-5}) = (1 - 0.8) \cdot 2.5 \cdot 10^6 = 488\,\text{Kb/s}$. These steady-state probabilities are equal to the probabilities of states in the previous case for $\rho > 1$. A similar picture is observed for $\rho < 1$. Figure 8 shows the calculation results for the case $\rho > 1$ before the jump and $\rho < 1$ after the jump. In contrast to

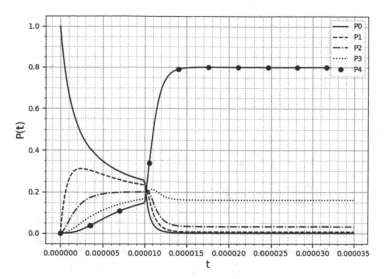

Fig. 7. Dependencies of the states probabilities of the buffer for $\rho = 1$ and jump of λ.

the previous cases, consider the decrease in intensity from $\lambda_0 = 8.3 \cdot 10^5$ packet/s to $\lambda_1 = 2.5 \cdot 10^5$ packet/s in the moment of time $t_1 = 1 \cdot 10^{-5}$ s. Obviously, the character of the probability of packet loss is changed at the moment of the jump, since the arrival rate of packets decreases sharply. It can be seen from the calculation results that in this case, the stationary mode is determined only by the characteristics of the system after the jump, and the arrival and service rates before the jump do not affect the stationary behavior of the system. For example for this case the probabilities of the system states in the stationary mode are $\pi_0 = P_0(t \to \infty) = 0.51$, $\pi_1 = P_1(t \to \infty) = 0.25$, $\pi_2 = P_2(t \to \infty) = 0.14$, $\pi_3 = P_3(t \to \infty) = 0.06$, $\pi_4 = P_4(t \to \infty) = 0.04$, the throughput is $\Lambda(t = 5.5 \cdot 10^{-5}) = (1 - 0.04) \cdot 2.5 \cdot 10^5 = 234$ Kb/s. These results practically coincide the results obtained in the Sect. 7.1 for the case $\rho < 1$. Moreover, if the jump occurs before the end of the transient mode, then the resulting transient time τ_{tr} is equal to the sum of the transient time of the system before the jump and the transient time after the jump ($\tau_{tr} \approx 5.5 \cdot 10^{-5}$ s). Thus, we have shown that the appearance of jumps in the intensity of the input flow during the transient mode leads to a decrease in the transient time, to a much faster process of buffer loading, and also to a significantly higher probability of packet loss regardless of the ratios of λ and μ before the jump. Therefore, the traffic jumps must be taken into account when performance metrics of telecommunication network are calculated.

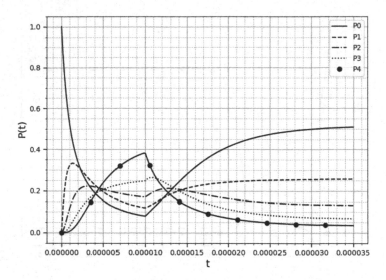

Fig. 8. Dependences of the states probabilities of the buffer for $\rho > 1$ and jump of λ.

6   Conclusion

In this paper, the transient behavior of the one of widely used queuing system such as the single-line system with a finite buffer $M|M|1|n$ is considered. For this, the accurate analytical method of the fundamental matrix of Kolmogorov equations was proposed. The method allows calculating the characteristics of queuing systems, both in transient and stationary modes. The proposed mathematical model allows analyzing the dependences of the system state probabilities on time. Also using this method it can be calculated the throughput of the system, the number of arrived and served packets for the time of transient mode and the transient time and other performance metrics. In addition, the presented approach makes it possible to solve the problems of the synthesis of such systems, since all parameters in the final expressions are presented in an explicit form. Also, the mathematical model of the queuing system in the presence of the arrival and service rate jumps on arbitrary value at an arbitrary moment of time has been proposed. The performance metrics of the telecommunication network model are calculated according to the proposed method. The numerical calculation results of the steady-state characteristics of the considered system coincide with obtained by the well-known methods.

References

1. Vishnevsky, V.M.: Theoretical Foundations for the Design of Computer Networks. Tekhnosfera, Moscow (2003)
2. Lakatos, L., Szeidl, L., Telek, M.: Introduction to Queuing Systems with Telecommunication Applications. Springer, Heidelberg (2013). https://doi.org/10.1007/978-1-4614-5317-8

3. Vishnevsky, V.M., Portnoy, S.L., Shakhnovich, I.V.: Encyclopedia of WiMAX. In: The Path to 4G. Tekhnosfera, Moscow (2010)
4. Giambene, G.: Encyclopedia of WiMAX. In: The Path to 4G. Tekhnosfera, Moscow (2010)
5. Zukerman, M.: Introduction to queueing theory and stochastic teletraffic models. arXiv arXiv:1307.2968 (2013)
6. Cohen, J.W.: The Single Server Queue, 2nd edn. North-Holland, Amsterdam (1982)
7. Prabhu, N.U.: Queues and Inventories. Wiley, New York (1965)
8. Harrison, P.G.: Transient behaviour of queueing networks. J. Appl. Probab. **18**(2), 482–490 (1981)
9. Abate, J., Whitt, W.: Transient behavior of the $M/M/1$ queue: starting at the origin. Queueing Syst. **2**, 41–65 (1987)
10. Whitt, W.: Decomposition approximations for time-dependent Markovian queueing networks. Oper. Res. Lett. **24**, 97–103 (1999)
11. Horváth, A., Angius, A.: Approximate transient analysis of queuing networks by decomposition based on time-inhomogeneous Markov arrival processes. EAI Endorsed Trans. IoT **1**(2) (2014). https://doi.org/10.4108/icst.valuetools
12. Kumar, R., Soodan, B.S.: Transient analysis of a single-server queuing system with correlated inputs and reneging. Reliab. Theor. Appl. **14**(1), 102–106 (2019)
13. Kumar, R., Arivudainambi, D.: Transient solution of an M/M/1 queue with catastrophes. Comput. Math. Appl. **40**, 1233–1240 (2000)
14. Sophia, S.: Transient analysis of a single server queue with chain sequence rates subject to catastrophes, failures and repairs. Int. J. Pure Appl. Math. **119**(15), 1969–1988 (2018)
15. Kumar, R., Krishnamoorthy, A., Pavai, M.S., Sadiq, B.S.: Transient analysis of a single server queue with catastrophes, failures and repairs. Queueing Syst. Theor. Appl. **56**(3–4), 133–141 (2007)
16. Rubino, G.: Transient analysis of Markovian queueing systems: a survey with focus on closed forms and uniformization, chap. 8. In: Advanced Trends in Queueing Theory 2, pp. 269–307 (2021)
17. Chan, R.H., Ma, K.C., Ching, W.K.: Boundary value methods for solving transient solutions of Markovian queueing networks. Appl. Math. Comput. **172**(2), 690–700 (2006)
18. Gantmacher, F.R.: The Theory of Matrices, vol. 1 and vol. 2. Chelsea Publishing Co. (1960)
19. Vytovtov, K., Barabanova, E., Vishnevskiy, V.: Accurate mathematical model of two-dimensional parametric systems based on 2×2 matrix. Commun. Comput. Inf. Sci. **1141**, 199–211 (2019)
20. Vytovtov, K., Barabanova, E.: Mathematical model of four-dimensional parametric systems based on block diagonal matrix with 2×2 blocks. Commun. Comput. Inf. Sci. **1141**, 139–155 (2019)
21. Barbashin, E.A.: Introduction to Stability Theory. Nauka, Moscow (1967)
22. Krivtsova, I., et al.: Implementing a broadcast storm attack on a mission-critical wireless sensor network. In: Mamatas, L., Matta, I., Papadimitriou, P., Koucheryavy, Y. (eds.) WWIC 2016. LNCS, vol. 9674, pp. 297–308. Springer, Cham (2016). https://doi.org/10.1007/978-3-319-33936-8_23

Analysis of Multi-server Loss Queueing System with the Batch Marked Markov Arrival Process

Chesoong Kim[1], Alexander Dudin[2,3]($\boxtimes$), Sergey Dudin[2], and Olga Dudina[2]

[1] Sangji University, Wonju, Kangwon 26339, Republic of Korea
dowoo@sangji.ac.kr
[2] Department of Applied Mathematics and Computer Science,
Belarusian State University, 220030 Minsk, Belarus
dudina@bsu.by
[3] Peoples' Friendship University of Russia (RUDN University),
6 Miklukho-Maklaya Street, Moscow 117198, Russia

Abstract. A multi-server queueing system with various types of customers arriving in accordance with a Batch Marked Markov Arrival Process ($BMMAP$) is considered. The service times have exponential distribution with the rate corresponding to the type of customer. The system does not have a buffer. The customer acceptance discipline is assumed to be partial admission. The operation of the system is described by a multi-dimensional continuous time Markov chain. The stationary distribution of the states of the chain and the key system indicators (such as loss probabilities of customers of different types and the sojourn time distribution) are computed. The results are illustrated numerically.

Keywords: Queueing system with heterogeneous customers · Batch Marked Markov Arrival Process · Loss probability · Sojourn time

1 Introduction

The queueing model considered in this paper is the essential direct generalization of the famous Erlang loss queue of $M/M/N/N$ type which was widely applied for performance evaluation and optimizing different telecommunication networks. This generalization is made into the following three important directions.

- The first direction consists of an account of the need for adequate modeling modern telecommunication networks in which the arrival flows exhibit the existence of dependence of the successive inter-arrival intervals and their high variability. This variability manifests itself via the existence of the alternating time intervals when customers arrive often with the intervals where customers come rarely or even do not come at all. The stationary Poisson process is the most popular in queueing and telegraphic theories arrival process model. However, this process fails to catch the inherent features of real-world

© Springer Nature Switzerland AG 2021
V. M. Vishnevskiy et al. (Eds.): DCCN 2021, LNCS 13144, pp. 182–195, 2021.
https://doi.org/10.1007/978-3-030-92507-9_16

arrival processes. The use of this model leads, as the rule, to the dramatic underestimation of the queue length, waiting time, and loss probabilities. This underestimation is especially significant in the case of the positive correlation of inter-arrival times.

As a good alternative to the model of the stationary Poisson process that accounts the effects of correlation and (or) high variability, the model of the Markov Arrival Process (MAP) was offered in the papers by the research group of the prominent professor Marcel Neuts (including D. Lucantoni, V. Ramaswami, S. Chakravarthy, Q.-M. He, etc.) from the University of Arizona, USA, and, independently and at the same time, by research group by the famous professor G.P. Basharin (including P.P. Bocharov, V. A. Naoumov, A.V. Pechinkin, K.E. Samouylov, etc.) from RUDN University, Russia. Brief surveys of the history and content of the related research can be found in, e.g., [1–3] and references therein. The MAP is completely defined by the finite state continuous-time underlying Markov chain the transition intensities of which without the generation of a customer are given by the matrix D_0 and transition intensities of which with the generation of a customer are given by the matrix D_1. The generator of this Markov chain is the sum of the matrices D_0 and D_1.

- The second direction consists of an account of the possibility of batch arrival of customers that is typical for lots of real systems where the arrived for service entity indeed is a group (batch) of individual customers. To take into account the possibility of batch arrivals, while keeping the opportunity of account of correlation, the model of the Batch Markov Arrival Process ($BMAP$) can be proposed by the mentioned above research groups. More information about the $BMAP$ and its properties as well as about the known results obtained in the analysis of queues with the $BMAP$, see, e.g., [1–4]. The $BMAP$ is an essential generalization of the MAP. It is defined by the finite state continuous-time underlying Markov process the transition intensities of which without the generation of a customer are given by the matrix D_0 and transition intensities of which that lead to the arrival of k customers are given by the matrix D_k, $k \geq 1$.

It is necessary to note that due to the finite number of servers the situation when the size of arriving batch exceeds the number of idle servers is possible. In such a situation, usually, three disciplines of admission (partial admission, complete rejection, and complete admission) are considered, see, e.g., [5]. All these disciplines have their pros and cons and are well studied in the literature. In this paper, we restrict ourselves to consideration of only the partial admission discipline.

- The third direction of generalization is the consideration of the heterogeneous arrival process. Arrival flows in many real systems are heterogeneous. This means that customers of different types have, generally speaking, different requirements for the main indicators of their service and different service time distributions. A more simple than considered in this paper arrival process is the so-called Marked Markov Arrival Process ($MMAP$) introduced by Q.-M. He in [6] as the transparent generalization of the MAP. The $MMAP$

is characterized by the finite state continuous-time underlying Markov chain the transition intensities of which without the generation of a customer are defined by the matrix D_0 and the transition intensities that accompanied by a type-k customer arrival are given by the matrix D_k, $k \geq 1$. The difference from the $BMAP$ is that here the entries of the matrices D_k, $k \geq 1$, are interpreted as intensities of type-k customers arrival while in the $BMAP$ definition they are the intensities of the arrival of a batch consisting of k customers.

In this paper, we assume the $BMMAP$ that generalizes in different directions the $BMAP$ and $MMAP$ arrival processes in which batches of customers of several types arrive. The underlying process of the $BMMAP$ is characterized by the set of matrices D_0, $D_r^{(k)}$ where the entries of the matrices $D_r^{(k)}$ define the intensities of transitions of the arrival of a batch consisting of k customers of type r. For more details see, e.g., [7].

The well-known property of the $M/M/N/N$ type queue, generalization of which we consider in this paper, is the following. The stationary distribution of the number of customers in this system at an arbitrary moment coincides with the analogous distribution in the $M/G/N/N$ type queue with the same arrival process and arbitrary distribution of service time having the same mean value as the exponentially distributed service time in the $M/M/N/N$ queue. In other words, this property means that the stationary distribution of the number of the number of busy servers in the $M/G/N/N$ system at an arbitrary moment depends on the average service time, but does not depend on the shape of the distribution function of the service time. This property is called in the literature the insensitivity (invariance) property. It was strictly proved by B.A. Sevastyanov, see, e.g. [8]. Namely due to the validity of this property, the results by A.K. Erlang obtained only for the $M/M/N/N$ type queue were successfully used for the design of many real-world systems including telephone networks in which, as a rule, the service time distribution is not exponential.

Note, that the invariance property was proven only for the queues with the stationary Poisson arrival process. But this property is not valid, generally speaking, for the system with the MAP. This is shown in [5] via analysis of the $BMAP/PH/N/N$ queue with the phase-type distribution of the service time. It is numerically illustrated in [5] that the stationary distribution of the number of customers in such a system depends not only on the average service time but also at least on the variance of the service time.

In this paper, we suppose that the service time distribution of an arbitrary customer is exponential with the rate depending on the type of the customer. The essential difficulty of the analysis of the model stems from the fact that if one tries to use the simple description of the Markov chain defining the behavior of the system by monitoring the type of a customer in service at each busy server, then the state space of the Markov chain may be very large. If the number of types of the customers equals to R, the number of servers equals to N, the cardinality of the state space of the underlying process of the $BMMAP$ is W, then the state space of the Markov chain has cardinality WT_N where $T_n = \frac{R^{n+1}-1}{R-1}$, $n \in \{1, \ldots, N\}$.

E.g., if $W = 2$, $R = 3$, $N = 11$, the cardinality is equal to 531 440. But in real systems, the number of servers, N, can be much larger than 11 and computations become practically impossible. In this paper, following an idea from [9], the Markov chain describing the behavior of the system is constructed in such a way that the stationary distribution of the chain can be successfully computed for essentially larger values of N.

The paper contains the following sections. The mathematical model of the system under study is presented in Sect. 2. Section 3 is devoted to the definition of the multi-dimensional Markov chain that describes the dynamics of the system and derivation of the infinitesimal generator of this chain. One of the possible numerically stable algorithms to compute the stationary probabilities of the states of the Markov chain is presented. In Sect. 4, expressions for computation of various non-trivial performance measures of the system in terms of the computed stationary probabilities of the states are given. These measures include, in particular, distributions of sojourn times and their averages. Section 5 contains the results of a brief numerical example. Some concluded remarks are presented in Sect. 6.

2 Model Description

We consider a queuing system having N identical servers processing R types of customers whose structure is presented in Fig. 1.

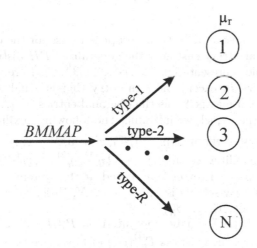

Fig. 1. Structure of the system.

The customers of R types arrive in the batches of different size at the system according to the $BMMAP$ which is defined by the irreducible continuous-time Markov chain ν_t, $t \geq 0$, with a state space $\{1, \ldots, W\}$ and the matrices

D_0, $D_r^{(k)}$, $r = \overline{1,R}$, $k = \overline{1,K_r}$. Here the parameter K_r, $K_r \geq 1$ defines the maximal size of a type r customers batch. The maximal batch size among all types of customers is assumed to be K where $K = \max\{K_r, r = \overline{1,R}\}$. The average intensity of type-r customer arrivals is denoted as λ_r and the total average intensity of customer arrivals is λ. For more information about the $BMMAP$, see, e.g., [12].

If the number of idle servers is greater than or equal to the size of an arriving batch, all the customers of this batch immediately start processing by the servers (service). If at a batch arrival moment the required number of servers is not available, the part of customers occupies the available servers, and the rest of customers, for which there are no idle servers, is lost. So, the partial admission discipline is considered.

The service time of r-type customer has an exponential distribution with the parameter μ_r, $r = \overline{1,R}$.

3 Process of the System States

The following regular irreducible continuous-time Markov chain

$$\xi_t = \{n_t, \nu_t, \eta_t^{(1)}, \eta_t^{(2)}, \ldots, \eta_t^{(R)}\}, \ t \geq 0,$$

describes the behavior of the considered system. Here, during the epoch t,

- n_t denotes the number of busy servers, $n_t = \overline{0,N}$;
- ν_t denotes the state of the underlying process of the $BMMAP$, $\nu_t = \overline{1,W}$;
- $\eta_t^{(r)}$ defines the number of r-type customers on service, $\eta_t^{(r)} = \overline{0,n_t}$, $r = \overline{1,R}$,
  $\sum_{r=1}^{R} \eta_t^{(r)} = n_t.$

To analyse the Markov chain ξ_t, we propose to use for the description of the service time of an arbitrary customer the generalized PH distribution, see [10], with the irreducible representation $(\boldsymbol{\beta}^1, \boldsymbol{\beta}^2, \ldots, \boldsymbol{\beta}^R, S)$. Here, all the entries of the vector $\boldsymbol{\beta}^r$ of size R, except the r-th entry that is equal to 1, are equal to zero and the diagonal matrix S has the diagonal entries $-\mu_1, \ldots, -\mu_R$.

For the use in the sequel, we introduce the following auxiliary matrices:

(a) The components of the matrix $P_n(\boldsymbol{\beta}^r)$, $n = \overline{0, N-1}$, $r = \overline{1,R}$, define the transition probabilities of the process $\{\eta_t^{(1)}, \eta_t^{(2)}, \ldots, \eta_t^{(R)}\}$ at the moment when a new type-r customer is accepted at the system at the epoch when the number of busy servers is n, $0 \leq n < N$. Here, the row vector $\boldsymbol{\beta}^r$ has the form $\boldsymbol{\beta}^r = (\beta_1^r, \beta_2^r, \ldots, \beta_R^r)$.
The matrices $P_n(\boldsymbol{\beta}^r)$ can be computed as $P_0(\boldsymbol{\beta}^r) = \boldsymbol{\beta}^r$ and $P_n(\boldsymbol{\beta}^r) = P_n^{(R-2)}(\boldsymbol{\beta}^r)$. Here, the matrices $P_n^{(l)}(\boldsymbol{\beta}^r)$ of block size $(n+1) \times (n+2)$, $n = \overline{1, N-1}$, are calculated using the following recursive formulas:

$$P_n^{(0)}(\boldsymbol{\beta}^r) = \begin{pmatrix} \beta_{R-1}^r & \beta_R^r & 0 & \cdots & 0 & 0 \\ 0 & \beta_{R-1}^r & \beta_R^r & \cdots & 0 & 0 \\ \vdots & \vdots & \vdots & \ddots & \vdots & \vdots \\ 0 & 0 & 0 & \cdots & \beta_{R-1}^r & \beta_R^r \end{pmatrix},$$

$$P_n^{(l)}(\boldsymbol{\beta}^r)$$

$$
= \begin{pmatrix}
\beta_{R-l-1}^r & \bar{\mathbf{r}}^{(l)} & \mathbf{0} & \mathbf{0} & \cdots & \mathbf{0} & \mathbf{0} \\
\mathbf{0}^T & \beta_{R-l-1}^r I & P_1^{(l-1)}(\boldsymbol{\beta}^r) & O & \cdots & O & O \\
\mathbf{0}^T & O & \beta_{R-l-1}^r I & P_2^{(l-1)}(\boldsymbol{\beta}^r) & \cdots & O & O \\
\vdots & \vdots & \vdots & \vdots & \ddots & \vdots & \vdots \\
\mathbf{0}^T & O & O & O & \cdots & \beta_{R-l-1}^r I & P_n^{(l-1)}(\boldsymbol{\beta}^r)
\end{pmatrix},
$$

$$l = \overline{1, R-2},$$

where the $\bar{\mathbf{r}}^{(l)}$ denotes the vector $\bar{\mathbf{r}}^{(l)} = (\beta_{R-l}^r, \beta_{R-l+1}^r, \ldots, \beta_R^r)$, $l = \overline{1, R-2}$.
Hereafter, I and O are an identity matrix and a zero matrix of an appropriate dimension, correspondingly;

(b) The entries of the matrix $B_n(\boldsymbol{\mu})$ define the intensities of transitions of the process $\{\eta_t^{(1)}, \eta_t^{(2)}, \ldots, \eta_t^{(R)}\}$ when some customer finishes its service. Here $\boldsymbol{\mu} = (\mu_1, \mu_2, \ldots, \mu_R)$.
The matrices $B_n(\boldsymbol{\mu})$, $n = \overline{1, N}$, are calculated as follows:

1. Compute recursively the matrices $B_n^{(l)}(\boldsymbol{\mu})$ by:

$$B_n^{(0)}(\boldsymbol{\mu}) = n\mu_R,$$

$$
B_n^{(l)}(\boldsymbol{\mu}) = \begin{pmatrix}
n\mu_{R-l} I & O & \cdots & O \\
B_1^{(l-1)}(\boldsymbol{\mu}) & (n-1)\mu_{R-l} I & \cdots & O \\
O & B_2^{(l-1)}(\boldsymbol{\mu}) & \cdots & O \\
\vdots & \vdots & \ddots & \vdots \\
O & O & \cdots & \mu_{R-l} I \\
O & O & \cdots & B_n^{(l-1)}(\boldsymbol{\mu})
\end{pmatrix}, \quad l = \overline{1, R-1}.
$$

2. Calculate the matrices $B_n(\boldsymbol{\mu})$ as $B_n(\boldsymbol{\mu}) = B_n^{(R-1)}(\boldsymbol{\mu})$.

Let:
the symbols $\otimes$ and $\oplus$ stand for the Kronecker product and sum of matrices, correspondingly, see [11];
$\mathbf{0}$ is a zero row vector and $\mathbf{e}$ is a column vector consisting of 1's;
$\hat{I}_n = -\text{diag}\{B_n(\boldsymbol{\mu})\mathbf{e}\}$, $n = \overline{1, N}$, where $\text{diag}\{\ldots\}$, i.e., $\hat{I}_n$ is the diagonal matrix with the diagonal elements given by the vector in the brackets;
Let also Q be the generator of the Markov chain ξ_t.

Theorem 1. The generator Q has the following form:

$$
Q = \begin{pmatrix}
Q_{0,0} & Q_{0,1} & Q_{0,2} & Q_{0,3} & \cdots & Q_{0,N} \\
Q_{1,0} & Q_{1,1} & Q_{1,2} & Q_{1,3} & \cdots & Q_{1,N} \\
O & Q_{2,1} & Q_{2,2} & Q_{2,3} & \cdots & Q_{2,N} \\
\vdots & \vdots & \vdots & \vdots & \ddots & \vdots \\
O & O & O & O & \cdots & Q_{N,N}
\end{pmatrix}.
$$

The non-zero matrices $Q_{i,j}$ are defined by:

$$Q_{0,0} = D_0,$$

$$Q_{n,n} = D_0 \oplus \hat{I}_n, \ n = \overline{1, N-1},$$

$$Q_{N,N} = D(1) \oplus \hat{I}_N,$$

$$Q_{n,n-1} = I_W \otimes B_n(\boldsymbol{\mu}), \ n = \overline{1, N},$$

$$Q_{0,n} = \sum_{r=1}^{R} X_n^r, \ n = \overline{1, \min\{N-1, K\}},$$

$$Q_{0,N} = \sum_{r=1}^{R} \sum_{k=N}^{K_r} D_r^{(k)} \otimes [P_0(\beta^r) P_1(\beta^r) \times \cdots \times P_{N-1}(\beta^r)],$$

$$Q_{n,n+k} = \sum_{r=1}^{R} Y_n^{r,k}, \ n = \overline{1, N}, \ k = \overline{1, \min\{N-n-1, K\}},$$

$$Q_{n,N} = \sum_{r=1}^{R} \sum_{k=N-n}^{K_r} D_r^{(k)} \otimes [P_n(\beta^r) P_{n+1}(\beta^r) \times \cdots \times P_{N-1}(\beta^r)], \ n = \overline{1, N},$$

where

$$X_n^r = \begin{cases} D_r^{(n)} \otimes [P_0(\beta^r) P_1(\beta^r) \times \cdots \times P_{n-1}(\beta^r)], & \text{if } n \leq K_r, \\ O, & \text{otherwise,} \end{cases}$$

$$Y_n^{r,k} = \begin{cases} D_r^{(k)} \otimes [P_n(\beta^r) P_{n+1}(\beta^r) \times \cdots \times P_{n+k-1}(\beta^r)], & \text{if } k \leq K_r, \\ O, & \text{otherwise.} \end{cases}$$

Let us enumerate the states of the Markov chain ξ_t in the reverse lexicographic order of the components $\eta_t^{(1)}, \ldots, \eta_t^{(R)}$ and the direct lexicographic order of the component ν_t, and form the vectors $\boldsymbol{\pi}_n$, $n = \overline{0, N}$, of the stationary probabilities.

The vectors $\boldsymbol{\pi}_n$, $n = \overline{0, N}$, can be found as the solution of the system

$$(\boldsymbol{\pi}_0, \boldsymbol{\pi}_1, \ldots, \boldsymbol{\pi}_N) Q = \mathbf{0}, \tag{1}$$

$$(\boldsymbol{\pi}_0, \boldsymbol{\pi}_1, \ldots, \boldsymbol{\pi}_N) \mathbf{e} = 1.$$

The size of the system (1) can be large and the algorithms that effectively account for the structure of the generator Q are recommended. Here, we apply the following algorithm, see [12].

Algorithm

Step 1. Calculate the matrices $P_{i,n}$ as

$$P_{i,N} = -Q_{i,N}(Q_{N,N})^{-1}, \ i = \overline{0, N-1},$$

$$P_{i,n} = -(Q_{i,n} + P_{i,n+1}Q_{n+1,n})(Q_{n,n} + P_{n,n+1}Q_{n+1,n})^{-1},$$
$$i = \overline{0, n-1}, \, n = N-1, N-2, \ldots, 1.$$

Step 2. Obtain the vector $\boldsymbol{\phi}_0$ as the unique solution of the system:

$$\boldsymbol{\phi}_0(Q_{0,0} + P_{0,1}Q_{1,0}) = \mathbf{0}, \quad \boldsymbol{\phi}_0\mathbf{e} = 1.$$

Step 3. Compute the vectors $\boldsymbol{\phi}_n$, $n = \overline{1, N}$, by the following formulas:

$$\boldsymbol{\phi}_n = \sum_{i=0}^{n-1} \boldsymbol{\phi}_i P_{i,n}, \, n = \overline{1, N}.$$

Step 4. Calculate the constant $c = \left(\sum\limits_{n=0}^{N} \boldsymbol{\phi}_n\mathbf{e} \right)^{-1}.$

Step 5. Obtain the stationary distribution vectors $\boldsymbol{\pi}_n$, $n = \overline{0, N}$, as

$$\boldsymbol{\pi}_n = c\boldsymbol{\phi}_n.$$

The end.

4 Performance Measures

After obtaining the vectors $\boldsymbol{\pi}_n$, we can compute a whole variety of stationary performance indicators of the queue. Formulas for calculation of some of them are given below.

The average number of customers in the system is

$$N_{customers} = \sum_{n=1}^{N} n\boldsymbol{\pi}_n\mathbf{e}.$$

The intensity of the output flow of the successfully serviced customers is

$$\lambda_{out} = \sum_{n=1}^{N} \boldsymbol{\pi}_n(I_W \otimes B_n(\boldsymbol{\mu}))\mathbf{e}.$$

The intensity of the output flow of type-r customers that successfully receive service is

$$\lambda_{out}^{(r)} = \sum_{n=1}^{N} \boldsymbol{\pi}_n(I_W \otimes B_n(\boldsymbol{\mu}_r))\mathbf{e} \quad \text{where} \quad \boldsymbol{\mu}_r = \underbrace{(0,0,\ldots,0,\mu_r,0,\ldots,0)}_{R}, \, r = \overline{1, R}.$$

The average number of type-r customers in the system is

$$N_{customers}^{(r)} = \frac{\lambda_{out}^{(r)}}{\mu_r}, \, r = \overline{1, R}.$$

The loss probability of an arbitrary customer is

$$
P_{loss} = 1 - \frac{\lambda_{out}}{\lambda} = \frac{1}{\lambda} \sum_{n=0}^{N} \pi_n \left[\left(\sum_{r=1}^{R} \sum_{k=N-n+1}^{K_r} (k-N+n) D_r^{(k)} \right) \otimes I_{T_n} \right] \mathbf{e}.
$$

The loss probability of an arbitrary customer of type-r is

$$
P_{loss}^{(r)} = 1 - \frac{\lambda_{out}^{(r)}}{\lambda_r} = \frac{1}{\lambda_r} \sum_{n=0}^{N} \pi_n \left[\left(\sum_{k=N-n+1}^{K_r} (k-N+n) D_r^{(k)} \right) \otimes I_{T_n} \right] \mathbf{e}, \; r = \overline{1, R}.
$$

The Laplace-Stieltjes transform $v(s)$ of the distribution of an arbitrary customer sojourn time in the system is defined by

$$
v(s) = P_{loss} + \frac{1}{\lambda} \sum_{n=0}^{N} \pi_n \sum_{r=1}^{R} \left[\sum_{k=1}^{\min\{K_r, N-n\}} k D_r^{(k)} \right.
$$

$$
\left. + \sum_{k=N-n+1}^{K_r} (N-n) D_r^{(k)} \right] \otimes I_{T_n} \mathbf{e} \frac{\mu_r}{\mu_r + s}.
$$

The Laplace-Stieltjes transform $v^{(not-lost)}(s)$ of the distribution of the sojourn time of an arbitrary customer, which is not lost, is computed by

$$
v(s)^{(not-lost)} = \frac{1}{\lambda(1 - P_{loss})} \sum_{n=0}^{N} \pi_n \sum_{r=1}^{R} \left[\sum_{k=1}^{\min\{K_r, N-n\}} k D_r^{(k)} \right.
$$

$$
\left. + \sum_{k=N-n+1}^{K_r} (N-n) D_r^{(k)} \right] \otimes I_{T_n} \mathbf{e} \frac{\mu_r}{\mu_r + s}.
$$

The average sojourn time v_1 of an arbitrary customer in the system is computed by

$$
v_1 = \frac{1}{\lambda} \sum_{n=0}^{N} \pi_n \sum_{r=1}^{R} \left[\sum_{k=1}^{\min\{K_r, N-n\}} k D_r^{(k)} + \sum_{k=N-n+1}^{K_r} (N-n) D_r^{(k)} \right] \otimes I_{T_n} \mathbf{e} \frac{1}{\mu_r}.
$$

The average sojourn time $v_1^{(not-lost)}$ of an arbitrary customer, which is not lost, is computed by

$$
v_1^{(not-lost)} = \frac{1}{\lambda(1 - P_{loss})} \sum_{n=0}^{N} \pi_n \sum_{r=1}^{R} \left[\sum_{k=1}^{\min\{K_r, N-n\}} k D_r^{(k)} \right.
$$

$$
\left. + \sum_{k=N-n+1}^{K_r} (N-n) D_r^{(k)} \right] \otimes I_{T_n} \mathbf{e} \frac{1}{\mu_r}.
$$

The distribution function $V(x)$ of an arbitrary customer sojourn time in the system is computed by

$$V(x) = P_{loss} + \frac{1}{\lambda} \sum_{n=0}^{N} \pi_n \sum_{r=1}^{R} \left[\sum_{k=1}^{\min\{K_r, N-n\}} kD_r^{(k)} \right.$$

$$+ \left. \sum_{k=N-n+1}^{K_r} (N-n)D_r^{(k)} \right] \otimes I_{T_n} e(1 - e^{-\mu_r x}).$$

The distribution function $V^{(not-lost)}(x)$ of the sojourn time of an arbitrary customer, which is not lost, is computed by

$$V^{(not-lost)}(x) = \frac{1}{\lambda(1 - P_{loss})} \sum_{n=0}^{N} \pi_n \sum_{r=1}^{R} \left[\sum_{k=1}^{\min\{K_r, N-n\}} kD_r^{(k)} \right.$$

$$+ \left. \sum_{k=N-n+1}^{K_r} (N-n)D_r^{(k)} \right] \otimes I_{T_n} e(1 - e^{-\mu_r x})$$

5   Numerical Example

In this brief numerical example, we assume that $R = 3$ types of customers arrive at the system. The arrival flow is the $BMMAP$ defined by the matrices

$$D_0 = \begin{pmatrix} -0.4769501 & 0 \\ 0 & -0.47839896 \end{pmatrix},$$

$$D_1^{(1)} = \begin{pmatrix} 0.00579524 & 0.00637479 \\ 0.0055055 & 0.00579526 \end{pmatrix}, \quad D_1^{(2)} = \begin{pmatrix} 0.0115906 & 0.00898266 \\ 0.011011 & 0.0121701 \end{pmatrix},$$

$$D_1^{(3)} = \begin{pmatrix} 0.0159369 & 0.0153574 \\ 0.0147779 & 0.017096 \end{pmatrix}, \quad D_1^{(4)} = \begin{pmatrix} 0.0173858 & 0.0176756 \\ 0.0179654 & 0.017096 \end{pmatrix},$$

$$D_2^{(1)} = \begin{pmatrix} 0.0292661 & 0.0582424 \\ 0.0875084 & 0.0866386 \end{pmatrix}, \quad D_2^{(2)} = \begin{pmatrix} 0.0434645 & 0.101417 \\ 0.0437542 & 0.101128 \end{pmatrix},$$

$$D_3^{(1)} = \begin{pmatrix} 0.0434645 & 0.0289763 \\ 0 & 0 \end{pmatrix}, \quad D_3^{(2)} = \begin{pmatrix} 0.0579526 & 0.00057951 \\ 0.0289763 & 0 \end{pmatrix},$$

$$D_3^{(3)} = \begin{pmatrix} 0.0144882 & 0 \\ 0.0289763 & 0 \end{pmatrix}.$$

The maximum batch size for the customers of type-1 is equal to four. For other types, the maximum is equal to two and three, correspondingly. The average arrival rate of type-1 customers is $\lambda_1 = 0.290482$, the average arrival rate of type-2 customers is $\lambda_2 = 0.420512$, and the average arrival rate of type-3 customers is $\lambda_3 = 0.189006$. The total average arrival rate is $\lambda = 0.9$. The service rate of type-1 customers is $\mu_1 = 0.1$, the service rate of type-2 customers is $\mu_2 = 0.1$, and the service rate of type-3 customers is $\mu_3 = 0.4$.

Let us change the number of servers N in the range $[1, 50]$ with step 1 and analyze the dependence of the main system performance indicators on the number of servers N.

Figure 2 shows the mean average number $N_{customers}^{(r)}$ of type-r, $r = 1, 2, 3$, customers in the system as the function of the number of servers N.

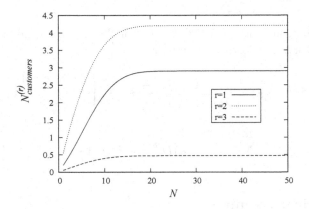

Fig. 2. Dependence of the average number $N_{customers}^{(r)}$ of type-r, $r = 1, 2, 3$, customers in the system on the number of servers N

The value of $N_{customers}^{(r)}$ increases with the growth of N because the larger value of N implies less chance to lose a type-r customer upon arrival due to the lack of servers. The values of $N_{customers}^{(r)}$ are ordered as

$$N_{customers}^{(2)} > N_{customers}^{(1)} > N_{customers}^{(3)}.$$

This order is easily explained, first of all, by the proportion of the intensities of the arrival of customers of the different types: $\lambda_2 > \lambda_1 > \lambda_3$. The larger arrival rate implies, generally speaking, the larger average number of the customers of the corresponding types that stay in the system.

Another essential factor defining the presented ordering of the values of $N_{customers}^{(r)}$ is the different maximum batch sizes for customers of different types. Type-2 customers arrive in batches with the maximum size equal to two (which is minimal among all types of customers) and have lower chances to be rejected because the available number of idle servers is less than the size of the arrived batch. This observation is supported by the shape of the curves in Figs. 3, 4 and 5.

Figure 3 shows the mean intensity $\lambda_{out}^{(r)}$ of the output flow of customers of type-r, $r = 1, 2, 3$, that are serviced successfully as the function of the number of servers N.

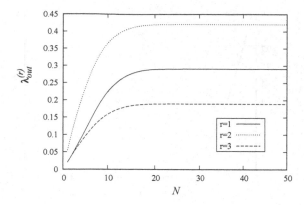

Fig. 3. Dependence of the average intensity of the output flow of the successfully serviced type-r, $r = 1, 2, 3$, customers $\lambda_{out}^{(r)}$ on the number of servers N

Figures 4 and 5 show the loss probability $P_{loss}^{(r)}$ of an arbitrary type-r, $r = 1, 2, 3$, customer and the total loss probability P_{loss} of an arbitrary customer as the functions of the number of servers N.

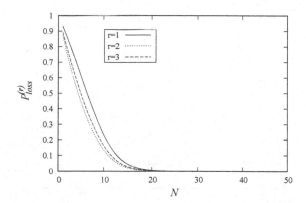

Fig. 4. Dependence of the loss probability $P_{loss}^{(r)}$ of an arbitrary type-r, $r = 1, 2, 3$, customer on the number of servers N

The presented results allow us to quantify the main performance indicators of the system what helps us to solve various optimization problems. In the simplest settings, the optimization problem can be formulated to define the minimum number of servers N that is necessary to guarantee that the loss probability of an arbitrary customer will not exceed the pre-assigned value ϵ.

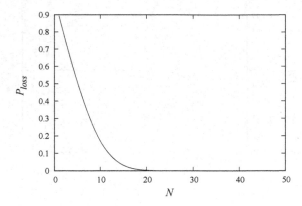

Fig. 5. Dependence of the total loss probability P_{loss} on the number of servers N

Let us fix $\epsilon = 10^{-5}$. In this case, from the Fig. 5 we can obtain that the optimal value of the number of servers $N^* = 30$. The shape of the sojourn time distribution function of an arbitrary non-lost customer $V^{(not-lost)}(x)$ in the system for this value of N is presented in Fig. 6.

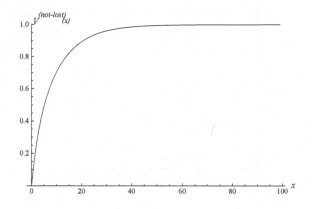

Fig. 6. Sojourn time distribution function of an arbitrary not-lost customer $V^{(not-lost)}(x)$ in the system

6   Conclusion

We have analyzed a multi-server queueing system with the correlated flow of the batches of heterogeneous customers and without an input buffer. We obtained the stationary distribution of the system states and the main performance indicators of the system including the probability of losses of different types of customers and Laplace-Stieltjes transform of sojourn time distribution. Numerical results are presented.

Disciplines of complete rejection and complete admission can be analyzed using the results of this paper. In principle, generalization to the case of the phase-type distribution of service times seems to be possible. But the only a relatively small number of servers can be dealt with in the numerical work for such a generalization.

Acknowledgments. This work was supported by the Basic Science Research Program through the National Research Foundation of Korea (NRF) funded by the Ministry of Education (Grant No. NRF-2020R1A2C1006999) and by the RUDN University Strategic Academic Leadership Program.

References

1. Chakravarthy, S.R.: The batch Markovian arrival process: a review and future work. In: Advances in Probability Theory and Stochastic Processes, pp. 21–29. Notable Publications Inc., New Jersey (2001)
2. Lucantoni, D.: New results on the single server queue with a batch Markovian arrival process. Commun. Stat. Stoch. Models **7**, 1–46 (1991)
3. Dudin, A.N., Klimenok, V.I., Vishnevsky, V.M.: The Theory of Queuing Systems with Correlated Flows. Springer, Cham (2020). https://doi.org/10.1007/978-3-030-32072-0
4. Vishnevskii, V.M., Dudin, A.N.: Queueing systems with correlated arrival flows and their applications to modeling telecommunication networks. Autom. Remote. Control. **78**(8), 1361–1403 (2017). https://doi.org/10.1134/S000511791708001X
5. Klimenok, V.I., Kim, C.S., Orlovsky, D.S., Dudin, A.N.: Lack of invariant property of Erlang $BMAP/PH/N/0$ model. Queueing Syst. **49**, 187–213 (2005)
6. He, Q.M.: Queues with marked customers. Adv. Appl. Probab. **28**, 567–587 (1996)
7. Al-Begain, K., Dudin, A.N., Mushko, V.V.: Novel queueing model for multimedia over downlink in 3.5 G wireless network. J. Commun. Softw. Syst. **2**, 68–80 (2006)
8. Kovalenko, I.N.: B.A. Sevastyanov's famous theorem. Proc. Steklov Inst. Math. **282**, 124–126 (2013). https://doi.org/10.1134/S0081543813060114
9. Ramaswami, V., Lucantoni, D.M.: Algorithms for the multi-server queue with phase type service. Stoch. Model. **1**, 393–417 (1984)
10. Dudin, A., Kim, C., Dudina, O., Dudin, S.: Multi-server queueing system with a generalized phase-type service time distribution as a model of call center with a call-back option. Ann. Oper. Res. **239**(2), 401–428 (2014). https://doi.org/10.1007/s10479-014-1626-2
11. Graham, A.: Kronecker Products and Matrix Calculus with Applications. Ellis Horwood, Cichester (1981)
12. Dudin, A., et al.: Analysis of single-server multi-class queue with unreliable service, batch correlated arrivals, customers impatience, and dynamical change of priorities. Mathematics **9**(11), 1–17 (2021)

Two Types of Single-Server Queueing Systems with Threshold-Based Renovation Mechanism

Viana C. C. Hilquias[1], I. S. Zaryadov[1,2](✉), and T. A. Milovanova[1]

[1] Department of Applied Probability and Informatics, Peoples' Friendship University of Russia (RUDN University), Miklukho-Maklaya Street 6, Moscow 117198, Russia
{1042195028,zaryadov-is,milovanova-ta}@rudn.ru
[2] Institute of Informatics Problems, FRC CSC RAS, IPI FRC CSC RAS, 44-2 Vavilova Street, Moscow 119333, Russia

Abstract. This work is devoted to the analysis of two $GI/M/1$ infinite capacity queuing systems with threshold control of the renovation mechanism, which is responsible for the probabilistic dropping of claims entered into the system upon service completions. In the system of the first type, the threshold value is the value of the queue length, upon passing which the renovation mechanism is activated. For a system of the second type, the threshold value not only activates the renovation, but also specifies the area in the queue where from the customers cannot be dropped. For both systems, the main stationary time-probabilistic characteristics are derived and also the results of simulation, which illustrate the performance of the queues, are presented.

Keywords: Queue management · Renovation mechanism · Threshold · Time-probabilistic characteristic · GPSS

1 Introduction

The modern active queue management (AQM) algorithms are required for network congestion reducing in Internet routers [1] and based on some rules of intelligent drop of incoming network packets inside a buffer, when it (buffer) becomes full or gets close to becoming full.

The main active queue management algorithms are algorithms of the RED (Random Early Detection or Random Exponential Marking) family (see, for example, [2–18]), or Blue [19] and stochastic fair Blue (SFB) algorithms, Explicit

This paper has been supported by the RUDN University Strategic Academic Leadership Program (Zaryadov I.S.—mathematical model development, Viana C. C. Hilquias—simulation model development). Also the publication has been funded by Russian Foundation for Basic Research (RFBR) according to the research project No. 19-07-00739 and No. 20-07-00804 (I. S. Zaryadov, T. A. Milovanova—numerical analysis based on the obtained analytical results).

Congestion Notification (ECN) [3], or Adaptive virtual queue algorithm [16–18], or controlled delay (CoDel) [20,21]).

So the congestion avoidance problem is considered to be actual [26–30], and to solve this important problem the IETF working group on "Active Queue Management and Packet Scheduling" [22] was created, and the main task of this working group is to research and standardize some more novel AQM algorithms. As shown above, a fairly large number of new AQM schemes have been proposed [17,23–31]. It should be noted that the analysis of the performance of most of the proposed algorithms is carried out by modeling (for example, [24,25,32]), and not by building analytical models based on the queuing theory. The number of works in which not only simulation, but also analytical results are presented is rather small. Among such works are the following [33–38].

In this paper, the mathematical model of active queue management ("RED-like") algorithm based on renovation mechanism with one threshold value is presented. We use the word "RED-like" because in the standard RED algorithms and its modifications (see [39,53–58,62]) the possible dropping of a packet occurs at the times of packets' arrivals and the control parameter(s) depend on the queue length (see [2] for classic RED). Unlike RED type AQMs the renovation is based on completely different idea: the possible dropping is synchronised with the times of service completion [39,53–55,58–63].

This paper presents some new results of the ongoing research of queuing systems with renovation as the AQM. We consider two types of renovation mechanism with one threshold value. The first type corresponds to the case when the threshold value determines the boundary in the queue, starting from which the renovation mechanism is activated. The second type covers the case when the threshold value specifies not only the moment of the renovation mechanism activation, but also the area in the queue, where from the arrived packets cannot be dropped.

The structure of the paper is as follows. Section 2 presents the main time-probabilistic characteristics for the queuing system under the first setting, Sect. 3 is devoted to the main time-probabilistic characteristics for the queuing system under the second setting. Simulation results are presented and analyzed in the Sect. 4. The last section concludes the paper with the short discussion.

2   The First Type System

First, we will consider the system of the first type, namely the $GI/M/1/\infty$ system depicted in the Fig. 1, with the implemented one threshold renovation mechanism.

The main idea of the renovation mechanism is, that a packet (at the time of the end of its service and before leaving the system) can either completely empty the queue (drop all other packets from the buffer) with some nonzero probability q, or will leave the system without discarding anything with additional probability $p = 1 - q$.

Thus, the queueing systems with renovation can be considered as something similar to queues with losses due to arrival of negative customers [40–42]) (or

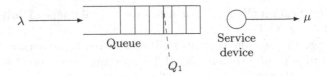

Fig. 1. Queuing system type 1

disasters [43–46]), when the incoming flow of signals cause the buffer to drop some or all the packets, or to queues with unreliable servers, which cause the packet dropping [47–52]).

For the considered system with the renovation mechanism, defined in [39, 53], and with threshold-based control of renovation probability q (this type of hysteretic load control is presented in [64,65])—the threshold value Q_1 sets the boundary value for the buffer and, if the current queue length is greater than Q_1 the renovation mechanism will be launched.

When deriving the main characteristics of the system, we will assume that packets entering the system are serviced in the order of arrival (the FCFS order).

2.1   The Stationary Probability Distribution

Let π_i be the steady-state for the embedded Markov chain probability that at the moment of a new request arrival exactly i other packets will be in the system. Then the steady-state probabilities will be obtained by (1):

$$\pi_0 = \sum_{i=0}^{\infty} \pi_i A_{i+1}^*, \quad \pi_i = \sum_{j=i-1}^{\infty} \pi_j A_{j-i+1}, \tag{1}$$

where

$$A_{i+1-j} = \int_0^{\infty} p^{i-Q_1} \frac{(\mu x)^{i+1-j}}{(i+1-j)!} e^{-\mu x} dA(x), 0 < j \le Q_1, \tag{2}$$

for the case, when the current queue length has not exceeded the threshold value, and

$$A_{i+1-j} = \int_0^{\infty} \frac{(p\mu x)^{i+1-j}}{(i+1-j)!} e^{-\mu x} dA(x), \tag{3}$$

if the threshold value Q_1 is exceeded. The probabilities A_j^* are determined by the formula:

$$A_{i+1}^* = 1 - \sum_{j=0}^{i+1} A_j, \quad i \ge 0. \tag{4}$$

It is not difficult to prove that $\pi_i = \pi_{Q_1+1} \cdot g^{i-Q_1-1}$, where $g = \alpha(\mu(1 - pg)), g \in (0,1)$, here $\alpha(s)$ is the Laplace-Stieltjes transform for the incoming flow distribution function.

2.2   The Service Probability and the Loss Probability for a Received Packet

Let $p^{(serv)}$ be the probability of an incoming packet (request) to be served and $p^{(loss)}$—the probability of a received in the system packet to be dropped by renovation mechanism.

The $p^{(serv)}$ probability is determined by (5)

$$p^{(serv)} = \pi_0 + \sum_{i=0}^{\infty} \pi_{i+1} \int_0^{\infty} p_{i,0}^{(serv)}(x)dx, \qquad (5)$$

where for $i + j \leq Q_1$

$$p_{i,j}^{(serv)}(x) = \frac{(\mu x)^{i+1}}{(i+1)!}e^{-\mu y}\overline{A}(x) + \int_0^y \sum_{k=0}^i \frac{(\mu y)^k}{k!}e^{-\mu y}dA(y)p_{i-k,j+1}^{(serv)}(x,y), \qquad (6)$$

and for $i + j > Q_1$

$$p_{i,j}^{(serv)}(x) = \overline{A}(x) \cdot \frac{(\mu x)^{i+1}}{(i+1)!}e^{-\mu x} \cdot p^{min(i+1,i+j+1-Q_1)}$$
$$+ \int_0^x \sum_{k=0}^i \frac{(\mu y)^k}{k!}e^{-\mu y} \cdot p_{k,j}^{(sevr)}dA(y)p_{i-k,j+1}^{(serv)}(x-y). \qquad (7)$$

Similarly, we determine the $p^{(loss)}$ by the Eq. (8)

$$p^{(loss)} = \sum_{i=1}^{\infty} \pi_i \int_0^{\infty} p_{i-1,0}^{(loss)}(x)dx, \qquad (8)$$

where

$$p_{i,j}^{(loss)}(x) = \int_0^x \sum_{k=0}^i \frac{(\mu y)^k}{k!}e^{-\mu y}dA(y)p_{i-k,j+1}^{(loss)}(x-y), \quad i+j \leq Q_1, \qquad (9)$$

$$p_{i,j}^{(loss)}(x) = \sum_{k=1}^{i+1+j-Q_1} qp^{k-1}\frac{(-\mu x)^k}{k!}e^{\mu x}\overline{A}(x)$$
$$+ \int_0^{\infty} \sum_{k=0}^i \frac{(\mu y)^k}{k!}e^{-\mu y} \cdot p_{k,j}^{(serv)}dA(y)p_{(i-k,j+1)}^{(loss)}, \quad i+j > Q_1. \qquad (10)$$

2.3   Time Characteristics for a Served Packet and a Dropped Packet

Let $W^{(serv)}(x)$ and $W^{(loss)}(x)$ be the distribution functions of the time spent in the queue by the served and dropped packets. Then

$$W^{(serv)}(x) = \frac{1}{p^{(serv)}} \sum_{i=0}^{\infty} W_{i,0}^{(serv)}(x)\pi_i, \tag{11}$$

where $W_{i,j}^{(serv)}(x)$ —the time distribution function for the served packet, if there are i other packets in the queue before the considered request and there are j other packets after it. For densities $w_{i,j}^{(serv)}(x) = \left(W_{i,j}^{(serv)}(x)\right)'$, we obtain:

$$w_{i,j}^{(serv)}(x) = \overline{A}(x) \cdot \frac{\mu^{i+1} x^i}{i!} e^{-\mu x} + \int_0^x \sum_{k=0}^i \frac{(\mu y)^k}{k!} e^{-\mu y} dA(y) \cdot w_{i-k,j+1}^{(serv)}(x-y), \quad i+j \le Q_1,$$

$$\tag{12}$$

$$w_{i,j}^{(serv)}(x) = \overline{A}(x) \cdot \frac{\mu^{i+1} x^i}{i!} e^{-\mu x} p_{i+1,j}^{(serv)}$$

$$+ \int_0^x \sum_{k=0}^i \frac{(\mu y)^k}{k!} e^{-\mu y} p_{k,j}^{(serv)} dA(y) \cdot w_{i-k,j+1}^{(serv)}(x-y), \quad i+j > Q_1, \tag{13}$$

$$p_{i+1,j}^{(serv)} = \begin{cases} p^{i+1}, & j \ge Q_1, \\ p^{i+j+1-Q_1}, & j < Q_1. \end{cases} \tag{14}$$

Similarly

$$W^{(loss)}(x) = \frac{1}{p^{(loss)}} \sum_{i=1}^{\infty} W_{i,0}^{(loss)}(x), \tag{15}$$

where $W_{i,j}^{(loss)}(x)$—the time distribution function for the dropped packet, if there are i other requests in the queue before the considered one and j other packets after it. For densities $w_{i,j}^{(loss)}(x) = \left(W_{i,j}^{(loss)}(x)\right)'_x$, we get:

$$w_{i,j}^{(loss)}(x) = \int_0^x \sum_{k=0}^i \frac{(\mu y)^k}{k!} e^{-\mu y} dA(y) w_{i-k,j+1}^{(loss)}(x-y), \quad i+j+1 \le Q_1, \tag{16}$$

$$w_{i,j}^{(loss)}(x) = \sum_{k=1}^{min(i,i+1+j-Q_1)} \frac{(\mu)^k x^{k-1}}{(k-1)!} e^{-\mu x} \cdot p^{k-1} \cdot q \cdot \overline{A}(x)$$

$$+ \int_0^x \sum_{k=0}^i \frac{(\mu y)^k}{k!} e^{-\mu y} \cdot p^{min(k,i+j+1-Q_1)} dA(y) w_{i-k,j+1}^{(loss)}(x-y), \quad i+j+1 \ge Q_1,$$

$$0 \le i \le Q_1 - 1, \tag{17}$$

$$w_{i,j}^{(loss)}(x) = \sum_{k=1}^{min(i,i+1+j-Q_1)} \frac{(\mu)^k x^{k-1}}{(k-1)!} e^{-\mu x} \cdot p^{k-1} \cdot q \cdot \overline{A}(x)$$

$$+ \int_0^x \sum_{k=0}^i \frac{(\mu y)^k}{k!} e^{-\mu y} \cdot p^{min(k,i+j+1-Q_1)} dA(y) w_{i-k,j+1}^{(loss)}(x-y), \quad i \geq Q_1. \quad (18)$$

3   The Second Type System

Now, we will investigate the $GI/M/1/\infty$ system, depicted in the Fig. 2, where the value Q_1 not only activates the renovation mechanism (if the current queue length becomes greater than the threshold value Q_1), but also sets a "safe" zone in the buffer, from which none of packets can be dropped.

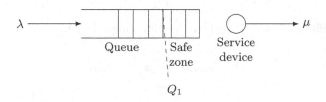

Fig. 2. Queuing system type 2

This queueing system is also investigated by using the Markov chain, embedded upon arrival times, and arriving packets are also served on a FCFS order.

3.1   The Stationary Probability Distribution

Just as in Sect. 2.1, let π_i be the steady-state (for the embedded Markov chain) probability that in the system there is exactly i packets at the time of a new request arrival, and the system for these steady-state probabilities has the form:

$$\pi_0 = \sum_{i=0}^{\infty} \pi_i \tilde{A}_{i+1}, \quad \pi_i = \sum_{j=i-1}^{\infty} \pi_j A_{j-i+1}, \quad (19)$$

where

$$A_{i+1-j} = \frac{-(1)^{i+1-j} \mu^{i+1-j}}{(i+1-j)!} \alpha^{(i+1-j)}(\mu), \quad 0 \leq i \leq Q_1 + 1, \quad (20)$$

$$A_{i+1-j} = \frac{(-p\mu)^{i+1-j}}{(i+1-j)!} \alpha^{i+1-j}(\mu), \quad i > Q_1 + 1, j \geq Q_1 + 1, \quad (21)$$

$$A_{i+1-Q_1} = \frac{(-\mu p)^{i+1-Q_1}}{(i+1-Q_1)!}\alpha^{(i+1-Q_1)}(\mu)$$

$$+ \sum_{k=1}^{i+1-Q_1} \frac{(-\mu)^k}{k!} \cdot p^{k-1}q\alpha^k(\mu), \quad i > Q_1+1, j = Q_1, \quad (22)$$

$$A_{i+1-j} = \int_0^\infty \int_0^x A_{i+1,Q_1}(y)dy A^*_{Q_1,j}(x-y)dA(x), \quad i > Q_1+1, 0 < j < Q_1, \quad (23)$$

where

$$A_{i,Q-1}(y) = \frac{(\mu y)^{i+1-Q_i}}{(i+1-Q_i)!}e^{-\mu y}p^{i+1-Q_i} + \sum_{k=1}^{i+1-Q_i}\frac{(\mu y)^k}{k!}e^{-\mu y}p^{k-1}q, \quad (24)$$

$$A^*_{Q_1,j}(x-y) = \frac{(\mu(x-y))^{Q_1-j}}{(Q_1-j)!}e^{-\mu(x-y)}. \quad (25)$$

The probabilities $\tilde{A}_i$ are defined by the following relation:

$$\tilde{A}_{i+1} = 1 - \sum_{j=0}^{i+1}A_j, \quad i \geq 0. \quad (26)$$

It is also not difficult to prove that $\pi_i = \pi_{Q_1+1} \cdot g^{i-Q_1-1}$ where $g = \alpha(\mu(1-pg)), g \in (0,1)$.

3.2   Service and Loss Probability for the Incoming Packet

Let $p^{(serv)}$ be the probability of an incoming packet to be served and $p^{(loss)}$—the probability of an incoming packet to be dropped (after entering the system). Then

$$p^{(serv)} = 1 - \pi_{Q_1+1} \cdot \frac{q}{(1-g)(1-pg)}, \quad p^{(loss)} = \pi_{Q_1+1}\frac{q}{(1-g)(1-gp)}, \quad (27)$$

where q—the drop probability.

3.3   Time Characteristics of Queuing System

In the terminology of the Sect. 2.3 we get

$$w_i^{(serv)}(x) = \frac{\mu^i x^{i-1}}{(i-1)!}e^{-\mu x}, \quad i = \overline{1, Q_1}, \quad (28)$$

$$w_i^{(serv)}(x) = p^i\frac{\mu^i x^{i-1}}{(i-1)!}e^{-\mu x}, \quad i > Q_1, \quad (29)$$

Then:

$$w^{(serv)}(x) = \frac{1}{p^{(serv)}} \left(\sum_{i=1}^{Q_1} \frac{\mu^i x^{i-1}}{(i-1)!} e^{-\mu x} \pi_i + \sum_{i=Q_1+1}^{\infty} p^{i-Q_1} \frac{\mu^i x^{i-1}}{(i-1)!} e^{-\mu x} \pi_i \right).$$

(30)

In terms of the Laplace-Stieltjes transformation, we get that

$$\omega^{serv}(s) = \frac{1}{p^{(serv)}} \left(\sum_{i=0}^{Q_1} \left(\frac{\mu}{\mu+s} \right)^i \pi_i + p \left(\frac{\mu}{\mu+s} \right)^{Q_1+1} \pi_{Q_1+1} \frac{\mu+s}{\mu+s-p\mu g} \right).$$

(31)

The average waiting time for a served packet is:

$$w^{(serv)} = \frac{1}{p^{(serv)}} \left(\sum_{i=0}^{Q_1} \frac{i}{\mu} \pi_i + p\pi_{Q_1+1} \left(\frac{Q_1+1}{\mu-p\mu+g} + \frac{p^2}{\mu(1-pg)^2} \right) \right). \quad (32)$$

For a dropped packet:

$$w^{(loss)}(x) = \frac{1}{p^{(loss)}} w_i^{(loss)}(x) \pi_i,$$

(33)

$$w_{Q_1+i}^{(loss)}(s) = \sum_{j=1}^{i} qp^{j-1} \left(\frac{\mu}{\mu+s} \right)^j, i \geq 1,$$

(34)

$$\omega^{(loss)}(s) = \frac{1}{p^{(loss)}} \sum_{i=0}^{\infty} w_i^{(loss)}(s) \pi_i = \frac{1}{p^{(loss)}} \cdot \frac{q\pi_{Q_1+1}}{1-g} \cdot \frac{\mu}{\mu+s-\mu pg}. \quad (35)$$

4   Simulation Results

This section presents the results of simulation for both systems, with the system of the first type denoted as **sys.1**, and the system of the second type—as **sys.2**.

The parameters of the first simulation are following: the threshold value $Q_1 = 30$, the arrival rate is 15 tasks per 1 unit of time, the rate of service is 13 task per 1 unit of time, and the simulation time is 100000 for different values of renovation probability q.

According to the GPSS simulation data presented in Table 1 the probability of dropping an accepted task for the first type system is higher than for a system of the second type, while the average waiting time or an undropped task is only 10% less than for a system of the second type. Thus, the second model is preferable to the first, unless the waiting time is critical.

Table 1. Simulation results for different values of drop probability q

Values of drop prob. q		0.0025	0.005	0.01	0.025	0.05	0.1
Generated tasks	sys.1	1502648	1499354	1502810	1501159	1501905	1499438
	sys.2	1500105	1500851	1498361	1501464	1501290	1500148
Serviced tasks	sys.1	1193383	1188527	1181532	1176373	1171087	1168985
	sys.2	1201271	1200236	1201447	1204133	1200470	1202273
Dropped tasks	sys.1	309218	310724	321254	324777	330785	330418
	sys.2	298712	300414	296886	297301	300793	297829
Probability of servicing	sys.1	0.7942	0.7927	0.7862	0.7836	0.7797	0.7796
	sys.2	0.8008	0.7997	0.8018	0.8020	0.7996	0.8014
Probability of dropping	sys.1	0.2058	0.2072	0.2138	0.2164	0.2202	0.2204
	sys.2	0.1991	0.2002	0.1981	0.1980	0.2004	0.1985
Average queue length	sys.1	108	58	35	21	17	14
	sys.2	130	81	55	39	34	31
Maximum queue length	sys.1	974	446	291	153	114	91
	sys.2	785	572	325	173	138	101
Average delay	sys.1	7.271	3.905	2.384	1.448	1.161	1.024
	sys.2	8.663	5.395	3.651	2.629	2.277	2.062

Table 2. Simulation results for different values of the threshold Q_1

Threshold Q_1 value		10	30	50	70	100
Generated tasks	sys.1	1500255	1502810	1502724	1502682	1500896
	sys.2	1499263	1498361	1499761	1500743	1498807
Serviced tasks	sys.1	1173281	1181532	1189273	1190678	1192209
	sys.2	1198212	1201447	1201250	1201295	1202179
Dropped tasks	sys.1	326959	321254	313381	308397	308609
	sys.2	300989	296886	298440	299369	296418
Probability of servicing	sys.1	0.7821	0.7862	0.7914	0.7924	0.7943
	sys.2	0.7992	0.8018	0.8010	0.8005	0.8021
Probability of dropping	sys.1	0.2179	0.2138	0.2085	0.2052	0.2056
	sys.2	0.2008	0.1981	0.1990	0.1995	0.1978
Average queue length	sys.1	29	35	42	50	65
	sys.2	35	55	75	95	125
Maximum queue length	sys.1	292	291	312	320	373
	sys.2	301	325	376	362	351
Average delay	sys.1	2.009	2.384	2.852	3.41	4.383
	sys.2	2.354	3.651	5.031	6.351	8.329

Table 3. Simulation results for different service intensities

Service intensity		8 B 1c	12 B 1c	15 B 1c	20 B 1c
Generated tasks	sys.1	1500254	1502810	1499461	1499738
	sys.2	1500114	1498361	1501682	1499738
Serviced tasks	sys.1	797746	1181532	1433571	1499738
	sys.2	799690	1201447	1464984	1499738
Dropped tasks	sys.1	702417	321254	65887	0
	sys.2	700287	296886	36674	0
Probability of servicing	sys.1	0.5317	0.7862	0.9561	1.0000
	sys.2	0.5331	0.8018	0.9756	1.0000
Probability of dropping	sys.1	0.4682	0.2138	0.0439	0.0000
	sys.2	0.4668	0.1981	0.0244	0.0000
Average queue length	sys.1	91	35	14	2
	sys.2	120	55	21	2
Maximum queue length	sys.1	915	291	113	36
	sys.2	980	325	134	36
Average delay	sys.1	6.106	2.384	0.992	0.2
	sys.2	8.003	3.651	1.466	0.2

The similar picture is observed according to the simulation results presented in Table 2, with a fixed value of the renovation probability q ($q = 0.01$) (the other simulation parameters are the same as for the Table 1) and with an increase of the threshold Q_1 value that controls the update mechanism and sets a safe zone in the buffer for a system of the second type.

The same picture (situation) can be seen from the simulation results presented in Table 3, where the results for different service intensities are presented (the other simulation parameters are the same as for the Table 1 and the Table 2).

5 Conclusion

In this paper we considered two $GI/M/1$ infinite capacity queuing systems with the renovation mechanism and a threshold policy. For each system the analytical expressions for the steady-state probability and time characteristics of the systems were obtained. The GPSS simulation results for both types of models were presented and compared. For the first type system, the probability of an accepted task to be dropped is greater than for a system of the second type, but the average waiting time is less than for a system of the second type. Thus, depending on which parameter of the system is more important (probability of service or average waiting time), a system of the first or second type should be preferred.

Further research tasks are to simplify the analytical expressions obtained for both systems, consider other service disciplines (for example LCFS), and also study a model with two threshold values that determine both the renovation (drop) probability and the safe zone in the queue. Also we will try to obtain results for queueing systems with general renovation (when an arbitrary number of tasks may be dropped with a given probability) mechanism based on threshold policy.

Even though the principle of active queue management by using the renovation mechanism is fundamentally different from the principle behind RED-type algorithms of active queue management, it can be noted that for certain values of the renovation mechanism parameters one can achieve at least the same level of system performance characteristics as for RED-like AQM [60].

References

1. Baker, F., Fairhurst, G.: IETF recommendations regarding active queue management. RFC 7567. Internet Engineering Task Force. https://tools.ietf.org/html/rfc7567. Accessed 27 May 2021
2. Floyd, S., Jacobson, V.: Random early detection gateways for congestion avoidance. IEEE/ACM Trans. Netw. **4**(1), 397–413 (1993). https://doi.org/10.1109/90.251892
3. Ramakrishnan, K., Floyd, S., Black, D.: The addition of explicit congestion notification (ECN) to IP. RFC 3168. Internet Engineering Task Force. https://tools.ietf.org/html/rfc3168. Accessed 27 May 2021
4. Floyd, S., Gummadi, R., Shenker, S.: Adaptive RED: an algorithm for increasing the robustness of RED's active queue management (2001). http://www.icir.org/floyd/papers/adaptiveRed.pdf
5. Floyd, S.: RED: discussions of setting parameters (1997). http://www.aciri.org/floyd/REDparameters.txt
6. Korolkova, A.V., Zaryadov, I.S.: The mathematical model of the traffic transfer process with a rate adjustable by RED. In: International Congress on Ultra Modern Telecommunications and Control Systems and Workshops (ICUMT), Moscow, Russia, pp. 1046–1050. IEEE (2010). https://doi.org/10.1109/ICUMT.2010.5676505
7. Jacobson, V., Nichols, K., Poduri, K.: RED in a different light. http://citeseerx.ist.psu.edu/viewdoc/summary?doi=10.1.1.22.9406. Accessed 29 May 2021
8. Class-based weighted fair queueing and weighted random early detection. http://www.cisco.com/c/en/us/td/docs/ios/12_0s/feature/guide/fswfq26.html. Accessed 27 May 2021
9. Floyd, S., Gummadi, R., Shenker, S.: Adaptive RED: an algorithm for increasing the robustness of RED's active queue management (2001). http://www.icir.org/floyd/papers/adaptiveRed.pdf. Accessed 1 Sept 2019
10. Changwang, Z., Jianping, Y., Zhiping, C., Weifeng, C.: RRED: robust RED algorithm to counter low-rate denial-of-service attacks. IEEE Commun. Lett. **14**(5), 489–491 (2010). https://doi.org/10.1109/LCOMM.2010.05.091407
11. Ott, T.J., Lakshman, T.V., Wong, L.H.: SRED: stabilized RED. In: Proceedings IEEE INFOCOM 1999, New York, NY, USA, vol. 3, pp. 1346–1355 IEEE (1999). https://doi.org/10.1109/INFCOM.1999.752153
12. Lin, D., Morris, R.: Dynamics of random early detection. Comput. Commun. Rev. **27**(4), 127–137 (1997)

13. Anjum, F.M., Tassiulas, L.: Balanced RED: an algorithm to achieve fairness in the Internet. Technical Research Report (1999). http://www.dtic.mil/dtic/tr/fulltext/u2/a439654.pdf
14. Aweya, J., Ouellette, M., Montuno, D.Y.: A control theoretic approach to active queue management. Comput. Netw. **36**, 203–235 (2001)
15. Sally Floyd website. http://www.icir.org/floyd/. Accessed 29 May 2021
16. Chrysostomoua, C., Pitsillidesa, A., Rossidesa, L., Polycarpoub, M., Sekercioglu, A.: Congestion control in differentiated services networks using fuzzy-RED. Control. Eng. Pract. **11**, 1153–1170 (2003)
17. Feng, W.-C.: Improving Internet congestion control and queue management algorithms. http://thefengs.com/wuchang/umich_diss.html. Accessed 29 May 2021
18. Al-Raddady, F., Woodward, M.: A new adaptive congestion control mechanism for the Internet based on RED. In: 21st International Conference on Advanced Information Networking and Applications, AINAW 2007 Workshops (2007)
19. Feng, W., Kandlur, D.D., Saha, D., Shin, K.G.: BLUE: a new class of active queue management algorithms. UM CSE-TR-387-99 (1999). https://www.cse.umich.edu/techreports/cse/99/CSE-TR-387-99.pdf
20. Nichols, K., Jacobson, V., McGregor, A., Iyengar, J.: Controlled delay active queue management. RFC 8289. Internet Engineering Task Force. https://tools.ietf.org/html/rfc8289. Accessed 29 Aug 2019
21. Hoeiland-Joergensen, T., McKenney, P., Taht, D., Gettys, J., Dumazet, E.: The flow queue CoDel packet scheduler and active queue management algorithm. Internet Engineering Task Force (2018). https://tools.ietf.org/html/rfc8290
22. IETF working group on active queue management and packet scheduling (AQM): description of the working group. http://tools.ietf.org/wg/aqm/charters. Accessed 29 Aug 2019
23. McKenney, P. E.: Stochastic fairness queueing. In: Proceedings of IEEE International Conference on Computer Communications, San Francisco, CA, USA, vol. 2, pp. 733–740. IEEE (1990). https://doi.org/10.1109/INFCOM.1990.91316
24. Adams, R.: Active queue management: a survey. IEEE Commun. Surv. Tut. **15**(3), 1425–1476 (2013)
25. Paul, A.K., Kawakami, H., Tachibana, A., Hasegawa, T.: An AQM based congestion control for eNB RLC in 4G/LTE network. In: 2016 IEEE Canadian Conference on Electrical and Computer Engineering (CCECE), Vancouver, BC, Canada, pp. 1–5. IEEE (2016). https://doi.org/10.1109/CCECE.2016.7726792
26. Irazabal, M., Lopez-Aguilera, E., Demirkol, I.: Active queue management as quality of service enabler for 5G networks. In: European Conference on Networks and Communications (EuCNC), Valencia, Spain, pp. 421–426. IEEE (2019). https://doi.org/10.1109/EuCNC.2019.8802027
27. Beshay, J.D., Nasrabadi, A.T., Prakash, R., Francini, A.: On active queue management in cellular networks. In: 2017 IEEE Conference on Computer Communications Workshops (INFOCOM WKSHPS), Atlanta, GA, USA, pp. 384–389. IEEE (2017). https://doi.org/10.1109/INFCOMW.2017.8116407
28. Pesántez-Romero, I.S., Pulla-Lojano, G.E., Guerrero-Vásquez, L.F., Coronel-González, E.J., Ordoñez-Ordoñez, J.O., Martinez-Ledesma, J.E.: Performance evaluation of hybrid queuing algorithms for QoS provision based on DiffServ architecture. In: Yang, X.-S., Sherratt, S., Dey, N., Joshi, A. (eds.) Proceedings of 6th International Congress on Information and Communication Technology. LNNS, vol. 216, pp. 333–345. Springer, Singapore (2022). https://doi.org/10.1007/978-981-16-1781-2_31

29. George, J., Santhosh, R.: Congestion control mechanism for unresponsive flows in internet through active queue management system (AQM). In: Shakya, S., Bestak, R., Palanisamy, R., Kamel, K.A. (eds.) Mobile Computing and Sustainable Informatics. LNDECT, vol. 68, pp. 765–777. Springer, Singapore (2022). https://doi.org/10.1007/978-981-16-1866-6_58

30. Singha, S., Jana, B., Jana, S., Mandal, N.K.: A novel congestion control algorithm using buffer occupancy RED. In: Das, A.K., Nayak, J., Naik, B., Dutta, S., Pelusi, D. (eds.) Computational Intelligence in Pattern Recognition. AISC, vol. 1349, pp. 519–528. Springer, Singapore (2022). https://doi.org/10.1007/978-981-16-2543-5_44

31. Korolkova, A.V., Kulyabov, D.S., Chernoivanov, A.I.: On the classification of RED algorithms. Bull. Peoples' Friendship Univ. Russ. Ser. Math. Inf. Sci. Phys. **3**, 34–46 (2009)

32. Korolkova, A.V., Velieva, T.R., Abaev, P.O., Sevastianov, L.A., Kulyabov, D.S.: Hybrid simulation of active traffic management. In: Proceedings of 30th European Conference on Modelling and Simulation (ECMS), Regensburg, Germany, pp. 692–697. ECMS (2016). https://doi.org/10.7148/2016-0685

33. Bonald, T., May, M., Bolot, J.: Analytic evaluation of RED performance. In: Proceedings IEEE INFOCOM 2000 Conference on Computer Communications, Tel Aviv, Israel, vol. 3, pp. 1415–1424. IEEE (2000). https://doi.org/10.1109/INFCOM.2000.832539

34. Chydzinski, A., Chrost, L.: Analysis of AQM queues with queue size based packet dropping. Int. J. Appl. Math. Comput. Sci. **21**(3), 567–577 (2011). https://doi.org/10.2478/v10006-011-0045-7

35. Chydzinski, A.: Improving quality of multimedia transmissions via dropping functions. In: Proceedings 2018 3rd International Conference on Smart and Sustainable Technologies (SpliTech), pp. 1–6 (2018)

36. Chydzinski, A.: On the transient queue with the dropping function. Entropy **22**, 825 (2020). https://doi.org/10.3390/e22080825

37. Barczyk, M., Chydzinski, A.: Experimental testing of the performance of packet dropping schemes. In: 2020 IEEE Symposium on Computers and Communications (ISCC), pp. 1–7, (2020). https://doi.org/10.1109/ISCC50000.2020.9219624

38. Konovalov, M.G., Razumchik, R.: Numerical analysis of improved access restriction algorithms in a $GI|G|1|N$ system. J. Commun. Technol. Electron. **63**(6), 616–625 (2018). https://doi.org/10.1134/S1064226918060141

39. Kreinin, A.: Queueing systems with renovation. J. Appl. Math. Stochast. Anal. **10**(4), 431–443 (1997). https://doi.org/10.1155/S1048953397000464

40. Gelenbe, E.: Product-form queueing networks with negative and positive customers. J. Appl. Probab. **28**(3), 656–663 (1991). https://doi.org/10.2307/3214499

41. Pechinkin, A.V., Razumchik, R.V.: The stationary distribution of the waiting time in a queueing system with negative customers and a bunker for superseded customers in the case of the LAST-LIFO-LIFO discipline. J. Commun. Technol. Electron. **57**(12), 1331–1339 (2012). https://doi.org/10.1134/S1064226912120054

42. Razumchik, R.V.: Analysis of finite capacity queue with negative customers and bunker for ousted customers using Chebyshev and Gegenbauer polynomials. Asia Pac. J. Oper. Res. **31**(04), 1450029 (2014). https://doi.org/10.1142/S0217595914500298

43. Semenova, O.V.: Multithreshold control of the $BMAP/G/1$ queuing system with MAP flow of Markovian disasters. Autom. Remote. Control. **68**(1), 95–108 (2007). https://doi.org/10.1134/S0005117907010092

44. Kim, C., Klimenok, V.I., Dudin, A.N.: Analysis of unreliable $BMAP|PH|N$ type queue with Markovian flow of breakdowns. Appl. Math. Comput. **314**, 154–172 (2017). https://doi.org/10.1016/j.amc.2017.06.035

45. Gudkova, I., et al.: Modeling and analyzing licensed shared access operation for 5G network as an inhomogeneous queue with catastrophes. In: International Congress on Ultra Modern Telecommunications and Control Systems and Workshops, Lisbon, Portugal, pp. 282–287. IEEE (2016). https://doi.org/10.1109/ICUMT.2016.7765372

46. Dudin, A.N., Klimenok, V.I., Vishnevsky, V.M.: Mathematical methods to study classical queuing systems. In: The Theory of Queuing Systems with Correlated Flows, pp. 1–61. Springer, Cham (2020). https://doi.org/10.1007/978-3-030-32072-0_1

47. Krishnamoorthy, A., Pramod, P.K., Chakravarthy, S.R.: Queues with interruptions: a survey. TOP **22**(1), 290–320 (2014). https://doi.org/10.1007/s11750-012-0256-6

48. Vishnevsky, V.M., Kozyrev, D.V., Semenova, O.V.: Redundant queuing system with unreliable servers. In: International Congress on Ultra Modern Telecommunications and Control Systems and Workshops, St. Petersburg, Russia, pp. 283–286. IEEE (2014). https://doi.org/10.1109/ICUMT.2014.7002116

49. Jain, M., Kaur, S., Singh, P.: Supplementary variable technique (SVT) for non-Markovian single server queue with service interruption (QSI). Oper. Res. Int. J. (2019). https://doi.org/10.1007/s12351-019-00519-8

50. Rykov, V.V., Kozyrev, D.V.: Analysis of renewable reliability systems by Markovization method. In: Rykov, V.V., Singpurwalla, N.D., Zubkov, A.M. (eds.) ACMPT 2017. LNCS, vol. 10684, pp. 210–220. Springer, Cham (2017). https://doi.org/10.1007/978-3-319-71504-9_19

51. Rykov, V., Kozyrev, D.: On the reliability function of a double redundant system with general repair time distribution. Appl. Stoch. Models Bus. Ind. **35**, 191–197 (2019). https://doi.org/10.1002/ASMB.2368

52. Nguyen, D.P., Kozyrev, D.: Reliability analysis of a multirotor flight module of a high-altitude telecommunications platform operating in a random environment. In: 2020 International Conference Engineering and Telecommunication (En&T), pp. 1–5 (2020). https://doi.org/10.1109/EnT50437.2020.9431312

53. Bocharov, P.P., Zaryadov, I.S.: Probability distribution in queueing systems with renovation. Bull. Peoples' Friendship Univ. Russ. Ser. Math. Inf. Sci. Phys. **1–2**, 15–25 (2007). (in Russian)

54. Zaryadov, I.S., Pechinkin, A.V.: Stationary time characteristics of the $GI/M/n/\infty$ system with some variants of the generalized renovation discipline. Autom. Remote. Control. **70**(12), 2085–2097 (2009). https://doi.org/10.1134/S0005117909120157

55. Zaryadov, I., Razumchik, R., Milovanova, T.: Stationary waiting time distribution in $G|M|n|r$ with random renovation policy. In: Vishnevskiy, V.M., Samouylov, K.E., Kozyrev, D.V. (eds.) DCCN 2016. CCIS, vol. 678, pp. 349–360. Springer, Cham (2016). https://doi.org/10.1007/978-3-319-51917-3_31

56. Konovalov, M., Razumchik, R.: Queueing systems with renovation vs. queues with RED. Supplementary material (2017). https://arxiv.org/abs/1709.01477

57. Konovalov, M., Razumchik, R.: Comparison of two active queue management schemes through the M/D/1/N queue. Informatika i ee Primeneniya (Inf. Appl.) **12**(4), 9–15 (2018). https://doi.org/10.14357/19922264180402

58. Bogdanova, E.V., Zaryadov, I.S., Milovanova, T.A., Korolkova, A.V., Kulyabov, D.S.: Characteristics of lost and served packets for retrial queueing system with general renovation and recurrent input flow. In: Vishnevskiy, V.M., Kozyrev, D.V. (eds.) DCCN 2018. CCIS, vol. 919, pp. 327–340. Springer, Cham (2018). https://doi.org/10.1007/978-3-319-99447-5_28

59. Zaryadov, I., Bogdanova, E., Milovanova, T., Matushenko, S., Pyatkina, D.: Stationary characteristics of the $GI|M|1$ queue with general renovation and feedback. In: 10th International Congress on Ultra Modern Telecommunications and Control Systems and Workshops (ICUMT), Moscow, Russia. IEEE (2019). Article no. 8631244. https://doi.org/10.1109/ICUMT.2018.8631244

60. Hilquias, V.C.C., et al.: The general renovation as the active queue management mechanism. some aspects and results. In: Vishnevskiy, V.M., Samouylov, K.E., Kozyrev, D.V. (eds.) DCCN 2019. CCIS, vol. 1141, pp. 488–502. Springer, Cham (2019). https://doi.org/10.1007/978-3-030-36625-4_39

61. Meykhanadzhyan, L.A., Zaryadov, I.S., Milovanova, T.A.: Stationary characteristics of the two-node Markovian tandem queueing system with general renovation. Sistemy i Sredstva Informatiki [Syst. Means Inform.] **30**(3), 14–31 (2020)

62. Gorbunova, A.V., Lebedev, A.V.: Queueing system with two input flows, preemptive priority, and stochastic dropping. Autom. Remote. Control. **81**(12), 2230–2243 (2020). https://doi.org/10.1134/S0005117920120073

63. Hilquias, C.C.V., Zaryadov, I.S.: Comparison of two single-server queueing systems with exponential service times and threshold-based renovation. In: CEUR Workshop Proceedings, vol. 2946, pp. 54–63 (2021)

64. Pechinkin, A.V., Razumchik, R.R., Zaryadov, I.S.: First passage times in $M_2^{[X]}|G|1|R$ queue with hysteretic overload control policy. In: AIP Conference Proceedings, Rhodes, vol. 1738. American Institute of Physics Inc. (2016). Article no. 220007. https://doi.org/10.1063/1.4952006

65. Razumchik, R.: Analysis of finite $MAP|PH|1$ queue with hysteretic control of arrivals. In: International Congress on Ultra Modern Telecommunications and Control Systems and Workshops, Lisbon. Portugal, pp. 288–293. IEEE (2016). Article no. 7765373. https://doi.org/10.1109/ICUMT.2016.7765373

Resource Queueing System $M/GI/\infty$ in a Random Environment

Nikita Krishtalev[ID], Ekaterina Lisovskaya$^{(\boxtimes)}$[ID], and Alexander Moiseev[ID]

Tomsk State University, 36 Lenin Avenue, Tomsk 634050, Russian Federation
krishtalevnik@gmail.com, ekaterina_lisovs@mail.ru, moiseev.tsu@gmail.com

Abstract. In this paper, we consider an infinite-service resource queueing system $M/GI/\infty$ operating in a random environment. When the environment changes its state, the service time and the occupied resource do not change for the customers already under service, however, for the new customers the arrival rate, the service time distribution, and resource requirements are changed. We apply the dynamic screening method and perform asymptotic analysis to find the approximation of the probability of the total amount of occupied resource.

Keywords: Infinite-server queue · Random environment · Resource queue · Asymptotic analysis

1 Introduction

Network slicing in 5G can be defined as a network configuration that allows multiple networks (virtualized and independent) to be created on top of a common physical infrastructure. This configuration has become an essential component of the overall 5G architectural landscape [17]. Each "slice" or portion of the network can be allocated based on the specific needs of the application, use case or customer.

Network slicing involves dividing the physical architecture of 5G into multiple virtual networks or layers. Each network layer includes control plane functions, user traffic plane functions, and a radio access network. Each layer has its own characteristics and is aimed at solving a particular business problem. 3GPP defines three standard network layers [1]:

- super-broadband access (eMBB, Enhanced Mobile Broadband) – users of the global Internet, CCTV cameras;
- ultra-reliability and low latency (URLLC, Ultra Reliable Low Latency Communication) – driverless transport, augmented and virtual reality;
- low enegry and low latency (IoT, Internet of Things) – millions of devices transmitting small amounts of data from time to time.

Figure 1 shows an example of a network slicing by traffic (service) types. In this paper, it is proposed to consider a resource queueing system operating in a

V. M. Vishnevskiy et al. (Eds.): DCCN 2021, LNCS 13144, pp. 211–225, 2021.
https://doi.org/10.1007/978-3-030-92507-9_18

random environment as a mathematical model of such a technology, assuming that service requests occur when the random environment is in the appropriate state. We analyze the system in the limiting condition of extremely frequently changing states of the environment, thus, this will insignificantly affect the main probabilistic and numerical characteristics. We assume that requests for the services form a Poisson process with constant intensity, depending on the type of service (i.e. the state of the random environment). The request service duration and the amount of the provided resource also depend on the type of service and do not change for requests that are in service when the random environment changes its state. We present a detailed description of the model in the Sect. 2.

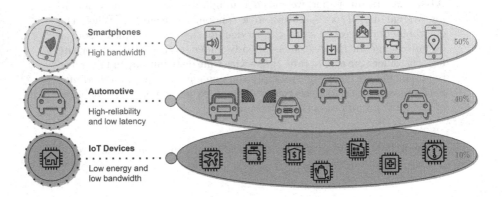

Fig. 1. Example of network slicing

There are many papers where queues in a random environment were studied. For example, [13] considers queueing system $M/M/C$, where the arrivals and service rates are modulated by a random environment for which the underlying process $C(t)$ is an irreducible continuous-time Markov chain. B. D'Auria [6] investigated an $M/M\infty$ queue whose parameters depend on an external random environment that is specified by a semi-Markovian process. Boxma and Kurkova [2] studied an $M/M/1$ queue with the special feature that the speed of the server alternates between two constant values. There are a lot more papers [3, 4, 7–10, 12], where authors consider queueing systems operating in a random environment.

Resource queueing systems have been analyzed extensively in recent years. For example, in [15], a model of a multi-server queueing system with losses caused by lack of resources necessary to service claims was considered. In [18], it was investigated a heterogeneous wireless network model in terms of a queueing system with limited resources and signals that trigger the resource reallocation process. In [19] Tikhonenko studied a queueing system with processor sharing and limited resources.

In [16], it was considered an $M/GI/\infty$ queueing system in a random environment. The dynamic screening method and asymptotic analysis were applied as well as we do in this paper. In [5], a mathematical model of an insurance company in the form of the infinite-server queueing system operating in a random environment was studied using the asymptotic analysis method. Paper [11] considers a non-Markovian infinite-server multi-resource queuing system. The result was found under the asymptotic condition of the growing intensity of the arrival process. All these papers consider an infinite-server queue in a random environment or with requirements for resources. Unlike them, we consider the $M/GI/\infty$ queueing model with both a random environment and requirements for resources in this paper.

The paper is organized as follows. In Sect. 2, the mathematical model is described and the goal of the study is formulated. In Sect. 3, the dynamic screening method is explained, moreover, balance equations are obtained and written using characteristic functions. The asymptotic analysis and final equations are derived in Sect. 4. Section 5 presents a numerical example and conclusions on the accuracy of the obtained approximation.

2 Mathematical Model

Consider a queueing system with an unlimited number of servers and an unlimited capacity of some resource that operates in a random environment (Fig. 2), such the functioning of the system depends on the environment state. The random environment is specified by a continuous-time Markov chain with a finite number of states $k \in \{1, \ldots, K\}$ and generator $\mathbf{Q} = \{q_{k\nu}\}$, $k, \nu = 1, \ldots, K$. When the process is in state k, the rate of the Poisson arrival process is equal to λ_k and the service time has distribution with CDF $B_k(x)$. We compose the arrivals rates into diagonal matrix $\mathbf{\Lambda} = \mathrm{diag}\{\lambda_k\}$, $k = 1, \ldots, K$. In addition, each arrival occupies a resource of a random size $v_k > 0$ with the CDF $G_k(y) = \mathrm{P}\{v_k < y\}$ which depends on the environment state.

When the environment state changes, the resource amounts and the service rates change only for new customers, as for customers already under service, these values stay the same. When a customer completes servicing, it leaves the system and releases the resource that it occupied during the capture. *Capture* is understood as the moment when the customer arrives, at which the resource is allocated.

A stochastic process $\{k(t), i(t), v(t)\}$ describes the system's state at time t as follows:

- the environment state at time t is denoted by $k(t) \in \{1, \ldots, K\}$,
- the number of customers in the system at time t is denoted by $i(t) \in \{0, 1, 2, \ldots\}$,
- the total amount of occupied resource at time t is denoted by $v(t) \geq 0$.

Our goal is to find the steady-state probability distribution of the total amount of the occupied resources $v(t)$.

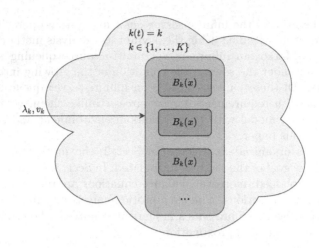

Fig. 2. Structure of the model

3   Dynamic Screening Method

We consider an infinite-server queue with non-exponential service times. This is the reason why we can not apply some classical methods directly here (for example, the method of supplementary variables). Otherwise, we should deal with the number of variables and equations that are unlimited and changing. To avoid the problem, we apply the dynamic screening method [14] whose modification for the resource queue is described below.

3.1   Method Description

Let at moment t_0 the system is empty. We fix a moment $T > t_0$ in the future. Let us draw two time axes (Fig. 3). The moments of customers arrivals are marked on the axis 0. We mark on axis 1 only those arrivals that before the moment T have not finished their service. We name the arrivals on axis 1 as "screened", and the entire point process on axis 1 is named as the "screened process".

Let us define function $S_k(t)$ that determines the dynamic screening probability on axis 1 as follows:

$$S_k(t) = 1 - B_k(T - t).$$

The customer arrived at the system at the moment $t < T$ does not finish service before moment T and occupies a certain amount of the resource with probability $S_k(t)$. On the other hand, the customer leaves the system and releases the resource occupied at the arriving with the probability $1 - S_k(t)$, hence it is not considered in the "screened" process. The values of $S_k(t)$ belong to the segment $[0, 1]$. In Fig. 3, colored areas depict different states of the random environment.

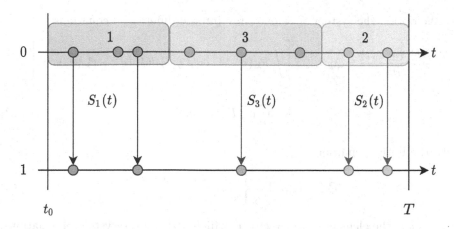

Fig. 3. Example of the customers arrivals screening

Let us denote the number of screened arrivals, which occurred during time $[t_0, t)$ by $n(t)$, and the total amount of resource occupied by the screened customers by $w(t)$.

Basing on the results obtained in [14] and [11], it is not hard to show that

$$P\{k(T) = k, i(T) = m, v(T) < x\}$$
$$= P\{k(T) = k, n(T) = m, w(T) < x\}, \quad k \in \{1, \dots, K\} \qquad (1)$$

for any non-negative values of m and x. So, we can first study the stochastic process $\{k(t), n(t), w(t)\}$ instead of process $\{k(t), i(t), v(t)\}$. After that, we can substitute $t = T$ into the final expressions and obtain the goal due to the moment T is chosen arbitrarily.

3.2 Balance Equations

We denote the probabilities

$$P_k(n, w, t) = P\{k(t) = k, n(t) = n, w(t) < w\}.$$

For $P_k(n, w, t)$, we write the following system according to the total probability law and dynamic screening method

$$P_k(n, w, t + \Delta t) = P_k(n, w, t)(1 - \lambda_k \Delta t)(1 + q_{kk} \Delta t)$$

$$+ \lambda_k \Delta t S_k(t) \int\limits_0^w P_k(n - 1, w - y, t) \, dG_k(y)$$

$$+ \lambda_k \Delta t (1 - S_k(t)) P_k(n, w, t) + \sum\limits_{\substack{v=1 \\ v \neq k}}^{K} q_{vk} \Delta t P_v(n, w, t) + o(\Delta t).$$

We obtain the system of Kolmogorov integro-differential equations

$$\frac{\partial P_k\left(n, w, t\right)}{\partial t} = \lambda_k S_k\left(t\right) \left[\int_0^w P_k\left(n-1, w-y, t\right) dG_k\left(y\right) - P_k\left(n, w, t\right)\right]$$

$$+ \sum_{\nu=1}^K q_{\nu k} P_\nu\left(n, w, t\right), \tag{2}$$

with the initial condition

$$P_k(n, dw, t_0) = \begin{cases} r_k \delta_0(dw), & \text{if } n = 0, \\ 0, & \text{else,} \end{cases} \tag{3}$$

where r_k is the element of row vector $\mathbf{r}$, which satisfies the system of equations:

$$\begin{cases} \mathbf{rQ = 0}, \\ \mathbf{re} = 1, \end{cases}$$

and $\mathbf{e}$ is a column vector that consists of ones.

3.3   Characteristic Functions

Let us define the partial characteristic functions as

$$h_k(u, v, t) = \sum_{n=0}^\infty e^{jun} \int_0^\infty e^{jvw} p_k(n, dw, t),$$

where $j = \sqrt{-1}$, and we rewrite the system (2)–(3) for functions $h_k(u, v, t)$

$$\frac{\partial h_k\left(u, v, t\right)}{\partial t} = \lambda_k S_k\left(t\right) \left(e^{ju} G_k^*(v) - 1\right) h_k(u, v, t)$$

$$+ \sum_{\nu=1}^K q_{\nu k} h_\nu\left(u, v, t\right), \quad k \in \{1, \ldots, K\},$$

with the initial condition

$$h_k(u, dv, t_0) = r_k \delta_0(dv)$$

where

$$G_k^*(v) = \int_0^\infty e^{jvy} dG_k(y).$$

Then, we rewrite the obtained system as matrix equation

$$\frac{\partial \mathbf{h}\left(u, v, t\right)}{\partial t} = \mathbf{h}\left(u, v, t\right) \left[\mathbf{\Lambda S}(t)\left(e^{ju}\mathbf{G}(v) - \mathbf{I}\right) + \mathbf{Q}\right], \tag{4}$$

with the initial condition

$$\mathbf{h}\left(u, dv, t_0\right) = \mathbf{r}\delta_0(dv) \tag{5}$$

where

$$\mathbf{h}\left(u, v, t\right) = \left[h_1\left(u, v, t\right), \ldots, h_K\left(u, v, t\right)\right],$$
$$\mathbf{S}\left(t\right) = \mathrm{diag}\{S_k(t)\}, \quad \mathbf{G}\left(v\right) = \mathrm{diag}\{G_k^*(v)\}$$

and $\mathbf{I}$ is identity matrix.

4 Asymptotic Analysis Method

Problem (4)–(5) seems as it can not be solved analytically in a direct way, so, we apply the method of asymptotic analysis to solve it.

4.1 Method Description

Asymptotic analysis method for queueing systems consists of analysis of the equations defining any characteristics of the system and allows to obtain probability distribution and numerical characteristics in the analytical form under some asymptotic condition. We will use the asymptotic analysis method under the condition of growing arrivals rate and extremely frequent changes in the environment state to solve the Eq. (4)–(5). We set

$$\tilde{\mathbf{\Lambda}} = N\mathbf{\Lambda}, \quad \tilde{\mathbf{Q}} = N\mathbf{Q}, \quad N \to \infty.$$

Then (4) can be rewrite as

$$\frac{1}{N}\frac{\partial \mathbf{h}\left(u, v, t\right)}{\partial t} = \mathbf{h}\left(u, v, t\right)\left[\mathbf{\Lambda S}\left(t\right)\left(e^{ju}\mathbf{G}\left(v\right) - \mathbf{I}\right) + \mathbf{Q}\right]. \tag{6}$$

4.2 First Order Asymptotic

In (6), (5), we substitute

$$\frac{1}{N} = \varepsilon, \quad u = \varepsilon x, \quad v = \varepsilon y, \quad \mathbf{h}\left(u, v, t\right) = \mathbf{f_1}\left(x, y, t, \varepsilon\right), \tag{7}$$

and then it can be rewritten as

$$\varepsilon\frac{\partial \mathbf{f_1}\left(x, y, t, \varepsilon\right)}{\partial t} = \mathbf{f_1}\left(x, y, t, \varepsilon\right)\left[\mathbf{\Lambda S}\left(t\right)\left(e^{j\varepsilon x}\mathbf{G}\left(\varepsilon y\right) - \mathbf{I}\right) + \mathbf{Q}\right]. \tag{8}$$

In (8) we set $\varepsilon \to 0$ and denote

$$\lim_{\varepsilon \to 0}\mathbf{f_1}(x, y, t, \varepsilon) = \mathbf{f_1}(x, y, t),$$

this yields
$$\mathbf{f_1}(x, y, t)\mathbf{Q} = \mathbf{0},$$
it then follows that $\mathbf{f_1}(x, y, t)$ can be represented as a product

$$\mathbf{f_1}(x, y, t) = \mathbf{r}\Phi_1(x, y, t), \tag{9}$$

where $\Phi_1(x, y, t)$ is a scalar function that satisfies the initial condition $\Phi_1(x, y, t_0) = 1$.

We multiply (8) by $\mathbf{e}$, divide by ε, set $\varepsilon \to 0$ and substitute (9):

$$\frac{\partial \Phi_1(x, y, t)}{\partial t} = \Phi_1(x, y, t)\,\mathbf{r}\mathbf{\Lambda}\mathbf{S}(t)\,(jx\mathbf{I} + jy\mathbf{A_1})\,\mathbf{e}, \tag{10}$$

where

$$\mathbf{A_1} = \mathrm{diag}\{a_k\}, \quad a_k = \int\limits_0^\infty y\,dG_k(y).$$

The solution of Eq. (10) is given as

$$\Phi_1(x, y, t) = \exp\left\{ jx \int\limits_{t_0}^t \mathbf{r}\mathbf{\Lambda}\mathbf{S}(\tau)\,\mathbf{e}\,d\tau + jy \int\limits_{t_0}^t \mathbf{r}\mathbf{\Lambda}\mathbf{S}(\tau)\,\mathbf{A_1}\mathbf{e}\,d\tau \right\}. \tag{11}$$

Denoting

$$m_1(t) \triangleq \int_{t_0}^t \mathbf{r}\mathbf{\Lambda}\mathbf{S}(\tau)\,\mathbf{e}\,d\tau, \quad m_2(t) \triangleq \int_{t_0}^t \mathbf{r}\mathbf{\Lambda}\mathbf{S}(\tau)\,\mathbf{A_1}\mathbf{e}\,d\tau,$$

and making substitution inverse to (9) and (7), we obtain the first order approximation

$$\mathbf{h}^{(1)}(u, v, t) = \mathbf{f_1}(x, y, t, \varepsilon) \approx \mathbf{f_1}(x, y, t) = \mathbf{r}\Phi_1(x, y, t)$$
$$= \mathbf{r}\exp\{juNm_1(t) + jvNm_2(t)\}.$$

4.3   Second Order Asymptotic

In (4), we make a substitution

$$\mathbf{h}(u, v, t) = \mathbf{h_2}(u, v, t)\exp\{juNm_1(t) + jvNm_2(t)\}, \tag{12}$$

we obtain

$$\frac{1}{N}\frac{\partial \mathbf{h_2}(u, v, t)}{\partial t} = \mathbf{h_2}(u, v, t)$$
$$\times \left[\mathbf{\Lambda}\mathbf{S}(t)\left(e^{ju}\mathbf{G}(v) - \mathbf{I}\right) + \mathbf{Q} - \mathbf{r}\mathbf{\Lambda}\mathbf{S}(t)\,\mathbf{e}\,(ju\mathbf{I} + jv\mathbf{A_1})\right]. \tag{13}$$

We substitute

$$\frac{1}{N} = \varepsilon^2, \quad u = \varepsilon x, \quad v = \varepsilon y, \quad \mathbf{h}_2(u, v, t) = \mathbf{f}_2(x, y, t, \varepsilon) \tag{14}$$

into (13), and we obtain

$$\varepsilon^2 \frac{\partial \mathbf{f}_2(x, y, t, \varepsilon)}{\partial t} = \mathbf{f}_2(x, y, t, \varepsilon) \tag{15}$$
$$\times \left[\mathbf{\Lambda S}(t) \left(e^{j\varepsilon x} \mathbf{G}(\varepsilon y) - \mathbf{I} \right) + \mathbf{Q} - \mathbf{r} \mathbf{\Lambda S}(t) \mathbf{e} \left(j\varepsilon x \mathbf{I} + j\varepsilon y \mathbf{A}_1 \right) \right].$$

As $\varepsilon \to 0$, denoting

$$\lim_{\varepsilon \to \infty} \mathbf{f}_2(x, y, t, \varepsilon) = \mathbf{f}_2(x, y, t),$$

we obtain

$$\mathbf{f}_2(x, y, t) \mathbf{Q} = \mathbf{0}.$$

This yields

$$\mathbf{f}_2(x, y, t) = \mathbf{r} \Phi_2(x, y, t), \tag{16}$$

where $\Phi_2(x, y, t)$ is a scalar function that satisfies the initial condition $\Phi_2(x, y, t_0) = 1$. We represent $\mathbf{f}_2(x, y, t)$ as the expansion

$$\mathbf{f}_2(x, y, t, \varepsilon) = \Phi_2(x, y, t) \left[\mathbf{r} + \mathbf{g}(t) \left(j\varepsilon x \mathbf{I} + j\varepsilon y \mathbf{A}_1 \right) \right] + O\left(\varepsilon^2\right), \tag{17}$$

where $\mathbf{g}(t)$ is a row vector.

We substitute the first degree Maclaurin expansion of e^z into Eq. (15)

$$\varepsilon^2 \frac{\partial \mathbf{f}_2(x, y, t, \varepsilon)}{\partial t} = \mathbf{f}_2(x, y, t, \varepsilon) \left[\mathbf{\Lambda S}(t) \left(\mathbf{I} - \mathbf{er} \right) \left(j\varepsilon x \mathbf{I} + j\varepsilon y \mathbf{A}_1 \right) + \mathbf{Q} \right].$$

Then we substitute (17) into the obtained equation

$$\Phi_2(x, y, t) \left[\mathbf{r} + \mathbf{g}(t) \left(j\varepsilon x \mathbf{I} + j\varepsilon y \mathbf{A}_1 \right) \right]$$
$$\times \left[\mathbf{\Lambda S}(t) \left(\mathbf{I} - \mathbf{er} \right) \left(j\varepsilon x \mathbf{I} + j\varepsilon y \mathbf{A}_1 \right) + \mathbf{Q} \right] = O\left(\varepsilon^2\right).$$

We divide both sides by ε and set $\varepsilon \to 0$

$$\Phi_2(x, y, t) \left[\mathbf{r} \mathbf{\Lambda S}(t) \left(\mathbf{I} - \mathbf{er} \right) + \mathbf{g}(t) \mathbf{Q} \right] \left[jx \mathbf{I} + jy \mathbf{A}_1 \right] = 0.$$

This yields

$$\mathbf{g}(t) \mathbf{Q} = \mathbf{r} \mathbf{\Lambda S}(t) \left(\mathbf{er} - \mathbf{I} \right). \tag{18}$$

Hence, vector $\mathbf{g}(t)$ is defined by the inhomogeneous linear system. The solution $\mathbf{g}(t)$ of the system (18) we can write as

$$\mathbf{g}(t) = c(t) \mathbf{r} + \mathbf{d}(t),$$

where $c(t)$ is an arbitrary scalar function and the row vector $\mathbf{d}(t)$ is any specific solution to system (18) satisfying a certain condition, for example:

$$\mathbf{d}(t)\mathbf{e} = 0.$$

The solution has the form

$$\mathbf{d}(t) = \mathbf{r}\mathbf{\Lambda}\mathbf{S}(t)\mathbf{G},$$

where

$$\begin{cases} \mathbf{GQ} = \mathbf{er} - \mathbf{I}, \\ \mathbf{Ge} = \mathbf{0}. \end{cases}$$

Finally, the solution of (18) has the form

$$\mathbf{g}(t) = c(t)\mathbf{r} + \mathbf{r}\mathbf{\Lambda}\mathbf{S}(t)\mathbf{G}.$$

Let us now derive the explicit expression for the function $\Phi_2(x, y, t)$. To do this, we approximate the exponential function in (15) with the second degree Maclaurin expansion and make substitution (17). This yields the equation

$$\varepsilon^2 \frac{\partial \Phi_2(x, y, t)}{\partial t}\mathbf{r} = \Phi_2(x, y, t)\left[\mathbf{r} + \mathbf{g}(t)(j\varepsilon x\mathbf{I} + j\varepsilon y\mathbf{A}_1)\right]$$

$$\times \left[j\varepsilon y\mathbf{\Lambda}\mathbf{S}(t)\mathbf{A}_1 + \frac{(j\varepsilon y)^2}{2}\mathbf{\Lambda}\mathbf{S}(t)\mathbf{A}_2 + j\varepsilon x\mathbf{\Lambda}\mathbf{S}(t) + j\varepsilon x j\varepsilon y\mathbf{\Lambda}\mathbf{S}(t)\mathbf{A}_1 \right.$$

$$\left. + \frac{(j\varepsilon x)^2}{2}\mathbf{\Lambda}\mathbf{S}(t) + \mathbf{Q} - j\varepsilon x\mathbf{r}\mathbf{\Lambda}\mathbf{S}(t)\mathbf{e}\mathbf{I} - j\varepsilon y\mathbf{r}\mathbf{\Lambda}\mathbf{S}(t)\mathbf{e}\mathbf{A}_1 \right],$$

where

$$\mathbf{A}_2 = \mathrm{diag}\{\alpha_k\}, \quad \alpha_k = \int\limits_0^\infty y^2 dG_k(y).$$

We then multiply both parts of the obtained equation by vector $\mathbf{e}$. Due to (18), we can write

$$\frac{\partial \Phi_2(x, y, t)}{\partial t} = \Phi_2(x, y, t)\left[\frac{(jx)^2}{2}[\mathbf{r} + 2\mathbf{g}(t)(\mathbf{I} - \mathbf{er})]\mathbf{\Lambda}\mathbf{S}(t)\mathbf{e} \right.$$

$$+ \frac{(jy)^2}{2}[\mathbf{r}\mathbf{A}_2 + 2\mathbf{g}(t)(\mathbf{I} - \mathbf{er})\mathbf{A}_1\mathbf{A}_1]\mathbf{\Lambda}\mathbf{S}(t)\mathbf{e}$$

$$\left. + jxjy[\mathbf{r} + 2\mathbf{g}(t)(\mathbf{I} - \mathbf{er})]\mathbf{A}_1\mathbf{\Lambda}\mathbf{S}(t)\mathbf{e} \right].$$

Denoting

$$K_{11}(t) \triangleq \int_{t_0}^{t} \left[\mathbf{r} + 2\mathbf{g}(\tau)(\mathbf{I} - \mathbf{er}) \right] \mathbf{\Lambda S}(\tau) \mathbf{e} d\tau,$$

$$K_{22}(t) \triangleq \int_{t_0}^{t} \left[\mathbf{r A_2} + 2\mathbf{g}(\tau)(\mathbf{I} - \mathbf{er}) \mathbf{A_1 A_1} \right] \mathbf{\Lambda S}(\tau) \mathbf{e} d\tau,$$

$$K_{12}(t) \triangleq \int_{t_0}^{t} \left[\mathbf{r} + 2\mathbf{g}(\tau)(\mathbf{I} - \mathbf{er}) \right] \mathbf{A_1 \Lambda S}(\tau) \mathbf{e} d\tau,$$

we obtain the solution

$$\Phi_2(x, y, t) = \exp\left\{ \frac{(jx)^2}{2} K_{11}(t) + \frac{(jy)^2}{2} K_{22}(t) + jxjy K_{12}(t) \right\}.$$

Making substitution inverse to (16), (14) and (12), we obtain the second order approximation

$$\mathbf{h}^{(2)}(u, v, t) \approx \mathbf{r} \exp\left\{ juN m_1(t) + jvN m_2(t) + \frac{(ju)^2}{2} NK_{11}(t) \right.$$

$$\left. + \frac{(jv)^2}{2} NK_{22}(t) + jujv NK_{12}(t) \right\}.$$

4.4   Main Result

Multiplying by vector $\mathbf{e}$, we obtain the approximation of the characteristic function of stochastic process $\{n(t), w(t)\}$

$$h^{(2)}(u, v, t) = \mathbf{h}^{(2)}(u, v, t) \mathbf{e},$$

and going to steady-state regime, we put $t = T$ and $t_0 \to -\infty$, using (1), we obtain the approximation for the characteristic function of process $\{i(t), v(t)\}$

$$h(u, v) \approx \exp\left\{ juN m_1 + jvN m_2 + \frac{(ju)^2}{2} NK_{11} + \frac{(jv)^2}{2} NK_{22} + jujv NK_{12} \right\},$$

where

$$m_1 = \mathbf{r\Lambda Be}, \quad m_2 = \mathbf{r\Lambda A_1 Be},$$

$$K_{11} = \mathbf{r\Lambda Be} + 2\mathbf{r\Lambda BG\Lambda e} - 2\mathbf{r\Lambda (M \circ G)\Lambda e},$$

$$K_{22} = \mathbf{rA_2 \Lambda Be} + 2\mathbf{rA_1 \Lambda BG A_1 \Lambda e} - 2\mathbf{rA_1 \Lambda (M \circ G) A_1 \Lambda e},$$

$$K_{12} = \mathbf{rA_1 \Lambda Be} + 2\mathbf{rA_1 \Lambda BG\Lambda e} - 2\mathbf{rA_1 \Lambda (M \circ G)\Lambda e},$$

$$\mathbf{B} = \mathrm{diag}\left\{ \int_0^{\infty} (1 - B_k(\tau)) d\tau \right\},$$

$$\mathbf{M} = \left[\int_0^{\infty} (1 - B_k(\tau))(1 - B_{k'}(\tau)) d\tau \right], \quad k, k' \in \{1, \dots, K\}.$$

Here notation $\mathbf{X} \circ \mathbf{Y}$ means Hadamard (element-wise) product of matrices $\mathbf{X}$ and $\mathbf{Y}$.

Finally, the steady-state probability distribution of two-dimensional stochastic process $\{i(t), v(t)\}$ can be approximated by the Gaussian distribution with means vector

$$\mathbf{m} = N\,[m_1 \quad m_2]$$

and covariance matrix

$$\mathbf{K} = N \begin{bmatrix} K_{11} & K_{12} \\ K_{12} & K_{22} \end{bmatrix}.$$

In particular, the stationary characteristic function of the distribution of the total amount of occupied resource $v(t)$ can be approximated as follows:

$$h\,(v) \approx \exp\left\{ jvNm_2 + \frac{(jv)^2}{2} NK_{22} \right\}. \tag{19}$$

So, this distribution can be approximated by the Gaussian one with mean Nm_2 and variance NK_{22}.

5   Numerical Example

Consider the following example. Let the random environment have three states $\{1, 2, 3\}$ and be defined by the generator

$$\widetilde{\mathbf{Q}} = N \cdot \begin{bmatrix} -3 & 1 & 2 \\ 1 & -2 & 1 \\ 2 & 2 & -4 \end{bmatrix}.$$

Let we have Poission arrivals with intensities

$$\widetilde{\mathbf{\Lambda}} = N \cdot \mathrm{diag}\{0.1; 1; 10\}.$$

Let service times have gamma distribution with the following shape and rate parameters:

$$\alpha_1 = 1.9, \; \beta_1 = 2; \quad \alpha_2 = 0.5, \; \beta_2 = 1; \quad \alpha_3 = 0.4, \; \beta_2 = 3,$$

and resource requirements, also, have gamma distribution with parameters

$$\bar{\alpha}_1 = \bar{\beta}_1 = 0.5; \quad \bar{\alpha}_2 = \bar{\beta}_2 = 1.5; \quad \bar{\alpha}_3 = \bar{\beta}_3 = 2.$$

Here indices mean the state of the random environment when the customer arrives.

Let us estimate the accuracy of obtained approximation (19) and find a lower bound of parameter N for the applicability of the proposed approximation. To do this, we carried out series of simulation experiments (in each of them 10^7 arrivals were generated) for increasing values of N and compared the asymptotic distributions with the empiric ones by using the Kolmogorov distance $\Delta = \sup\limits_x |F(x) - \tilde{F}(x)|$ as an accuracy measure. Here $F(x)$ is the cumulative distribution function (CDF) built on the basis of simulation results, and $\tilde{F}(x)$ is the CDF based on Gaussian approximation (19).

Table 1 presents the Kolmogorov distances between the asymptotic and empirical distribution functions of the total amount of resources occupied in the system. We see that the approximation accuracy increases with growing of parameter N, which is also illustrated by Fig. 4.

Table 1. Kolmogorov distances for the distribution of the total amount of occupied resource between ones basing on the approximation and the simulation results for various values of parameter N

N	5	10	15	20	25	50	100
Δ	0.085	0.061	0.051	0.045	0.041	0.033	0.023

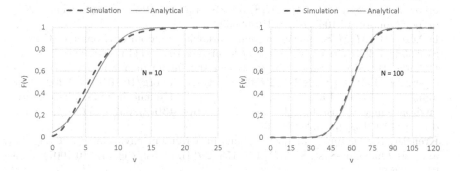

Fig. 4. Comparison of the approximation and simulation results for the distribution function of the total amount of occupied resource

If we suppose that the error $\Delta \leq 0.05$ is acceptable, we may conclude that the Gaussian approximation can be applicable for the cases $N > 15$.

6   Conclusion

In the paper, we have studied a resource queueing system $M/GI/\infty$ operating in a random environment. We have considered the case when the service time and the occupied resource do not change for the customers already under service when the random environment state changes. We have applied the dynamic screening method and the asymptotic analysis method to find the approximation of the probability of the total amount of occupied resource. It has been obtained that this distribution can be approximated by Gaussian distribution under the condition of growing arrival rate and extremely frequent changes of the states of the random environment. The parameters of the corresponding Gaussian distribution have been obtained.

References

1. 3GPP: Study on Enhancement of Network Slicing (Release 16). 3GPP TS 23.740 V16.0.0 (December 2018)
2. Boxma, O.J., Kurkova, I.A.: The M/M/1 queue in a heavy-tailed random environment. Stat. Neerl. **54**(2), 221–236 (2000). https://doi.org/10.1111/1467-9574.00138
3. Choudhury, G.L., Mandelbaum, A., Reiman, M.I., Whitt, W.: Fluid and diffusion limits for queues in slowly changing environments. Commun. Stat. Stoch. Models **13**(1), 121–146 (1997). https://doi.org/10.1080/15326349708807417
4. Cordeiro, J.D., Kharoufeh, J.P.: The unreliable M/M/1 retrial queue in a random environment. Stoch. Model. **28**(1), 29–48 (2012). https://doi.org/10.1080/15326349.2011.614478
5. Dammer, D.: Research of mathematical model of insurance company in the form of queueing system in a random environment. In: Dudin, A., Nazarov, A., Kirpichnikov, A. (eds.) ITMM 2017. CCIS, vol. 800, pp. 204–214. Springer, Cham (2017). https://doi.org/10.1007/978-3-319-68069-9_17
6. D'Auria, B.: $M/M/\infty$ queues in semi-Markovian random environment. Queueing Syst. **58**(3), 221–237 (2008). https://doi.org/10.1007/s11134-008-9068-7
7. D'Auria, B.: Stochastic decomposition of the queue in a random environment. Oper. Res. Lett. **35**(6), 805–812 (2007). https://doi.org/10.1016/j.orl.2007.02.007
8. Jiang, T., Liu, L., Li, J.: Analysis of the M/G/1 queue in multi-phase random environment with disasters. J. Math. Anal. Appl. **430**(2), 857–873 (2015). https://doi.org/10.1016/j.jmaa.2015.05.028
9. Krenzler, R., Daduna, H.: Loss systems in a random environment: steady state analysis. Queueing Syst. **80**(1-2), 127–153 (2014). https://doi.org/10.1007/s11134-014-9426-6
10. Krishnamoorthy, A., Jaya, S., Lakshmy, B.: Queues with interruption in random environment. Ann. Oper. Res. **233**(1), 201–219 (2015). https://doi.org/10.1007/s10479-015-1931-4
11. Lisovskaya, E.Y., Moiseev, A.N., Moiseeva, S.P., Pagano, M.: Modeling of mathematical processing of physics experimental data in the form of a non-Markovian multi-resource queuing system. Russ. Phys. J. **61**(12), 2188–2196 (2019). https://doi.org/10.1007/s11182-019-01655-6
12. Liu, Y., Honnappa, H., Tindel, S., Yip, N.K.: Infinite server queues in a random fast oscillatory environment. Queueing Syst. **98**(1-2), 145–179 (2021). https://doi.org/10.1007/s11134-021-09704-z
13. Liu, Z., Yu, S.: The M/M/C queueing system in a random environment. J. Math. Anal. Appl. **436**(1), 556–567 (2016). https://doi.org/10.1016/j.jmaa.2015.11.074
14. Moiseev, A., Nazarov, A.: Queueing network $MAP - (GI/\infty)^K$ with high-rate arrivals. Eur. J. Oper. Res. **254**(1), 161–168 (2016). https://doi.org/10.1016/j.ejor.2016.04.011
15. Naumov, V.A., Samuilov, K.E., Samuilov, A.K.: On the total amount of resources occupied by serviced customers. Autom. Remote. Control. **77**(8), 1419–1427 (2016). https://doi.org/10.1134/s0005117916080087

16. Nazarov, A., Baymeeva, G.: The $M/G/\infty$ queue in random environment. In: Dudin, A., Nazarov, A., Yakupov, R., Gortsev, A. (eds.) ITMM 2014. CCIS, vol. 487, pp. 312–324. Springer, Cham (2014). https://doi.org/10.1007/978-3-319-13671-4_36

17. Samdanis, K., Costa-Perez, X., Sciancalepore, V.: From network sharing to multi-tenancy: the 5g network slice broker. IEEE Commun. Mag. **54**(7), 32–39 (2016). https://doi.org/10.1109/MCOM.2016.7514161

18. Sopin, E., Vikhrova, O., Samouylov, K.: LTE network model with signals and random resource requirements. In: 2017 9th International Congress on Ultra Modern Telecommunications and Control Systems and Workshops (ICUMT) (2017). https://doi.org/10.1109/icumt.2017.8255155

19. Tikhonenko, O.M.: Queuing system with processor sharing and limited resources. Autom. Remote. Control. **71**(5), 803–815 (2010). https://doi.org/10.1134/s0005117910050073

Numerical Analysis of a Retrial System with Unreliable Servers Based on Laplace Domain Description

András Mészáros[1,2]($\boxtimes$), Evsey Morozov[3,4,5], Taisia Morozova[3,6],
and Miklós Telek[1,2]

[1] Department of Networked Systems and Services, Budapest University
of Technology and Economics, Budapest, Hungary
{meszarosa,telek}@hit.bme.hu
[2] MTA-BME Information Systems Research Group, ELKH, Budapest, Hungary
[3] Petrozavodsk State University, Lenin Street 33, Petrozavodsk 185910, Russia
[4] Institute of Applied Mathematical Research of the Karelian Research Centre
of RAS, Pushkinskaya Street 11, Petrozavodsk 185910, Russia
emorozov@karelia.ru
[5] Moscow Center for Fundamental and Applied Mathematics, Moscow State
University, Moscow 119991, Russia
[6] Uppsala University, Uppsala, Sweden

Abstract. In this paper, we consider a retrial queuing system with unreliable servers and analyze the distribution of the stationary generalized service time which includes also the unavailable periods (setup times) occurring during service of the customer. We consider three service interruption disciplines: preemptive resume (PRS), preemptive repeat different (PRD), and preemptive repeat identical (PRI); and provide the stationary distribution of the generalized service time and the remaining generalized service time for these disciplines in Laplace transform (LT) domain.

The main focus of the paper is on the numerical analysis based on LT domain descriptions, which we evaluate for various numerical examples.

Keywords: Numerical Inverse Laplace transform · Preemptive resume · Preemptive repeat different · Preemptive repeat identical

1 Introduction

Recent developments in Numerical Inverse Laplace transform (NILT) provide an efficient tool for the analysis of stochastic models [1,4]. In this work, we investigate the applicability of the NILT approach by evaluating the generalized

This work is partially supported by the OTKA K-123914 project and the Artificial Intelligence National Laboratory Programme of Hungary and by the Russian Science Foundation, Project No. 21-71-10135, https://rscf.ru/en/project/21-71-10135/.

service time distribution and its remaining time distribution (also referred to as equilibrium distribution [5, p. 437] [7, p. 432, 469]) of a retrial queuing system with an unreliable server.

The system behaves as follows. Incoming customers queue up in an infinite buffer. The single server serves the customers in FIFO order. The server is subject to breakdown. In case of a server breakdown, the server gets back to operational after an independent, identically distributed (i.i.d.) setup time. If the server was busy at breakdown, it continues the service of the interrupted customer when it gets back to operational according to one of the following three preemption policies: preemptive resume (PRS), preemptive repeat different (PRD), and preemptive repeat identical (PRI). With the PRS policy, the server continues the service of the interrupted customer from the point it was interrupted. With the PRD policy, after an interruption the server restarts the service of the interrupted customer with i.i.d. service time. With the PRI policy the server restarts the service of the interrupted customer, and the service time of the customer in the current operational period of the server is identical to the one of the previous operational period.

We selected this stochastic model, because depending on the applied discipline at server failure the complexity of the Laplace transform (LT) description varies a lot, consequently the applicability of the NILT analysis raises more and more severe numerical and computational complexity issues.

Similarly, the probability distributions used for the (non-generalized) service time and the downtime of the server can significantly affect the computational characteristics of the model. In our numerical experiment we consider a set of distributions with differing degrees of complexity in LT domain description.

Various performance measures of this model have been investigated in preceding papers [2,6]. For example, the LT description of the generalized service time distribution with PRS and PRD policies are available in [2]. But none of the preceding papers considered the NILT based numerical analysis of the generalized service time distribution.

In this paper we extend the LT domain description with the PRI case, but the main focus of the paper is the investigation of the NILT based numerical analysis in case of different preemption policies and service and setup time distributions.

2   Analysis of the Generalized Service Time Distribution

The generalized service time, G, is the time from the instant the server starts the service of a customer until it completes the service of that costumer considering the potential breakdown and setup cycles of the server and the applied preemption policy.

The CDF, PDF, and the LT of the (breakdown free) service time, S, are denoted by $F(x) = Pr(S < x)$, $f(x) = dF(x)/dx$, and $f^*(s) = E(e^{-sS})$, respectively. Similarly, the CDF, PDF, and the LT of the setup time, R, and the generalized service time G are denoted by $R(x)$, $r(x)$, $r^*(s)$ and $G(x)$, $g(x)$, $g^*(s)$, respectively.

In this work, we assume that the server breaks down with constant rate ν. That is, when the server is operational, the time to the next breakdown is exponentially distributed with parameter ν (independent of the time of the last breakdown), thus the time of breakdown B has PDF $b(x) = \nu e^{-\nu x}$.

2.1   Preemptive Repeat Different – PRD

Theorem 1 [2]. *In case of PRD preemption policy, the LT of the generalized service time is*

$$g^*(s) = \frac{(s+\nu)f^*(s+\nu)}{(s+\nu) - \nu(1 - f^*(s+\nu))r^*(s)}. \tag{1}$$

Proof.

$$(G|B = h, S = x) = \begin{cases} x & h > x \\ h + R + G & h < x \end{cases}$$

$$E(e^{-sG}|B = h, S = x) = \begin{cases} e^{-sx} & h > x \\ e^{-sh} \underbrace{E(e^{-sR})}_{r^*(s)} \underbrace{E(e^{-sG})}_{g^*(s)} & h < x \end{cases}$$

$$g^*(s) = E(e^{-sG}) = \int_h b(h) \int_x f(x) E(e^{-sG}|B = h, S = x) dx dh$$

$$= \int_{h=0}^{\infty} b(h) \left(\int_{x=0}^{h} f(x)e^{-sx}dx + \int_{x=h}^{\infty} f(x)e^{-sh}r^*(s)g^*(s)dx \right) dh$$

$$= \int_{h=0}^{\infty} \nu e^{-\nu h} \left(\int_{x=0}^{h} f(x)e^{-sx}dx + (1 - F(h))e^{-sh}r^*(s)g^*(s) \right) dh$$

$$= \int_{x=0}^{\infty} f(x)e^{-sx} \underbrace{\int_{h=x}^{\infty} \nu e^{-\nu h}dh}_{e^{-\nu h}} dx + \nu \underbrace{\int_{h=0}^{\infty} e^{-(s+\nu)h}(1 - F(h))dh}_{\frac{1-f^*(s+\nu)}{s+\nu}} r^*(s)g^*(s)$$

That is,

$$g^*(s) = f^*(s+\nu) + \nu\frac{1 - f^*(s+\nu)}{s+\nu}r^*(s)g^*(s),$$

from which

$$g^*(s) = \frac{f^*(s+\nu)}{1 - \nu\frac{1-f^*(s+\nu)}{s+\nu}r^*(s)} = \frac{(s+\nu)f^*(s+\nu)}{(s+\nu) - \nu(1 - f^*(s+\nu))r^*(s)}$$

2.2   Preemptive Resume – PRS

Theorem 2 [2]. *In case of PRS preemption policy, the LT of the generalized service time is*

$$g^*(s) = f^*(s + \nu - \nu r^*(s)). \tag{2}$$

Proof. The number of interruptions during the service time $S = x$ is N_x. N_x is Poisson $(x\nu)$ distributed, i.e., $Pr(N_x = i) = \frac{(x\nu)^i}{i!} e^{-x\nu}$

$$(G|S = x, N_x = i) = x + \sum_{j=1}^{i} R_j$$

$$E(e^{-sG}|S = x, N_x = i) = e^{-sx}(r^*(s))^i$$

$$g^*(s) = \int_x \sum_{i=0}^{\infty} Pr(N_x = i) f(x) E(e^{-sG}|S = x, N_x = i) dx$$

$$= \int_x \sum_{i=0}^{\infty} \frac{(x\nu)^i}{i!} e^{-x\nu} f(x) e^{-sx} (r^*(s))^i dx$$

$$= \int_x f(x) e^{-(s+\nu)x} \underbrace{\sum_{i=0}^{\infty} \frac{(x\nu r^*(s))^i}{i!}}_{e^{x\nu r^*(s)}} dx$$

$$= f^*(s + \nu - \nu r^*(s))$$

2.3   Preemptive Repeat Identical – PRI

Theorem 3. *In case of PRI preemption policy, the mean and the LT of the generalized service time are*

$$E(G) = \left(E(R) + \frac{1}{\nu}\right)\left(f^*(-\nu) - 1\right) \tag{3}$$

and

$$g^*(s) = \frac{(s+\nu)}{(s+\nu) - \nu r^*(s)} \cdot \sum_{j=0}^{\infty} \left(\frac{-\nu r^*(s)}{(s+\nu) - \nu r^*(s)}\right)^j f^*((j+1)(s+\nu)). \tag{4}$$

Before proving the theorem we need the following lemma.

Lemma 1.

$$\sum_{i=j}^{\infty} \binom{i}{j} a^i = a^j (1-a)^{-j-1}$$

Proof (Lemma 1). Using

$$\frac{d}{da}(1-a)^{-j-1} = (j+1)(1-a)^{-j-2},$$

$$\frac{d^n}{da^n}(1-a)^{-j+1} = (j+1)\ldots(j+n)(1-a)^{-j-n-1} = \frac{(j+n)!}{j!}(1-a)^{-j-n-1},$$

the Taylor series of $(1-a)^{-j-1}$ is

$$(1-a)^{-j-1} = \sum_{n=0}^{\infty} \frac{(n+j)!}{n!j!}a^n = \sum_{n=0}^{\infty} \binom{j+n}{n}a^n = \sum_{n=0}^{\infty} \binom{j+n}{j}a^n$$

from which

$$\sum_{i=j}^{\infty} \binom{i}{j}a^i = a^j \sum_{i=j}^{\infty} \binom{i}{j}a^{i-j} = a^j \sum_{n=0}^{\infty} \binom{n+j}{j}a^n = a^j(1-a)^{-j-1}.$$

Proof (Theorem 3). Let N_x be the number of interruptions if the service time is $S = x$. N_x is Geometrical(p) distributed, i.e., $Pr(N_x = i) = p(1-p)^i$, with $p = e^{-x\nu}$ and $E(N_x) = \frac{1-p}{p} = \frac{1-e^{-x\nu}}{e^{-x\nu}}$.

$$(G|S = x, N_x = i) = x + \sum_{j=1}^{i} (B_j(x) + R_j), \tag{5}$$

where the interruption time, $B(x)$, is truncated exponentially distributed, i.e., for $0 < h < x$

$$Pr(B(x) < h) = \frac{1-e^{-h\nu}}{1-e^{-x\nu}} \quad \text{and} \quad \frac{d}{dh}Pr(B(x) < h) = \frac{\nu e^{-h\nu}}{1-e^{-x\nu}}.$$

Consequently,

$$E(B(x)) = \int_{h=0}^{x} h\frac{\nu e^{-h\nu}}{1-e^{-x\nu}}dh = \frac{1}{\nu} - \frac{xe^{-x\nu}}{1-e^{-x\nu}}$$

and

$$i^*(s,x) = E(e^{-sB(x)}) = \int_{h=0}^{x} e^{-sh}\frac{\nu e^{-h\nu}}{1-e^{-x\nu}}dh = \frac{\nu}{1-e^{-x\nu}}\int_{h=0}^{x} e^{-(s+\nu)h}dh$$

$$= \frac{\nu}{1-e^{-x\nu}} \cdot \frac{1-e^{-(s+\nu)x}}{s+\nu} = \frac{\nu(1-e^{-(s+\nu)x})}{(s+\nu)(1-e^{-x\nu})}$$

From (5) we get

$$E(G|S = x, N_x = i) = x + \Big(E(R) + E(B(x))\Big)i,$$

and

$$E(e^{-sG}|S = x, N_x = i) = e^{-sx}(r^*(s))^i(i^*(s,x))^i.$$

From these we get

$$E(G) = \int_x f(x) \sum_{i=0}^{\infty} Pr(N_x = i)E(G|S = x, N_x = i)dx$$

$$= \int_x f(x) \sum_{i=0}^{\infty} Pr(N_x = i)\Big(x + (E(R) + E(B(x)))i\Big)dx$$

$$= \int_x f(x)\Big(x + (E(R) + E(B(x)))E(N_x)\Big)dx$$

$$= E(S) + \int_x f(x)\Big(E(R) + \frac{1}{\nu} - \frac{xe^{-x\nu}}{1 - e^{-x\nu}}\Big)\frac{1 - e^{-x\nu}}{e^{-x\nu}}dx$$

$$E(G) = E(S) + \int_x f(x)\Big(E(R) + \frac{1}{\nu} - \frac{xe^{-x\nu}}{1 - e^{-x\nu}}\Big)\frac{1 - e^{-x\nu}}{e^{-x\nu}}dx$$

$$= E(S) + \Big(E(R) + \frac{1}{\nu}\Big)\int_x f(x)\frac{1 - e^{-x\nu}}{e^{-x\nu}}dx - \underbrace{\int_x f(x)xdx}_{E(S)}$$

$$= \Big(E(R) + \frac{1}{\nu}\Big)\int_x f(x)(e^{x\nu} - 1)dx$$

$$= \Big(E(R) + \frac{1}{\nu}\Big)\Big(f^*(-\nu) - 1\Big)$$

and

$$g^*(s) = \int_x f(x) \sum_{i=0}^{\infty} Pr(N_x = i)E(e^{-sG}|S = x, N_x = i)dx$$

$$= \int_x \sum_{i=0}^{\infty} e^{-x\nu}(1 - e^{-x\nu})^i f(x)e^{-sx}(i^*(s,x))^i(r^*(s))^i dx$$

$$= \int_x \sum_{i=0}^{\infty} e^{-x\nu}(1 - e^{-x\nu})^i f(x)e^{-sx}\Big(\frac{\nu(1 - e^{-(s+\nu)x})}{(s + \nu)(1 - e^{-x\nu})}\Big)^i (r^*(s))^i dx$$

$$= \int_x f(x)e^{-(s+\nu)x} \sum_{i=0}^{\infty} \Big(\frac{\nu r^*(s)}{s + \nu}\Big)^i (1 - e^{-(s+\nu)x})^i dx$$

$$= \int_x f(x)e^{-(s+\nu)x} \sum_{i=0}^{\infty} \Big(\frac{\nu r^*(s)}{s + \nu}\Big)^i \sum_{j=0}^{i} \binom{i}{j}(-e^{-(s+\nu)x})^j dx$$

$$= \int_x f(x)e^{-(s+\nu)x} \sum_{i=0}^{\infty} \Big(\frac{\nu r^*(s)}{s + \nu}\Big)^i \sum_{j=0}^{i} \binom{i}{j}(-1)^j e^{-j(s+\nu)x} dx$$

Further on

$$g^*(s) = \int_x f(x)e^{-(s+\nu)x} \sum_{j=0}^{\infty}(-1)^j e^{-j(s+\nu)x} \underbrace{\sum_{i=j}^{\infty}\binom{i}{j}\left(\frac{\nu r^*(s)}{s+\nu}\right)^i}_{}\,dx$$

$$= \int_x f(x)\sum_{j=0}^{\infty}(-1)^j e^{-(j+1)(s+\nu)x}\frac{\left(\frac{\nu r^*(s)}{s+\nu}\right)^j}{\left(1-\frac{\nu r^*(s)}{s+\nu}\right)^{j+1}}\,dx$$

$$= \sum_{j=0}^{\infty}(-1)^j\frac{(s+\nu)\,(\nu r^*(s))^j}{((s+\nu)-\nu r^*(s))^{j+1}}\int_x f(x)e^{-(j+1)(s+\nu)x}\,dx$$

$$= \sum_{j=0}^{\infty}\frac{(s+\nu)\,(-\nu r^*(s))^j}{((s+\nu)-\nu r^*(s))^{j+1}}f^*((j+1)(s+\nu))$$

$$= \frac{(s+\nu)}{(s+\nu)-\nu r^*(s)}\cdot\sum_{j=0}^{\infty}\left(\frac{-\nu r^*(s)}{(s+\nu)-\nu r^*(s)}\right)^j f^*((j+1)(s+\nu)),$$

where we used Lemma 1 to rewrite the expression highlighted by the brace under it.

The infinite summation in (4), provides the complete description of the generalized service time distribution, but in order to compute $g^*(s)$ based on (4), we need to truncate the summation at a given threshold.

The region of convergence for the $f^*(s) = \int_{x=0}^{\infty} e^{-sx}f(x)dx$ integral is always of the form $\{s : Re(s) > a\}$ (possibly including some points of the boundary line $\{Re(s) = a\}$), or empty ($a = \infty$), or the entire complex plane ($a = -\infty$). The real constant a is known as the abscissa of absolute convergence.

Corollary 1. *With PRI policy, the mean generalized service time in (3) is finite when the abscissa of absolute convergence of the service time distribution is less than $-\nu$, that is,*

$$f^*(-\nu) = \int_0^{\infty} f(x)e^{x\nu}dx < \infty.$$

2.4   Remaining Time Distribution

The LT domain description of the remaining time distribution of the generalized service time when the server is busy, $h^*(s)$, and for an arriving customer, $\hat{h}^*(s)$, are

$$h^*(s) = \frac{1-g^*(s)}{sE(G)}, \quad \hat{h}^*(s) = (1-p_{busy}) + p_{busy}\frac{1-g^*(s)}{sE(G)},$$

where $E(G)$ is the mean of the generalized service time, that is $E(G) = -\frac{d}{ds}g^*(s)|_{s=0}$ and $p_{busy} = \lambda_{arr}E(G)$ is the probability that an arriving customer finds the server busy, with λ_{arr} being the arrival rate of customers.

3   NILT Using Abate-Whitt Framework Methods

In this work we restrict our attention to NILT methods in the Abate-Whitt framework [1]. In this framework, the order N approximate of $f(t)$ at point $t = T$ is obtained based on $f^*(s) = \int_t f(t)e^{-st}dt$ as

$$f(T) \approx f_N(T) := \sum_{n=1}^{N} \frac{\eta_n}{T} f^* \left(\frac{\beta_n}{T} \right), \tag{6}$$

where the coefficients η_n and β_n are determined by the order (N) and the NILT method (e.g., Euler [1], Gaver [3], Talbot [8], CME [4]) and they are independent of the function $f^*(s)$. We assume that the η_n and β_n coefficients are available with negligible computational cost, since, at worst, they can be calculated and stored in advance. This way, the computational complexity of computing $f_N(T)$ based on (6) is approximately N times the computational cost of evaluating $f^*(s)$ at potentially complex points.

4   Laplace Transform of Positive Distributions

We consider the following list of service time and setup time distributions.

- Weibull distribution with density $f(t) = \alpha\lambda(\lambda t)^{\alpha-1}e^{-(\lambda t)^\alpha}$:
  The complexity of $f^*(s) = \int_t f(t)e^{-st}dt$ depends on α.
  - When α is irrational, $f^*(s)$ does not have a closed form expression.
  - When α is rational, $f^*(s)$ can be described with generalized hypergeometric functions. The complexity of the hypergeometric function depends on α. The following two cases result in the simplest LT expressions
    * heavy tailed case ($\alpha = 1/2$): $f^*(s) = \frac{\sqrt{\pi\lambda}}{2\sqrt{s}}e^{\frac{\lambda}{4s}}Erfc\left(\frac{\sqrt{\lambda/s}}{2}\right)$,
    * light tailed case ($\alpha = 2$): $f^*(s) = 1 - \frac{s\sqrt{\pi}}{2\lambda}e^{\frac{s^2}{4\lambda^2}}Erfc\left(\frac{s}{2\lambda}\right)$,

    where $Erfc$ is the complementary error function defined as $Erfc(z) = \frac{2}{\sqrt{\pi}}\int_{t=z}^{\infty}e^{-t^2}dt$. The hypergeometric function and its special case, the $Erfc$ function, are integral functions and the computational complexity of the evaluation of these functions depends on their implementation. For $\alpha < 1$ the abscissa of absolute convergence of the Weibull distributed density is $a = 0$ and for $\alpha > 1$ it is $a = -\infty$.
- Gamma distribution with density $f(t) = \frac{\lambda^\alpha t^{\alpha-1}e^{-\lambda t}}{\Gamma(\alpha)}$:
  $f^*(s)$ has the analytic form

$$f^*(s) = (1 + s/\lambda)^{-\alpha},$$

which can be calculated with low computational cost. The abscissa of absolute convergence is $a = -\lambda$. The special case when α is a positive integer gives the Erlang distribution and when $\alpha = 1$ gives the exponential distribution.

- Pareto distribution with density $f(t) = \alpha(t+1)^{-(\alpha+1)}$ and support on $(0, \infty)$: $f^*(s)$ can be expressed with the use of the exponential integral function $E_x(s) = \int_1^\infty t^{-x}e^{-st}dt$ as

$$f^*(s) = \alpha e^s E_{\alpha+1}(s).$$

The abscissa of absolute convergence is $a = 0$.

- Lognormal distribution with density $f(t) = \frac{1}{\sigma t\sqrt{2\pi}}e^{\frac{-(\log(t)-\mu)^2}{2\sigma^2}}$:

$f^*(s)$ has no closed form. $f^*(s) = \int_t f(t)e^{-st}dt$ needs to be evaluated. The abscissa of absolute convergence is $a = 0$.

5   Numerical Experiments

The previous sections introduced LT domain distributions with various complexity. We investigate their numerical behaviour in this section. We used the order $N = 60$ CME method [4] in all cases.

5.1   Weibull Distributed Service Time

In the following we consider the case of PRS, PRD, and PRI preemption with light and heavy tailed Weibull distributed service times.

Figure 1 plots the PDF of the generalized service time distribution and the remaining generalized service time distribution for the three preemption policies - calculated from $h^*(s)$ - with light and heavy tailed Weibull service time distribution (defined in Sect. 4 with $\alpha = 2$ and $1/2$, respectively). The setup time is exponentially distributed with parameter 4 ($r^*(s) = \frac{4}{s+4}$) and the server break down rate is $\nu = 2$. The shape of the service time distribution has a profound effect on the shape of the generalized service time distribution, but this effect vanishes in the remaining generalized service time distribution. In the heavy tailed case, $a = 0$ and the mean generalized service time with PRI policy is infinite, that is why we cannot plot the PDF of the remaining generalized service time, while the PDF of the generalized service time is still computable. In the light tailed case, $a = -\infty$ and the mean generalized service time with PRI policy is finite. In this case both densities of the PRI policy are available.

5.2   Computational Complexity

There are two main factors that affect the computational cost of the calculation of the PDF of the generalized service time $(g(t))$ via NILT: the order of the NILT (N) and the complexity of the Laplace domain function of the generalized service time $(g^*(s))$. To obtain the time domain function $(g(t))$ in a single point, $g^*(s)$ has to be evaluated in N points. Assuming that the evaluation of $g^*(s)$ dominates the complexity of NILT and that the complexity of the evaluation of $g^*(s)$ is independent of the point it is evaluated in, the computational complexity of the NILT is N times the computational complexity of evaluating $g^*(s)$.

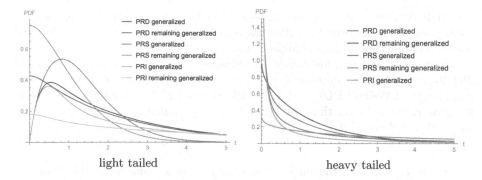

Fig. 1. Density function of the generalized service time distribution and the remaining generalized service time distribution with PRD, PRS and PRI policies, when the service time is light ($\lambda = 1, \alpha = 2$) and heavy ($\lambda = 1, \alpha = 1/2$) tailed Weibull distributed, the setup time is exponentially distributed with parameter 4 and the failure rate of the server is $\nu = 2$.

To evaluate $g^*(s)$ in one point, in the PRD and the PRS case we have to evaluate $r^*(s)$ in one point and $f^*(s)$ in one point (Theorem 1 and 2), in the PRI case, if we truncate the infinite sum at $j = k - 1$, then we have to evaluate $r^*(s)$ in one point, and $f^*(s)$ in k points (Theorem 3). In the following we only discuss the computational cost of the evaluation of a Laplace transform PDF in a single point (e.g., the evaluation of $f^*(s)$ or $r^*(s)$), from which the cost of NILT of $g^*(s)$ can be easily calculated. E.g., according to (1), 60 $f^*(s)$ evaluations and 60 $r^*(s)$ evaluations are needed to obtain $g(t)$ in a single point in the PRD case with $N = 60$.

The computational complexity of the evaluation of a Laplace transform PDF (say $f^*(s)$) is a nuanced question, a detailed discussion of which is out of the scope of this paper. To give a practical perspective, we investigated the computational time of the evaluation of such PDFs using Wolfram Mathematica for the distributions listed in Sect. 4. For these we measured the average evaluation time of $f^*(s)$ using 100 random complex s values. The evaluation times in seconds can be seen in Table 1. Unlike other distributions, the lognormal distribution does not have a closed form Laplace transform, therefore in this case numerical integration is needed, which is considered as part of the evaluation of $f^*(s)$.

In accordance with the expectations, the Laplace transform of the gamma distribution can be evaluated extremely fast. The Laplace transforms of the other distributions do not have closed forms, but both the hypergeometric function (whose special case is the $Erfc$ function) in the Laplace transform of the Weibull distribution and the exponential integral function in the Laplace transform of the Pareto distribution can be calculated efficiently, in general. More precisely, the order of the hyper geometric function in case of the Weibull distribution depends on $\max\{a, b\}$ for the rational $\alpha = a/b$, where a and b are relative primes.

For large a or b (integer) parameters the hypergeometric function is of high order and the evaluation of $f^*(s)$ can become quite complex, which explains the high computational cost for $\alpha = 11/100$ in Table 1. Finally, the Laplace transform of the lognormal distribution requires numerical integration, thus the related computational time is higher than most other cases.

Comparing the result for Weibull distribution with $\alpha = 11/100$ and lognormal distribution (based on numerical integration) suggests, that the computational complexity of Weibull PDF with $\alpha = 11/100$ using high order hyper geometric function could be larger than the evaluation of the numerical integral according to $f^*(s) = \int_t f(t)e^{-st}dt$.

Table 1. Evaluation time of $f^*(s)$ in a single point for different distributions

	Weibull			Gamma	Pareto	Lognormal
	$\alpha = 1/2$	$\alpha = 2$	$\alpha = 11/100$	$\alpha = 5/2, \lambda = 1$	$\alpha = 2$	$\alpha = 2, \lambda = 1$
Time	$7.34 \cdot 10^{-4}$	$1.56 \cdot 10^{-4}$	$6.40 \cdot 10^{-1}$	$<10^{-6}$	$1.56 \cdot 10^{-5}$	$4.68 \cdot 10^{-3}$

5.3   Accuracy of the NILT Results

The distribution of the service time and generalized service time cannot be calculated analytically for the more complex functions of this paper. Therefore, to verify the accuracy of NILT, we implemented the models with PRD, PRS, and PRI preemption using a discrete event simulator. The model specific features of the applied simulation tool are as follows:

- Based on the the PASTA property, the *remaining generalized service time* is measured at independent Poisson instants that arrive at a constant rate.
- Utilizing that the stationary distributions of the *elapsed time* and the *remaining time* are identical, the simulation collects statistics on the *elapsed time* for implementation convenience.
- In case of the heavy-tailed Weibull distributed service time with PRI policy, the mean generalized service time has an infinite mean, which requires a special simulation approach of the generalized service time, which utilizes the fact that we are interested in the CDF until a known upper bound. Consequently, the simulation follows the life of customers only until their system time reaches the upper bound.

We ran simulations for PRD, PRS, and PRI preemption using Weibull distributed service time distribution and exponential setup and server break down distributions using the same parameters as in Sect. 5.1. We compared the CDF of the generalized service times as well as the remaining generalized service times obtained using simulation and the NILT of the corresponding formulas. We obtained the empirical CDF (ECDF) curves as the average of 200 simulation runs for each interruption mode with 1000 served customer in each run and also calculated their 95% confidence intervals. The results are presented in Fig. 2 and Fig. 3. To approximate the infinite sum for the PRI preemption, we used the first 21 terms ($j_{max} = 20$). The figures show that the simulation and the NILT give almost identical results. Because the confidence intervals are highly tight, we did not plot them as they would not be informative. We state, however, that the mean length of the intervals vary from 0.003 to 0.004, with the CDF obtained using NILT always lying within the interval bounds. These results verify that the NILT based approach is safely applicable for the complex functions discussed in this paper.

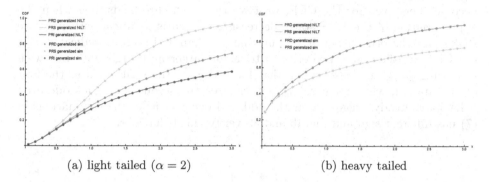

(a) light tailed ($\alpha = 2$)   (b) heavy tailed

Fig. 2. NILT and simulation of generalized service time, for light tailed Weibull distributed service time

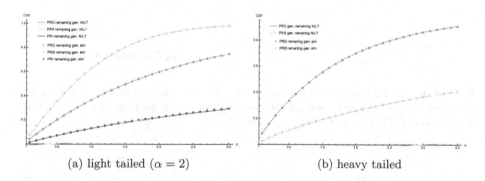

(a) light tailed ($\alpha = 2$)   (b) heavy tailed

Fig. 3. NILT and simulation of remaining generalized service time, for light tailed Weibull distributed service time

5.4   Truncation of the Infinite Summation in the PRI Case

According to Theorem 3, the LT of the generalized service time with PRI policy is obtained as a result of the infinite summation in (4). In practice, we approximate the LT as

$$
g^*_{j_{max}}(s) = \frac{(s+\nu)}{(s+\nu) - \nu r^*(s)} \cdot \sum_{j=0}^{j_{max}} \left(\frac{-\nu r^*(s)}{(s+\nu) - \nu r^*(s)} \right)^j f^*((j+1)(s+\nu)), \quad (7)
$$

i.e., we truncate the infinite sum at $j = j_{max}$.

Figure 4a and 5a demonstrate the behaviour of the obtained finite approximation of the PDF and the CDF of the generalized service time for various values of j_{max}. These show that the initial part of the PDF and CDF can be approximated well using lower j_{max}, but higher j_{max} is needed for a good approximation of their tail. It is hard to determine an exact threshold when the error of the PDF and the CDF become significant (e.g., higher than a predefined ϵ value) for a given j_{max}. For the PDF the only certain threshold is when the approximation becomes negative. For the CDF, we have two such thresholds: one when the approximation of the CDF starts decreasing (which is identical with negative PDF) and the other when it becomes larger than 1. Figure 5a indicates that the CDF satisfies these two error criteria for larger and larger intervals with increasing j_{max}. It is also visible that for odd j_{max} the CDF violates the first error criterion, while for even j_{max} it violates the second one. This odd-even behaviour already suggests that the odd and even terms of the summations in (7) has different sign and Fig. 4b and 5b verify this behaviour.

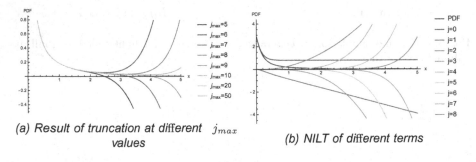

(a) Result of truncation at different j_{max} values

(b) NILT of different terms

Fig. 4. Density function of the generalized service time distribution with PRI policy, when the service time and the setup time are exponentially distributed with parameter 3 and 4 and the failure rate of the server is $\nu = 2$.

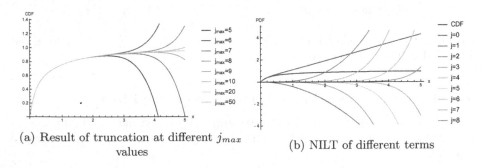

(a) Result of truncation at different j_{max} values

(b) NILT of different terms

Fig. 5. Cumulative density function of the generalized service time distribution with the same settings as in Fig. 4.

6   Conclusion

In this work, we investigated the applicability of numerical inverse Laplace transformation based analysis of complex queueing systems, where various distributions and preemption policies characterize the complexity of the Laplace transform expression of the measure of interest.

Based on a wide set numerical experimentation, we conclude, that the NILT is generally stable, and the computational complexity of the analysis comes from the evaluation of the LT transform function. There are positive distributions, often considered in qeueuing models (e.g. the lognormal distribution) for which no closed form transform domain description is available, but the Laplace domain expression can be evaluate via numerical integration.

References

1. Abate, J., Whitt, W.: A unified framework for numerically inverting laplace transforms. INFORMS J. Comput. **18**(4), 408–421 (2006)
2. Dimitriou, I., Morozov, E., Morozova, T.: Multiclass retrial system with coupled orbits and service interruptions: verification of stability conditions. In: Conference of Open Innovations, pp. 75–81 (2019)
3. Gaver, D.P.: Observing stochastic processes and approximate transform inversion. Oper. Res. **14**, 444–459 (1966)
4. Horváth, I., Horváth, G., Almousa, S.A.D., Telek, M.: Numerical inverse Laplace transformation using concentrated matrix exponential distributions. Perf. Eval. (2019). https://doi.org/10.1016/j.peva.2019.102067
5. Kulkarni, V.: Modeling and Analysis of Stochastic Systems. Chapman & Hall, London (1995)
6. Morozov, E., Morozova, T.: On stationary remaining service time in queueing systems. In: Proceedings of SMARTY: International Conference Stochastic Modeling and Applied Research of Technology, Petrozavodsk, Russia, pp. pp. 1–12 (2020)
7. Ross, S.: Introduction to Probability Models, 8th edn. Academic Press, San Diego (2003)
8. Talbot, A.: The accurate numerical inversion of laplace transforms. IMA J. Appl. Math. **23**(1), 97–120 (1979). https://doi.org/10.1093/imamat/23.1.97

Scaling Limits of a Tandem Retrial Queue with Common Orbit and Poisson Arrival Process

Anatoly Nazarov[1] , Svetlana Paul[1(✉)] , Tuan Phung-Duc[2] ,
and Mariya Morozova[1]

[1] National Research Tomsk State University, 36 Lenina Avenue,
634050 Tomsk, Russia
[2] University of Tsukuba, 1-1-1 Tennodai, Tsukuba, Ibaraki 305-8573, Japan
tuan@sk.tsukuba.ac.jp

Abstract. In this paper, we consider an asymptotic analysis for the stationary queue length of a tandem queueing system with one orbit, Poisson arrival process of incoming calls and two sequentially connected servers. Under the condition that the average delay time of calls in the orbit is extremely large, we obtain the asymptotic probability distribution of the number of calls there. It turns out that the scaled version of the number of calls in the orbit follow the Gaussian distribution. Then we evaluate the applicability of the asymptotic results by simulation.

Keywords: Tandem RQ-system with two sequentially connected servers · Asymptotic analysis method · Gaussian approximation

1 Introduction

In queuing theory there exists a special class of systems, in which the following situation is characterized: if a call finds the server busy, instead of queueing before the server it goes into the orbit, from there, after some random time, it tries to get onto the server again. Such models with orbits are called retrial queuing systems or RQ-systems [1–5].

On the other hand, tandem queuing systems represent a connection between one node queue and queuing networks: such systems can be considered as queuing networks with a linear topology [6]. Furthermore, tandem RQ-systems can be used to simulate the processing process, in which incoming requests are serviced sequentially at several stages. The need for sequential services arises in processing requests in call-centers [7–9], in controlling the data flow between elements of a multi-agent robotic system [10], etc. Tandem queuing networks are extensively studied. If the buffer is full, then the request is lost in such systems [11]. In contrast to this, we study tandem systems with an orbit of infinite capacity. Studies in this area have already been carried out by several authors. In [14], a model with a correlated flow of arrivals and the operation of the second station is described by a Markov chain.

© Springer Nature Switzerland AG 2021
V. M. Vishnevskiy et al. (Eds.): DCCN 2021, LNCS 13144, pp. 240–250, 2021.
https://doi.org/10.1007/978-3-030-92507-9_20

Retrial tandem queues are studied by several authors. Arvachenkov and Yechiali [2,3] study the model with constant retrial ate and they obtained some analytic and approximate results. However, as far as we know, tandem queues with classical retrial policy, i.e., the retrial rate is proportional to the number of customers in the obit, are less studied in the literature. We are aware of only one the related work in this line by Phung-Duc [13] in which the author studied a tandem retrial queue where only blocked customers at the first server join the orbit while those are blocked at the second one are lost. In this model, explicit joint distribution of the queue length and the state of the servers is obtained. It should be noted that the loss at the second server makes the model simple and allows the explicit solution.

In contrast to this, the underlying Markov chain of the model in the current paper is non-homogeneous because the retrial rate is proportional to the number of customers in the orbit. As a result, the model can be formulated using a level-dependent quasi-birth-and-death process where the level is the number of customers in the orbit and the phase represents the states of the servers. However, it is well-known that level-dependent QBD does not have analytical solution in general and in our model. Thus, our aim in this paper is to obtain an explicit form for the distribution of the number of customers in the orbit under some asymptotic condition. To this end, we study the model in a special regime, i.e., the case in which the retrial rate is extremely small. Under this regime, the number of customers in the orbit explodes. However, after some appropriate scaling, the scaled version of the number of customers in the orbit follows a proper distribution. The main tool to derive our results is the method of asymptotic analysis [12] under the condition of a large delay of calls in the orbit. We also validate the accuracy of the analytical results by comparing them with simulations.

The rest of our paper is organized as follows. In Sect. 2, we present the model in details. Section 3 presents the Kolmogorov's equations for the model while Sect. 4 shows the asymptotic analysis. In Sect. 5, we utilize the asymptotic results to build an approximation an validate it by comparing with simulation. Section 6 concludes our paper.

2   Mathematical Model and Problem Statement

We consider a retrial queueing system with Poisson arrival process of incoming calls with rate λ and two sequentially connected servers (see Fig. 1). Upon the arrival of a call, if the first server is free, the call occupies it. The call is served for a random time exponentially distributed with parameter μ_1 and then tries to go to the second server. If the second server is free, the call moves to it for a random time exponentially distributed with parameter μ_2. When a call arrives, if the first server is busy, the call instantly goes to the orbit, stays there for an exponentially distributed time with parameter σ and then tries to occupy the first server again. If after completing the service at the first server if the call finds that the second server is busy, it instantly goes to the same orbit, where, after random exponentially distributed delay with parameter σ, tries to move to the first server for service again.

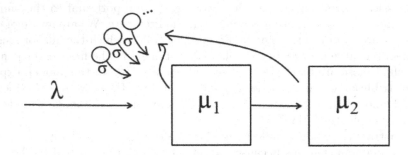

Fig. 1. Tandem RQ-system.

Let us denote:

Process $N_1(t)$ - the state of the first server at time t: 0, if the server is free; 1, if the server is busy;

Process $N_2(t)$ - the state of the second server at time t: 0, if the server is free; 1, if the server is busy;

Process $I(t)$ - the number of calls in the orbit at the time t.

The goal of the study is to obtain the stationary probability distribution of the number of calls in the orbit $I(t)$ and the probability distribution of servers' states in the considered system.

3   Derivation of Differential Kolmogorov Equations

We define probabilities

$$P_{n_1 n_2}(i, t) = P\{N_1(t) = n_1, N_2(t) = n_2, I(t) = i\}; n_1 = 0, 1; n_2 = 0, 1. \quad (1)$$

The three-dimensional process $\{N_1(t), N_2(t), I(t)\}$ is a Markov chain. For probability distribution (1) we can write the system of differential Kolmogorov equations:

$$\frac{\partial P_{00}(i, t)}{\partial t} = -(\lambda + i\sigma)P_{00}(i, t) + \mu_2 P_{01}(i, t),$$

$$\frac{\partial P_{10}(i, t)}{\partial t} = \lambda P_{00}(i, t) + (i + 1)\sigma P_{00}(i + 1, t) - (\lambda + \mu_1)P_{10}(i, t)$$

$$+\lambda P_{10}(i - 1, t) + \mu_2 P_{11}(i, t),$$

$$\frac{\partial P_{01}(i, t)}{\partial t} = \mu_1 P_{10}(i, t) - (\lambda + i\sigma + \mu_2)P_{01}(i, t) + \mu_1 P_{11}(i - 1, t),$$

$$\frac{\partial P_{11}(i, t)}{\partial t} = \lambda P_{01}(i, t) + (i + 1)\sigma P_{01}(i + 1, t) - (\lambda + \mu_1 + \mu_2)P_{11}(i, t)$$

$$+\lambda P_{11}(i - 1, t).$$

$$(2)$$

We introduce partial characteristic functions, denoting $j = \sqrt{-1}$

$$H_{n_1 n_2}(u, t) = \sum_{i=0}^{\infty} e^{jui} P_{n_1 n_2}(i, t). \tag{3}$$

Rewriting system (2) in the following form

$$\frac{\partial H_{00}(u, t)}{\partial t} = -\lambda H_{00}(u, t) + j\sigma \frac{\partial H_{00}(u, t)}{\partial u} \mu_2 H_{01}(u, t),$$

$$\frac{\partial H_{10}(u, t)}{\partial t} = \lambda H_{00}(u, t) - j\sigma e^{-ju} \frac{\partial H_{00}(u, t)}{\partial u}$$

$$-(\lambda + \mu_1 - \lambda e^{ju}) H_{10}(u, t) + \mu_2 H_{11}(u, t),$$

$$\frac{\partial H_{01}(u, t)}{\partial t} = \mu_1 H_{10}(u, t) - (\lambda + \mu_2) H_{01}(u, t) \tag{4}$$

$$+ j\sigma \frac{\partial H_{01}(u, t)}{\partial u} + \mu_1 e^{ju} H_{11}(u, t),$$

$$\frac{\partial H_{11}(u, t)}{\partial t} = \lambda H_{01}(u, t) - j\sigma e^{-ju} \frac{\partial H_{01}(u, t)}{\partial u}$$

$$-(\lambda + \mu_1 + \mu_2 - \lambda e^{ju}) H_{11}(u, t).$$

Denote matrices

$$\mathbf{A} = \begin{bmatrix} -\lambda & \lambda & 0 & 0 \\ 0 & -(\lambda + \mu_1) & \mu_1 & 0 \\ \mu_2 & 0 & -(\lambda + \mu_2) & \lambda \\ 0 & \mu_2 & 0 & -(\lambda + \mu_1 + \mu_2) \end{bmatrix},$$

$$\mathbf{B} = \begin{bmatrix} 0 & 0 & 0 & 0 \\ 0 & \lambda & 0 & 0 \\ 0 & 0 & 0 & 0 \\ 0 & 0 & \mu_2 & \lambda \end{bmatrix}, \mathbf{I}_0 = \begin{bmatrix} 1 & 0 & 0 & 0 \\ 0 & 0 & 0 & 0 \\ 0 & 0 & 1 & 0 \\ 0 & 0 & 0 & 0 \end{bmatrix}, \mathbf{I}_1 = \begin{bmatrix} 0 & 1 & 0 & 0 \\ 0 & 0 & 0 & 0 \\ 0 & 0 & 0 & 1 \\ 0 & 0 & 0 & 0 \end{bmatrix}. \tag{5}$$

Let us write the system (4) in the matrix form

$$\frac{\partial \mathbf{H}(u, t)}{\partial t} = \mathbf{H}(u, t)\{\mathbf{A} + e^{ju}\mathbf{B}\} + ju \frac{\partial \mathbf{H}(u, t)}{\partial u}\{\mathbf{I}_0 - e^{-ju}\mathbf{I}_1\}. \tag{6}$$

Multiplying equations of system (6) an identity column vector $\mathbf{e}$, we get scalar equation and add it to the system (6) in order to have

$$\frac{\partial \mathbf{H}(u, t)}{\partial t} = \mathbf{H}(u, t)\{\mathbf{A} + e^{ju}\mathbf{B}\} + ju \frac{\partial \mathbf{H}(u, t)}{\partial u}\{\mathbf{I}_0 - e^{-ju}\mathbf{I}_1\},$$

$$\frac{\partial \mathbf{H}(u, t)}{\partial t}\mathbf{e} = (e^{ju} - 1)\{\mathbf{H}(u, t)\mathbf{B} + j\sigma e^{-ju} \frac{\partial \mathbf{H}(u, t)}{\partial u}\mathbf{I}_1\}\mathbf{e}. \tag{7}$$

This system of equations is the basis in further research. We will solve it by an asymptotic method under the asymptotic condition $\sigma \to 0$.

4   Research of the Tandem RQ-System by the Method of Asymptotic Analysis

We will solve the Eq. (7) by a method of asymptotic analysis under the asymptotic condition of unlimitedly increasing the average delay of calls in the orbit, i.e., $1/\sigma \to \infty$. Under the the steady-state regime, the system of Eq. (7) is written as follows.

$$\mathbf{H}(u)\{\mathbf{A} + e^{ju}\mathbf{B}\} + ju\mathbf{H}'(u)\{\mathbf{I}_0 - e^{-ju}\mathbf{I}_1\} = 0,$$
$$\{\mathbf{H}(u)\mathbf{B} + j\sigma e^{-ju}\mathbf{H}'(u)\mathbf{I}_1\}\mathbf{e} = 0. \tag{8}$$

4.1   The First Order Asymptotic

Denote $\sigma = \epsilon$ and perform the following substitution in (8)

$$u = \epsilon w, \mathbf{H}(u) = \mathbf{F}(w, \epsilon). \tag{9}$$

We obtain

$$\mathbf{F}(w, \epsilon)\{\mathbf{A} + e^{j\epsilon w}\mathbf{B}\} + j\frac{\partial \mathbf{F}(w, \epsilon)}{\partial w}\{\mathbf{I}_0 - e^{-j\epsilon w}\mathbf{I}_1\} = 0,$$
$$\{\mathbf{F}(w, \epsilon)\mathbf{B} + je^{-j\epsilon w}\frac{\partial \mathbf{F}(w, \epsilon)}{\partial w}\mathbf{I}_1\}\mathbf{e} = 0. \tag{10}$$

Theorem 1. *Under the asymptotic condition $\sigma \to 0$, the following equality is true*

$$\lim_{\sigma \to 0} E e^{jw\sigma i(t)} = e^{jw\kappa_1}, \tag{11}$$

where κ_1 is a solution of the scalar equation

$$\mathbf{r}(\kappa_1)\{\mathbf{B} - \kappa_1\mathbf{I}_1\}\mathbf{e} = 0, \tag{12}$$

and vector $\mathbf{r}(\kappa_1)$ satisfies the normality condition

$$\mathbf{r}(\kappa_1)\mathbf{e} = 1, \tag{13}$$

and is a solution of matrix equation

$$\mathbf{r}(\kappa_1)\{(\mathbf{A} + \mathbf{B}) - \kappa_1(\mathbf{I}_0 - \mathbf{I}_1)\} = 0. \tag{14}$$

Proof. Let us take the limit $\epsilon \to 0$ in the system (10) and get the system for $\mathbf{F}(w) = \lim_{\epsilon \to 0} \mathbf{F}(w, \epsilon)$:

$$\mathbf{F}(w)(\mathbf{A} + \mathbf{B}) + j\mathbf{F}'(w)(\mathbf{I}_0 - \mathbf{I}_1) = 0,$$
$$(\mathbf{F}(w)\mathbf{B} + j\mathbf{F}'(w)\mathbf{I}_1)\mathbf{e} = 0. \tag{15}$$

We find the solution of this system in the form

$$\mathbf{F}(w) = \mathbf{r}\Phi(w), \tag{16}$$

where row vector $\mathbf{r}$ defines two-dimensional probability distribution of the states of servers (n_1, n_2), the sum of the elements of which is equal to one, according to the normalization condition.

Substituting the Eq. (16) in the system (15), we obtain

$$\mathbf{r}(\mathbf{A} + \mathbf{B}) + j\mathbf{r}\frac{\Phi'(w)}{\Phi(w)}(\mathbf{I}_0 - \mathbf{I}_1) = 0,$$

$$\mathbf{r}\left\{\mathbf{B} + j\frac{\Phi'(w)}{\Phi(w)}\mathbf{I}_1\right\}\mathbf{e} = 0. \tag{17}$$

Because the ratio $\frac{\Phi'(w)}{\Phi(w)}$ not depends on w, the scalar function $\Phi(w)$ has the form

$$\Phi(w) = e^{jw\kappa_1}, \tag{18}$$

then $j\frac{\Phi'(w)}{\Phi(w)} = -\kappa_1$. Let us substitute this equation to the system (17)

$$\mathbf{r}(\mathbf{A} + \mathbf{B}) - \mathbf{r}\kappa_1(\mathbf{I}_0 - \mathbf{I}_1) = 0,$$

$$\mathbf{r}(\mathbf{B} - \kappa_1\mathbf{I}_1)\mathbf{e} = 0. \tag{19}$$

Solving this system, we find the probability distribution of states of servers $\mathbf{r}$ and κ_1.

The first order asymptotic only defines the mean asymptotic value κ_1/σ of the number of calls in the orbit in prelimit situation of nonzero values of σ. For more detailed information of the number $I(t)$ of calls in the orbit, let us consider the second order asymptotic.

4.2   The Second Order Asymptotic

Substituting the following equation in the system (8)

$$\mathbf{H}(u) = \exp\left(j\frac{u}{\sigma}\kappa_1\right)\mathbf{H}^{(2)}(u), \tag{20}$$

we obtain

$$\mathbf{H}^{(2)}(u)\{\mathbf{A} + e^{ju}\mathbf{B} - \kappa_1(\mathbf{I}_0 - e^{-ju}\mathbf{I}_1)\} + j\sigma\frac{d\mathbf{H}^{(2)}(u)}{du}\{\mathbf{I}_0 - e^{-ju}\mathbf{I}_1\} = 0,$$

$$\mathbf{H}^{(2)}(u)(\mathbf{B} - e^{-ju}\kappa_1\mathbf{I}_1)\mathbf{e} + j\sigma e^{-ju}\frac{d\mathbf{H}^{(2)}(u)}{du}\mathbf{I}_1\mathbf{e} = 0. \tag{21}$$

Denote $\sigma = \epsilon^2$ and perform the following substitution in (21)

$$u = \epsilon w, \mathbf{H}^{(2)}(u) = \mathbf{F}^{(2)}(w, \epsilon), \tag{22}$$

and obtain the system

$$\mathbf{F}^{(2)}(w, \epsilon)\{\mathbf{A} + e^{j\epsilon w}\mathbf{B} - \kappa_1(\mathbf{I}_0 - e^{-j\epsilon w}\mathbf{I}_1)\} + j\epsilon\frac{\partial \mathbf{F}^{(2)}(w, \epsilon)}{\partial w}\{\mathbf{I}_0 - e^{-j\epsilon w}\mathbf{I}_1\} = 0,$$

$$\mathbf{F}^{(2)}(w, \epsilon)(\mathbf{B} - e^{-j\epsilon w}\kappa_1\mathbf{I}_1)\mathbf{e} + j\epsilon e^{-j\epsilon w}\frac{\partial \mathbf{F}^{(2)}(w, \epsilon)}{\partial w}\mathbf{I}_1\mathbf{e} = 0. \tag{23}$$

Theorem 2. *In the context of Theorem 1 the following equation is true*

$$\lim_{\sigma \to 0} E e^{jw\sqrt{\sigma}\left(i(t)-\frac{\kappa_1}{\sigma}\right)} = e^{\frac{(jw)^2}{2}\kappa_2}, \tag{24}$$

where κ_2 is a solution of the scalar equation

$$\mathbf{g}(\kappa_2)(\mathbf{B} - \kappa_1\mathbf{I}_1)\mathbf{e} = \mathbf{r}\mathbf{I}_1(\kappa_2 - \kappa_1)\mathbf{e}, \tag{25}$$

and vector $\mathbf{g}(\kappa_2)$ is a solution of the system

$$\mathbf{g}(\kappa_2)\{\mathbf{A} + \mathbf{B} - \kappa_1(\mathbf{I}_0 - \mathbf{I}_1)\} = \mathbf{r}(\kappa_2\mathbf{I}_0 - \kappa_2\mathbf{I}_1 - \mathbf{B} + \kappa_1\mathbf{I}_1), \\ \mathbf{g}(\kappa_2)\mathbf{e} = 0. \tag{26}$$

Proof. Let us substitute the following expansion into the system (23)

$$\mathbf{F}^{(2)}(w,\epsilon) = \Phi_2(w)(\mathbf{r} + j\epsilon w\mathbf{f}) + O(\epsilon^2), \tag{27}$$

where $\mathbf{r} = \begin{bmatrix} r_{00} & r_{10} & r_{01} & r_{11} \end{bmatrix}$ and $\mathbf{f} = \begin{bmatrix} f_{00} & f_{10} & f_{01} & f_{11} \end{bmatrix}$, we obtain

$$\Phi_2(w)(\mathbf{r} + j\epsilon w\mathbf{f})\left\{\mathbf{A} + e^{j\epsilon w}\mathbf{B} - \kappa_1(\mathbf{I}_0 - e^{-j\epsilon w}\mathbf{I}_1)\right\} \\ +j\epsilon(\Phi_2'(w)(\mathbf{r} + j\epsilon w\mathbf{f}) + \Phi_2(w)j\epsilon\mathbf{f})(\mathbf{I}_0 - e^{-j\epsilon w}\mathbf{I}_1) = O(\epsilon^2), \\ \Phi_2(w)(\mathbf{r} + j\epsilon w\mathbf{f})(\mathbf{B} - e^{-j\epsilon w}\kappa_1\mathbf{I}_1)\mathbf{e} \\ +j\epsilon e^{-j\epsilon w}(\Phi_2'(w)(\mathbf{r} + j\epsilon w\mathbf{f}) + \Phi_2(w)j\epsilon\mathbf{f})\mathbf{I}_1\mathbf{e} = O(\epsilon^2). \tag{28}$$

Rewrite the system (28) in the following form:

$$\Phi_2(w)(\mathbf{r} + j\epsilon w\mathbf{f})\left\{\mathbf{A} + e^{j\epsilon w}\mathbf{B} - \kappa_1(\mathbf{I}_0 - e^{-j\epsilon w}\mathbf{I}_1)\right\} \\ +j\epsilon(\Phi_2'(w)(\mathbf{r}(\mathbf{I}_0 - e^{-j\epsilon w}\mathbf{I}_1) = O(\epsilon^2), \\ \Phi_2(w)(\mathbf{r} + j\epsilon w\mathbf{f})(\mathbf{B} - e^{-j\epsilon w}\kappa_1\mathbf{I}_1)\mathbf{e} + j\epsilon e^{-j\epsilon w}\Phi_2'(w)\mathbf{r}\mathbf{I}_1\mathbf{e} = O(\epsilon^2). \tag{29}$$

Let us expand the exponent in a series

$$\Phi_2(w)(\mathbf{r} + j\epsilon w\mathbf{f})\left\{\mathbf{A} + (1 + j\epsilon w)\mathbf{B} - \kappa_1(\mathbf{I}_0 - (1 - -j\epsilon w)\mathbf{I}_1)\right\} \\ +j\epsilon(\Phi_2'(w)(\mathbf{r}(\mathbf{I}_0 - (1 - j\epsilon w)\mathbf{I}_1) = O(\epsilon^2), \\ \Phi_2(w)(\mathbf{r} + j\epsilon w\mathbf{f})(\mathbf{B} - (1 - j\epsilon w)\kappa_1\mathbf{I}_1)\mathbf{e} + j\epsilon(1 - j\epsilon w)\Phi_2'(w)\mathbf{r}\mathbf{I}_1\mathbf{e} = O(\epsilon^2). \tag{30}$$

Open the parentheses and group the terms for ϵ^0 and ϵ^1

$$\Phi_2(w)\mathbf{r}\left\{\mathbf{A} + \mathbf{B} - \kappa_1(\mathbf{I}_0 - \mathbf{I}_1)\right\} \\ +\Phi_2(w)j\epsilon w\left\{\mathbf{r}\mathbf{B} - \mathbf{r}\kappa_1\mathbf{I}_1 + \mathbf{f}\mathbf{A} + \mathbf{f}\mathbf{B} - \mathbf{f}\kappa_1(\mathbf{I}_0 - \mathbf{I}_1)\right\} \\ +j\epsilon\Phi_2'(w)\mathbf{r}(\mathbf{I}_0 - \mathbf{I}_1) = O(\epsilon^2), \\ \Phi_2(w)\mathbf{r}(\mathbf{B} - \kappa_1\mathbf{I}_1)\mathbf{e} + \Phi_2(w)j\epsilon w(\mathbf{r}\kappa_1\mathbf{I}_1 + \mathbf{f}\mathbf{B} - \mathbf{f}\kappa_1\mathbf{I}_1)\mathbf{e} \\ +j\epsilon\Phi_2'(w)\mathbf{r}\mathbf{I}_1\mathbf{e} = O(\epsilon^2). \tag{31}$$

Taking into account the system (19), rewrite the system (31) in the form

$$\mathbf{rB} - \mathbf{r}\kappa_1\mathbf{I}_1 + \mathbf{fA} + \mathbf{fB} - \mathbf{f}\kappa_1(\mathbf{I}_0 - \mathbf{I}_1) + \frac{\Phi_2'(w)}{w\Phi_2(w)}\mathbf{r}(\mathbf{I}_0 - \mathbf{I}_1) = 0,$$

$$(\mathbf{r}\kappa_1\mathbf{I}_1 + \mathbf{fB} - \mathbf{f}\kappa_1\mathbf{I}_1)\mathbf{e} + \frac{\Phi_2'(w)}{w\Phi_2(w)}\mathbf{r}\mathbf{I}_1\mathbf{e} = 0. \tag{32}$$

Because the ratio $\frac{d\Phi_2'(w)/dw}{w\Phi_2(w)}$ not depends on w, the scalar function $\Phi_2(w)$ has the form

$$\Phi_2(w) = e^{\left\{\frac{(jw)^2}{2}\kappa_2\right\}}, \tag{33}$$

then $\frac{\Phi_2'(w)}{w\Phi_2(w)} = -\kappa_2$. Let us substitute this equation to the system (32)

$$\mathbf{f}\{\mathbf{A} + \mathbf{B} - \kappa_1(\mathbf{I}_0 - \mathbf{I}_1)\} = \mathbf{r}(\kappa_2\mathbf{I}_0 - \kappa_2\mathbf{I}_1 - \mathbf{B} + \kappa_1\mathbf{I}_1),$$

$$\mathbf{f}(\mathbf{B} - \kappa_1\mathbf{I}_1)\mathbf{e} = \mathbf{r}\mathbf{I}_1(\kappa_2 - \kappa_1)\mathbf{e}. \tag{34}$$

The system (34) is an inhomogeneous system of linear algebraic equations for $\mathbf{f}$. Since the determinant of the matrix of coefficients of the system is equal to 0, and the rank of the extended matrix is equal to the rank of the matrix of coefficients, the system is consistent and has many solutions.

Let us consider the inhomogeneous system of Eqs. (34) and homogeneous system of Eqs. (19). If we compare them, we can see that system (19) is a homogeneous system for system (34). In this case, we can write the solution to system (34) in the form

$$\mathbf{f} = C\mathbf{r} + \mathbf{g}, \tag{35}$$

where C is a constant, $\mathbf{r}$ is the stationary distribution of the probabilities of the states of the servers and the row vector $\mathbf{g}$ is a particular solution of the inhomogeneous system (34), to which we will assign the condition $\mathbf{ge} = 0$.

Substituting the expression (35) in the system (34), we obtain

$$\mathbf{g}\{\mathbf{A} + \mathbf{B} - \kappa_1(\mathbf{I}_0 - \mathbf{I}_1)\} = \mathbf{r}(\kappa_2\mathbf{I}_0 - \kappa_2\mathbf{I}_1 - \mathbf{B} + \kappa_1\mathbf{I}_1),$$

$$\mathbf{g}(\mathbf{B} - \kappa_1\mathbf{I}_1)\mathbf{e} = \mathbf{r}\mathbf{I}_1(\kappa_2 - \kappa_1)\mathbf{e}. \tag{36}$$

The solution of this system of inhomogeneous equations allows us to find a parameter κ_2, that determines the variance of the number of claims in the orbit as κ_2/σ.

The second order asymptotic shows that the asymptotic probability distribution of the number $I(t)$ of calls in the orbit is Gaussian with mean asymptotic κ_1/σ and dispersion as κ_2/σ.

5   Approximation Accuracy and its Application Area

Now we could build a Gaussian approximation

$$P^{(2)}(i) = (L(i + 0.5) - L(i - 0.5))(1 - L(-0.5))^{-1}, \tag{37}$$

where $L(x)$ is the normal distribution function with parameters κ_1/σ and κ_2/σ.

Approximation accuracy $P^{(2)}(i)$ will be defined by using Kolmogorov range.

$$\Delta = \max_{k \geq 0} \left| \sum_{i=0}^{k} \left(P_i^{(2)} - P_i \right) \right|, \tag{38}$$

where P_i is the probability probability distribution of the number of claims in the orbit, obtained by the simulation.

The table contains values for this range for various values of σ and ρ (system load):

$$\rho = \frac{\lambda(\mu_1 + \mu_2)}{\mu_1 \mu_2}. \tag{39}$$

We consider $\mu_1 = 1$ and $\mu_2 = 2$ for all experiments (Table 1).

Table 1. Kolmogorov range.

σ	$\rho = 0.5$	$\rho = 0.6$	$\rho = 0.7$	$\rho = 0.8$	$\rho = 0.9$
0.5	0.142	0.125	0.112	0.146	0.198
0.1	0.071	0.049	0.055	0.071	0.097
0.05	0.034	0.039	0.04	0.036	0.074
0.02	0.022	0.024	0.026	0.031	0.049

In Figs. 2, 3 and 4, the solid line shows the approximation of $P_i^{(2)}$, the dashed line - the probability distribution of the number of claims in the orbit, obtained by the simulation (P_i).

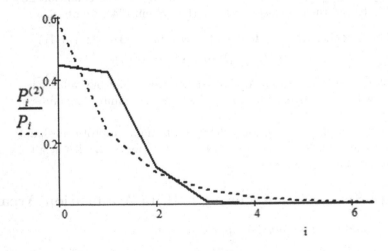

Fig. 2. The probability distribution of the number of claims in the orbit $\sigma = 0.5$, $\rho = 0.5$.

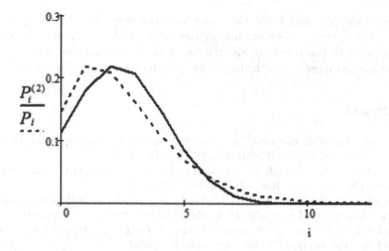

Fig. 3. The probability distribution of the number of claims in the orbit $\sigma = 0.1$, $\rho = 0.5$.

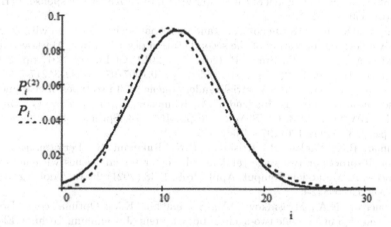

Fig. 4. The probability distribution of the number of claims in the orbit $\sigma = 0.02$, $\rho = 0.5$.

It can be seen from the table that the accuracy of the approximations increases with decreasing parameters ρ and σ. The Gaussian approximation is applicable for values of $\sigma < 0.02$, where the relative error, in the form of the Kolmogorov distance, does not exceed 0.05.

6 Conclusion

In this paper, we consider the tandem retrial queueing system with Poisson arrival process. Using the method of asymptotic analysis under the asymptotic condition of the long delay in the orbit, we obtain mean asymptotic κ_1/σ and dispersion as κ_2/σ and build the Gaussian approximation for the probability distribution of the number of calls in the orbit in the considered RQ-system. Comparing with the results of simulation, it is shown that the accuracy of the approximations increases with decreasing parameters σ and the system load.

References

1. Artalejo, J.R., Gómez-Corral, A.: Retrial Queueing Systems. Springer, Heidelberg (2008). https://doi.org/10.1007/978-3-540-78725-9
2. Avrachenkov, K., Yechiali, U.: On tandem blocking queues with a common retrial queue. Comput. Oper. Res. **37**(7), 1174–1180 (2010)
3. Avrachenkov, K., Yechiali, U.: Retrial networks with finite buffers and their application to internet data traffic. Prob. Eng. Inf. Sci. **22**(4), 519–536 (2008)
4. Basharin, G.P.: Analysis of queues in computer networks: theory and calculation methods. The science. Ch. ed. phys.-mat. lit. (1989)
5. Falin, G., Templeton, J.G.: Retrial Queues, vol. 75. CRC Press, Boca Raton (1997)
6. Kim, C., Dudin, A., Dudin, S., Dudina, O.: Tandem queueing system with impatient customers as a model of call center with interactive voice response. Perf. Eval. **70**(6), 440–453 (2013)
7. Kim, C., Dudin, A., Klimenok, V.: Tandem retrial queueing system with correlated arrival flow and operation of the second station described by a Markov chain. In: Kwiecień, A., Gaj, P., Stera, P. (eds.) CN 2012. CCIS, vol. 291, pp. 370–382. Springer, Heidelberg (2012). https://doi.org/10.1007/978-3-642-31217-5_39
8. Klimenok, V., Savko, R.: A retrial tandem queue with two types of customers and reservation of channels. In: Dudin, A., Klimenok, V., Tsarenkov, G., Dudin, S. (eds.) BWWQT 2013. CCIS, vol. 356, pp. 105–114. Springer, Heidelberg (2013). https://doi.org/10.1007/978-3-642-35980-4_12
9. Kumar, B.K., Sankar, R., Krishnan, R.N., Rukmani, R.: Performance analysis of multi-processor two-stage tandem call center retrial queues with non-reliable processors. Methodol. Comput. Appl. Prob. 1–48 (2021). https://doi.org/10.1007/s11009-020-09842-6
10. Kuznetsov, N.A., Myasnikov, D.V., Semenikhin, K.V.: Optimal control of data transmission in a mobile two-agent robotic system. J. Commun. Technol. Electron. **61**(12), 1456–1465 (2016). https://doi.org/10.1134/S1064226916120159
11. Meester, L.E., Shanthikumar, J.G.: Concavity of the throughput of tandem queueing systems with finite buffer storage space. Adv. Appl. Prob. **22**(3), 764–767 (1990)
12. Nazarov, A.A., Moiseeva, S.P.: The method of asymptotic analysis in queuing theory. Publishing house of Scientific and technical literature (2006)
13. Phung-Duc, T.: An explicit solution for a tandem queue with retrials and losses. Oper. Res. **12**(2), 189–207 (2012)
14. Vishnevsky, V.M., Larionov, A.A., Semenova, O.V.: Performance evaluation of the high-speed wireless tandem network using centimeter and millimeter-wave channels. Probl. Upravlen. **4**, 50–56 (2013)

On Regenerative Estimation of Extremal Index in Queueing Systems

Irina Peshkova[1]($\boxtimes$)(iD), Evsey Morozov[1,2,3](iD), and Maria Maltseva[1](iD)

[1] Petrozavodsk State University, Lenin Street 33, Petrozavodsk 185910, Russia
`iaminova@petrsu.ru`
[2] Institute of Applied Mathematical Research of the Karelian Research Centre
of RAS, Pushkinskaya Street 11, Petrozavodsk 185910, Russia
`emorozov@karelia.ru`
[3] Moscow Center for Fundamental and Applied Mathematics,
Moscow State University, Moscow 119991, Russia

Abstract. In this paper, the extremal behavior of the waiting time and queue size in some queueing systems is considered, based on the regenerative property of these systems. The basic properties of the extremal index of a stationary sequence are discussed, which further relate to the maximum waiting time and queue size processes. Simulation results are presented with the aim to construct estimate and to evaluate the sensitivity of the extremal index to the level of the exceedance of the process of a threshold. We discuss the additional conditions placed on the normalizing sequences to compare extremal indexes of two different stationary sequences of random variables. This analysis is then applied to compare the waiting time extremal indexes in two $M/G/1$ systems with the ordered service times having Pareto distribution.

Keywords: Extremal index · Performance analysis · Queueing system · Regenerative simulation

1 Introduction

The *extreme value theory* models the occurrence of the rare events such as high (low) values of a given process over some period of time [4,5,8]. Within the finance context, extreme events present themselves spectacularly whenever major stock market crashes. In credit insurance, mortgage–backed securities, catastrophic insurance futures, monitoring of catastrophic environmental events, applied problems of signal processing and communication network performance it is crucial to understand the risk of occurrence of extreme events and there is

The research of E. Morozov is supported by Russian Foundation for Basic Research, projects No. 19-07-00303. The research of I. Peshkova, M. Maltseva has been prepared with the support of Russian Science Foundation according to the research project No.21-71-10135 https://rscf.ru/en/project/21-71-10135/.

© Springer Nature Switzerland AG 2021
V. M. Vishnevskiy et al. (Eds.): DCCN 2021, LNCS 13144, pp. 251–264, 2021.
https://doi.org/10.1007/978-3-030-92507-9_21

an increasing need for modelling of such events to make decisions on the basis of how the extreme values will behave.

It is often observed that extremes cluster in time. Such short-range clustering is also accommodated by extreme value theory via the so-called extremal index. The extremal index is an important parameter which determines the limiting distribution of extreme values of the dependent stationary sequences as a measure of the clustering tendency of high–threshold exceedances u_n. An alternative characterisation of the extremal index is that it measures the mean cluster size in the point process of exceedance times over a high threshold. This suggests that a suitable way to estimate the extremal index is to identify clusters of the high level exceedances, and to calculate the mean size of those clusters, see for instance, [12,14,15].

The problem of studying the extreme values of the performance measures of queueing systems has been repeatedly considered, see for example, [1,3,6,7,13]. The extreme value theory for independent and *one-dependent* regeneration processes is developed in [13]. The limiting behavior of the maximum waiting time, maximum virtual waiting time and the maximum queue size in $GI/G/1$ system is the main objective of the paper [7]. The algorithm of computing the extremal index of the stationary waiting time of a stable $G/G/1$ system with distribution belonging to the domain of attraction of Gumbel distribution is given in [6].

When service time in the stable $GI/G/1$ system has exponential tail asymptotics, then there is the solution $\gamma \neq 0$ of equation $\mathsf{E}e^{\gamma(S-T)} = 1$ and the limiting distribution of maximum of waiting times is Gumbel. (Here S denotes a generic service time, and T denote a generic interarrival times.) In many cases the extremal index can be obtained explicitly or iteratively, see for instance, [6,7]. When the service times are heavy-tailed, for example have Pareto distribution, then $\gamma = 0$, but in this case we can use the tail asymptotics of the waiting times and queue sizes given in [2].

The main contribution of this research is to demonstrate that, combining the extreme value theory and the *regenerative simulation* it is possible to successfully execute a confidence estimation of the extremal index of the maximum waiting time and maximum queue size in the queueing systems. Moreover under an additional condition on the normalizing sequences of the maximum (described in Statement 1) we can compare the extremal indexes of the distributions with different parameters. Ordering of the service times in two queueing systems allows to compare the extremal indexes of the waiting time, and it is shown for the $M/G/1$ systems with Pareto service times (see Statement 2).

The structure of the paper is as follows. In Sect. 2, we give some basic definitions and results from the extreme value theory. Also, in this section, Statement 1 is given, which states that under additional conditions on the normalizing sequences, and stochastic ordering of random variables, their extremal indexes are ordered in an opposite way. In Sect. 3, we recall a few basic notions from the regenerative theory and then consider classical regeneration and estimation of the extremal index based on the regeneration cycles. The conditions leading to the explicit form of the extremal index for the waiting time are discussed as well. The limiting distribution of the maximum waiting time for the $M/G/1$ system with Pareto service time, based on the known tail asymptotics, is derived. This analysis is further applied in Sect. 4 to compare the estimates of extremal indexes with stochastically ordered service times. Some numerical simulation results for different levels of exceedances $\{u_n\}$ for the waiting time and the queue size processes in the $M/G/1$ systems are presented in Sect. 5.

2   Extremal Index

Consider the sequence of independent and identically distributed (iid) random variables (r.v.'s) $\{X_n, n \geq 1\}$ with a common distribution function (d.f.) F, and denote the maximum $M_n = \max(X_1, \ldots, X_n)$. It is evident, that d.f. of M_n satisfies

$$P(M_n \leq x) = F^n(x).$$

It is known [8] that if there exist the constants b_n, $a_n > 0$, $n \geq 1$, such, that for a non-degenerate d.f. G the following relation holds:

$$P((M_n - b_n)/a_n \leq x) \to G(x), \quad n \to \infty, \tag{1}$$

then d.f. F belongs to the maximum domain of attraction of d.f. G, denoted as $F \in MDA(G)$.

The distributions G satisfying assumption (1) are called extreme value distributions and have the following general form [4]:

$$P(X \leq x) =: H(x) = \begin{cases} \exp\left(-(1 + \eta \cdot \frac{x - \nu}{\sigma})^{-1/\eta}\right) & \eta \neq 0; \\ \exp\left(-\exp(-\frac{x - \nu}{\sigma})\right) & \eta = 0. \end{cases} \tag{2}$$

where $1 + \eta \cdot \frac{x - \nu}{\sigma} > 0$. We note that if $\eta > 0$ then H is Frechet distribution, if $\eta < 0$ then H is Weibull distribution, and if $\eta = 0$ then H is Gumbel distribution.

For a fixed x, consider the linear normalised sequence

$$u_n(x) = a_n x + b_n, \quad n \geq 1.$$

If there exists such sequence $\{u_n := u_n(x)\}$ that $n\overline{F}(u_n) \to \tau(x)$ as $n \to \infty$ (where $\overline{F} = 1 - F$ is the tail distribution), then

$$P(M_n \leq u_n(x)) \to e^{-\tau(x)}, \text{ as } n \to \infty, \ 0 < \tau(x) < \infty, \tag{3}$$

and conversely, (3) implies $n\overline{F}(u_n) \to \tau(x)$ [8]. At that, function $\tau(x) = e^{-x}$ for Gumbel distribution, $\tau(x) = x^{-\eta}$ for Frechet distribution and $\tau(x) = (-x)^\eta$ for Weibull distribution. It is easy to find the sequence $\{u_n(x)\}$ for some distributions. For example, for exponential distribution $\overline{F}(x) = e^{-\lambda x}$,

$$u_n(x) = \frac{1}{\lambda}(x + \log n) \tag{4}$$

and maximum M_n has Gumbel limit distribution. Pareto distribution

$$F(x) = 1 - \left(\frac{x_0}{x_0 + x}\right)^\xi, \ \xi > 0, \ x_0 > 0, \ x \geq 0, \tag{5}$$

(denoted by $Pareto(\xi, x_0)$) belongs to the maximum domain of attraction of Frechet distribution with the sequence

$$u_n(x) = x_0 n^{1/\xi} x - x_0.$$

In the case of the dependent *strictly stationary sequences* $\{X_n\}$, some additional conditions are required (on the mixing of the r.v.'s, see [8], Theorem 3.7.1), to ensure the asymptotic extremal behaviour, and relation (3) then becomes

$$\mathsf{P}(M_n \leq u_n(x)) \to e^{-\theta\tau(x)}, \ \text{as } n \to \infty, \ 0 < \tau(x) < \infty, \tag{6}$$

where parameter $\theta \in [0,1]$ is called the *extremal index* of the sequence $\{X_n\}$. On the other hand, consider the iid sequence $\{\hat{X}_n\}$ (called associated with original dependent sequence $\{X_n\}$) such that $\hat{X}$ has the same (marginal) distribution F as any X_n. Then it follows that

$$\lim_{n\to\infty} \mathsf{P}(\hat{M}_n \leq u_n(x)) = e^{-\tau(x)},$$

where $\hat{M}_n = \max(\hat{X}_1, \ldots, \hat{X}_n)$. Thus, the normalised and centred sequences M_n and $\hat{M}_n$ have the limit laws of the same type, with different $\theta > 0$.

Next statement allows to compare extremal indexes of two stationary sequences $\{X_n\}$ and $\{Y_n\}$ with different distributions. Let

$$M_n^X = \max(X_1, \ldots, X_n), \ M_n^Y = \max(Y_1, \ldots, Y_n),$$

where X_i have a common distribution F and Y_i have a common distribution G. (The proof of this statement will be given in a future work.) We say that r.v. X is stochastically less than a r.v. Y, and denote it $X \leq_{st} Y$, if the tail distributions are ordered as $\overline{F}(x) \leq \overline{G}(x)$ for all x.

Statement 1. *If there exist such sequences $\{u_n(x) = a_n x + b_n\}$ and $\{u_n'(x) = a_n' x + b_n'\}$ that*

$$n\overline{F}(u_n) \to \tau(x), \ n\overline{G}(u_n') \to \tau'(x),$$

as $n \to \infty$, $u_n(x) \geq u_n'(x)$ for all x, $n \geq 1$, and $\overline{F}(x) \leq \overline{G}(x)$ for all x, then

$$\theta_X \geq \theta_Y, \tag{7}$$

where θ_X and θ_Y are the extremal indexes of the sequences $\{X_n\}$ and $\{Y_n\}$, respectively.

Note, that condition $u_n(x) \geq u_n'(x)$ is met if the following relation holds

$$x \geq \frac{b_n' - b_n}{a_n - a_n'} := g_n, \ \text{for all } n \geq 1. \tag{8}$$

The inequality (8) satisfies for all $x \geq 0$ if $\sup_n g_n \leq 0$.

Remark 1. Let X and Y have exponential distributions with parameters λ and λ', respectively. If $\lambda \geq \lambda'$, then r.v. X is stochastically less than r.v. Y. Using relation (4) it is easy to find that $g_n = -\log n < 0$, $n \geq 1$.

Remark 2. Consider two r.v.'s X and Y having Pareto d.f., $Pareto(\xi, x_0)$ and $Pareto(\xi', x_0')$, respectively. If $\xi \geq \xi'$ and $x_0' > x_0 \geq 1$, then X is stochastically less than Y and the inequality (8) holds, therefore the ordering (7) is fulfilled as well.

To estimate the extremal index θ, the so-called block method can be used [4]. Notice that convergence (6), together with $n\overline{F}(u_n) \to \tau$, imply the basic relation

$$\theta = \lim_{n \to \infty} \frac{\log P(M_n \leq u_n)}{n \log F(u_n)}. \tag{9}$$

Now we assume that $n = m\,h$, and divide sequence $X_1, \ldots, X_n$ into m blocks of the size h. To apply the block method, we need to calculate the maximum of the sequence over each block, that is consider the sequence

$$M_{n,k} = \max(X_{(i-1)h+1}, \ldots, X_{ih}), \ i = 1, \ldots, m.$$

Denote $N(u_n) = \#(i \leq n : X_i > u_n)$, the number of the exceedances of the level u_n by the sequence $X_1, \ldots, X_n$, and let

$$m(u_n) = \#(k \leq n, M_{n,k} > u_n),$$

be the number of blocks with one or more exceedances. Then

$$\hat{\theta} = \frac{1}{h} \frac{\log\left(1 - \frac{m(u_n)}{m}\right)}{\log\left(1 - \frac{N(u_n)}{n}\right)}, \tag{10}$$

is the consistent estimator of the extremal index θ [4].

3   Regeneration and Extreme Values in $GI/G/1$ Queueing System

We consider the $GI/G/1$ queueing system with renewal input with iid interarrival times $T_i = t_{i+1} - t_i$, $i \geq 0$, iid service time $\{S_i\}$, and let W_i be the waiting time of the i-th customer. (Recall that, to denote a generic element of an iid sequence, we omit sequential index.) Denote by $\lambda = 1/ET$ the input rate and by $\mu = 1/ES$ the service rate, and let $\rho = \lambda/\mu$. It is well-known that this system regenerates when an arrival meets the system idle [9], and the waiting times sequence satisfies the well-known Lindley recursion:

$$W_{n+1} = (W_n + S_n - T_n)^+, \quad n \geq 1, \tag{11}$$

where we assume that $W_1 = 0$ (zero delayed process), and $(\cdot)^+ = \max(0, \cdot)$. Note that recursion (11) defines the remaining work at instants $\{t_i^-\}$, that is just before the arrivals. If $\rho < 1$ and distribution of T is *non-lattice*, then there exists the stationary distribution of waiting time, that is $W_n \Rightarrow W$ (see for instance, [1]).

For a *stationary, regenerative sequences* satisfying some additional requirements, it has been shown in [7,13] that the normalized process of level exceedances converges in distribution to a compound Poisson process. Moreover, this work provides also explicit expression for the extremal index θ. The main idea of the proof is to use the regeneration cycles instead of the blocks of the identical size. More exactly, consider a regenerative sequence $\{Z_n\}$ with regeneration instants β_k, and denote by M_{Y_k} its maximum in the kth regeneration cycle, that is, $M_{Y_k} = \sup\{Z_n, \beta_{k-1} \leq n < \beta_k\}$, and denote by $\mathsf{E}\alpha$ the mean cycle length. Then it is shown in [13] that

$$\sup_x |\mathsf{P}(M_n \leq x) - \mathsf{P}(M_{Y_1} \leq x)^{n/\mathsf{E}\alpha}| \to 0 \text{ as } n \to \infty$$

and the stationary regenerative sequence $\{Z_n\}$ has extremal index θ if and only if there exists the limit

$$\theta = \lim_{n \to \infty} \frac{\mathsf{P}(M_{Y_1} > u_n)/\mathsf{E}\alpha}{\mathsf{P}(Z_1 > u_n)} \tag{12}$$

for some exceedance sequence u_n with $n(1 - F(u_n)) \to \tau > 0$. By (10), the estimate of the extremal index based on the regeneration cycles can be constructed as follows:

$$\hat{\theta} = \frac{1}{\mathsf{E}\alpha} \cdot \frac{\log\left(1 - \frac{\hat{m}(u_n)}{m}\right)}{\log\left(1 - \frac{\hat{N}(u_n)}{n}\right)}, \tag{13}$$

where

$$\hat{m}(u_n) = \sum_{i=1}^{m} I_{\{M_{Y_i} > u_n\}}, \quad \hat{N}(u_n) = \sum_{i=1}^{n} I_{\{Z_i > u_n\}},$$

$I_{\{A\}}$ is the indicator function of an event A, and m is the number of regeneration cycles used in simulation.

Consider a random walk $Y_n = X_1 + \cdots + X_n$, $n \geq 1$ ($Y_0 = 0$) with iid non-lattice $X_i = S_i - T_i$ (with generic element X) which have a non-degenerate distribution F. Also assume that there exists the number γ such that

$$\mathsf{E}e^{\gamma X} = 1, \quad \mathsf{E}[Xe^{\gamma X}] = \mu_\gamma < \infty. \tag{14}$$

It is well-known that if $\rho < 1$ and (14) has solution $\gamma \neq 0$, then $\sup Y_n$ has an exponential tail [7], that is,

$$\mathsf{P}(\sup_{n \geq 0} Y_n > x) \sim Ke^{-\gamma x}, \quad x \to \infty.$$

(Relation $a \sim b$ means that $a/b \to 1$.) Further, define the maximum waiting time $W_n^* = \max\{W_i : 0 \le i \le n\}$. If $\rho \le 1$, then [7]

$$\lim_{n \to \infty} \mathsf{P}(\gamma W_n^* - \log(bn) \le x) = \Lambda^{1/\mathsf{E}\alpha}(x), \tag{15}$$

where $\Lambda(x) = \exp(-e^{-x})$ is Gumbel distribution and b is a constant. Moreover, W_n^* grows (in distribution) asymptotically like $\log n^{1/\gamma}$ [7]:

$$\frac{W_n^*}{\log n^{1/\gamma}} \Rightarrow 1.$$

It was proved in [1], that for the $GI/M/1$ queueing system under stability condition $\rho < 1$, the Eq. (14) has a unique positive solution γ. For stationary $M/G/1$ queue, Eq. (14) has the following form:

$$\mathsf{E}e^{\gamma S} = 1 + \frac{\gamma}{\lambda}. \tag{16}$$

Further, if Eq. (16) can be solved (explicitly or numerically), then the extremal index θ can be calculated from the following expression [6]:

$$\theta = \frac{\gamma(1 - \rho)}{\gamma + \lambda}. \tag{17}$$

For instance, for exponential, uniform, Erlang and deterministic service times, solution of Eq. (16) can be obtained either explicitly or through a simple iteration procedure [6]. For the simplest $M/M/1$ queue, it follows that $b = \rho$ and $\gamma = \mu - \lambda$, implying $\theta = (1 - \rho)^2$. In this case we can compare the numerical simulation results with the exact solution. It turns out that the regenerative approach shows a better accuracy compared with the block method.

If the service times have *regularly varying* tail distributions then (14) does not hold, and moreover, after a suitable normalization, W^* has a Frechet limit distribution [1]. For example, if the service times have Pareto distribution

$$B(x) = 1 - \left(\frac{x_0}{x_0 + x}\right)^\xi, \ \xi > 0, \ x_0 > 0, \ x \ge 0,$$

that is belongs to the class of heavy-tailed distributions, then the moment generation function satisfies

$$M_S(\gamma) := \int_0^\infty e^{\gamma x} dB(x) = \xi(-x_0\gamma)^\xi e^{-\gamma x} \Gamma(-\xi, -\gamma x_0),$$

where $\Gamma(\xi, x) = \int_x^\infty e^{-t} t^{\xi-1} dt$ is the incomplete Gamma function, in which case Eq. (16) has not solution $\gamma \ne 0$.

It is known that in heavy-tailed case (in particular, when service times are subexponential), the waiting time distribution has the following tail asymptotic [2]

$$\mathsf{P}(W > x) \sim \frac{\rho}{1 - \rho} \mathsf{P}(S_e > x), \ x \to \infty, \tag{18}$$

where the stationary remaining renewal time S_e (in the renewal process generated by the service times) has the equilibrium density $\overline{B}(x)/ES$. For Pareto distribution (5) we have $\rho = \lambda x_0/(\xi - 1)$, $\xi > 1$, and

$$B_e(x) := \mathsf{P}(S_e \leq x) = \frac{\xi - 1}{x_0} \int\limits_0^x \left(\frac{x_0}{x_0 + y} \right)^\xi dy = 1 - \left(\frac{x_0}{x_0 + x} \right)^{\xi - 1}.$$

Therefore, the tail distribution of the waiting time is

$$\mathsf{P}(W > x) \sim \frac{\lambda x_0}{\xi - 1 - \lambda x_0} \left(\frac{x_0}{x_0 + x} \right)^{\xi - 1}.$$

Let $u_n(x) = x_0 n^{1/(\xi - 1)} x - x_0$, then, as $n \to \infty$,

$$n\mathsf{P}(W > u_n(x)) \to \frac{\lambda x_0}{\xi - 1 - \lambda x_0} x^{-\xi + 1}. \tag{19}$$

Thus $W_n^* = \max(W_1, \ldots, W_n)$ has Frechet-type distribution with parameter $\xi - 1$ for n large.

4   Ordering of Waiting Times Extremal Indexes

In this section we consider an example of comparison the extremal indexes of waiting times of two queueing systems. First we recall some known monotonicity results of queueing systems with stochastically ordered service times.

We consider two $GI/G/1$ queueing systems, denoted by $\Sigma^{(1)}$ and $\Sigma^{(2)}$. (In what follows, the superscript (i) relates to the system i.) The service discipline is assumed to be First-In-First-Out. We denote by $S_n^{(i)}$ the service time of customer n, and by $t_n^{(i)}$ his arrival instant. The sequence of the iid interarrival times $T_n^{(i)} = t_{n+1}^{(i)} - t_n^{(i)}$, $n \geq 1$, and the sequence of the iid service times $\{S_n^{(i)}, n \geq 1\}$ are assumed to be independent, $i = 1, 2$. Denote by $S^{(i)}$ the generic service time, and by $T^{(i)}$ the generic interarrival time, $\mathsf{E}T^{(i)} = 1/\lambda_{T^{(i)}}$, $i = 1, 2$. Now we compare the steady-state queue-size processes in the systems $\Sigma^{(1)}$ and $\Sigma^{(2)}$. At the arrival instant of customer n in the system $\Sigma^{(i)}$, denote by $\nu_n^{(i)}$ the number of customers, by $Q_n^{(i)}$ the *queue size* and by $W_n^{(i)}$ the *waiting time* of this customer. Denote, when exists, the limits (in distribution)

$$Q_n^{(i)} \Rightarrow Q^{(i)}, \ \nu_n^{(i)} \Rightarrow \nu^{(i)}, \ W_n^{(i)} \Rightarrow W^{(i)}, \ n \to \infty, \ i = 1, 2.$$

These limits exists, in particular, when the interarrival times $T^{(i)}$, $i = 1, 2$ are *non-lattice* and $\rho_i = \lambda_{T^{(i)}} \mathsf{E}S^{(i)} < 1$ [1]. Assume that the following stochastic relations hold:

$$\nu_1^{(1)} =_{st} \nu_1^{(2)} = 0, \ T^{(1)} =_{st} T^{(2)}, \ S^{(1)} \leq_{st} S^{(2)}. \tag{20}$$

Then [16],

$$Q_n^{(1)} \leq_{st} Q_n^{(2)}, \ W_n^{(1)} \leq_{st} W_n^{(2)}, \ n \geq 1. \tag{21}$$

Now we consider two $M/G/1$ systems with the same input $\lambda_{T^{(1)}} = \lambda_{T^{(2)}} = \lambda_T$. Let service times have Pareto distribution $Pareto(\xi, x_0)$ and $Pareto(\xi', x_0')$ in systems $\Sigma^{(1)}$ and $\Sigma^{(2)}$, respectively. Under condition $\xi \geq \xi'$ and $x_0' > x_0 \geq 1$, service times are stochastically ordered, so $S^{(1)} \leq_{st} S^{(2)}$, therefore inequalities (21) hold, including stochastic ordering of waiting times, $W_n^{(1)} \leq_{st} W_n^{(2)}$. Now define the maximum waiting times in both systems,

$$W_n^{(1)*} = \max(W_1^{(1)}, \ldots, W_n^{(1)}), \ W_n^{(2)*} = \max(W_1^{(2)}, \ldots, W_n^{(2)}).$$

We know that the limit distributions of $W_n^{(1)*}$ and $W_n^{(2)*}$ have form (19) with parameters ξ, x_o and ξ', x_0', respectively. Discussion above implies the following result.

Statement 2. *If, in the described above systems $\rho_i < 1$, $i = 1, 2$ and the following conditions on the parameters of the Pareto service time are satisfied*

$$\xi \geq \xi' > 1, \quad x_0' > x_0 \geq 1,$$

then the extremal indexes of the waiting times in these systems are ordered as

$$\theta_{W^{(1)}} \geq \theta_{W^{(2)}}.$$

5 Simulation Results

In this section we present the numerical examples of simulation of waiting time and queue size extremal indexes based on expressions (10) and (13) with focus on the system with Pareto service time, and using the results on the sensitivity of the extremal index on the level of exceedances.

Figure 1 shows the estimates of waiting time extremal index calculated by the block method, the regenerative method and explicit expression, for the $M/M/1$ queueing system with input rate $\lambda_T = 0.2$, and service rate $\lambda = 0.8$. It is seen that the regenerative estimate is closer to explicit result (relative error less than 1.7%).

Figure 2 shows simulation results of $M/G/1$ system with input rate $\lambda_T = 1.87$ and $Pareto(2.1, 0.5)$ service time, implying $\rho = 0.85$. The block method and the regeneration method are used to estimate extremal index of the waiting times. In both cases the estimate is very closed to zero, and this is typical for the heavy-tailed distributions.

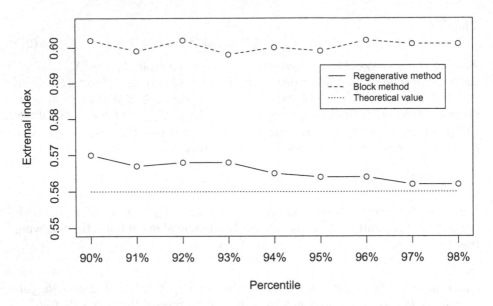

Fig. 1. Comparison of extremal indexes of waiting time calculated by block method, regenerative method and via explicit expression; input rate $\lambda_T = 0.2$, service rate $\lambda_1 = 0.8$.

Figure 3 shows the extremal indexes of waiting times in two $M/G/1$ systems with the same input rate $\lambda_T = 1.87$. The service time have Pareto distribution with parameters $\xi_1 = 2.3$ and $\xi_2 = 2.1$, respectively. These parameters guarantees that systems are stable and service times are stochastically ordered, $S^{(1)} \leq_{st} S^{(2)}$. By Statements 1, 2, the extremal indexes of waiting times are ordered as

$$\theta_{W^{(1)}} \geq \theta_{W^{(2)}},$$

and it is confirmed by simulation.

Now we consider three $M/G/1$ systems: $\Sigma^{(1)}$ with exponential service time with rate $\lambda = 4.3$, $\Sigma^{(2)}$ with Pareto service time with parameters $\xi = 2.1, x_0 = 0.5$ and $\Sigma^{(3)}$ with exponential-Pareto service time having mixture distribution with parameters $x_0 = 0.5$, $\lambda = 4.3$, $\xi = 2.1$ and mixing proportion $p = 0.5$. (For more detail on Exponential-Pareto distribution, see [10].) The input rate is $\lambda_T = 1.87$ in all three systems. As we see on Fig. 4, the estimates of the extremal indexes of waiting times are ordered in the following way:

$$\hat{\theta}_{W^{(1)}} \geq \hat{\theta}_{W^{(3)}} \geq \hat{\theta}_{W^{(2)}}. \tag{22}$$

Relation (22) is intuitive because the service times in these systems are stochastically ordered in the following order

$$S^{(1)} \leq_{st} S^{(3)} \leq_{st} S^{(2)}.$$

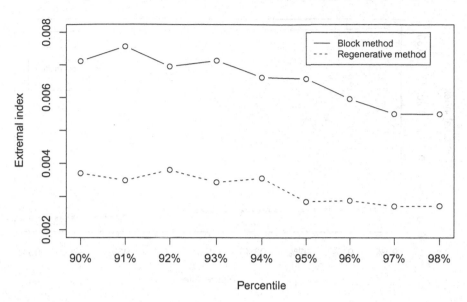

Fig. 2. Extremal index for waiting times in $M/G/1$ with Pareto service time and parameters $\lambda_T = 1.87$, $\xi = 2.1$, $x_0 = 0.5$.

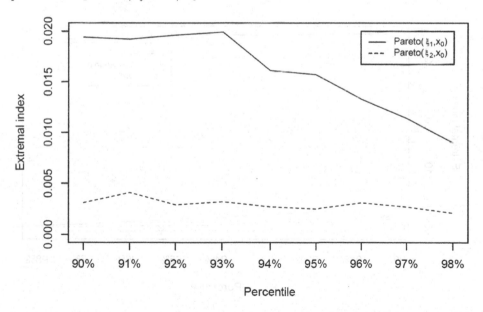

Fig. 3. Extremal indexes for waiting times in two $M/G/1$ systems with Pareto service time and parameters $\lambda_T = 1.87$, $\xi_1 = 2.3, \xi_2 = 2.1$, $x_0 = 0.5$.

Note that the ordering of the performance measures of the queueing systems with mixture service times is discussed in [11]. Figure 5 demonstrates simulation results of the extremal indexes of the queue sizes for all three systems described above.

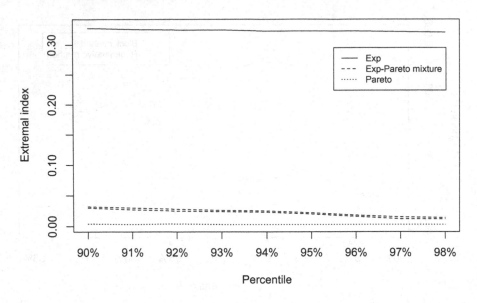

Fig. 4. Extremal index of waiting time. Input rate $\lambda_T = 1.87$, $p = 0.5, x_0 = 0.5$, $\lambda = 4.3$, $\xi = 2.1$.

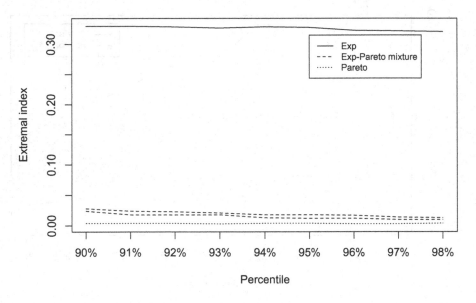

Fig. 5. Extremal index of queue size. Input rate $\lambda_\tau = 1.87$, $p = 0.5, x_0 = 0.5$, $\lambda = 4.3$, $\xi = 2.1$.

6   Conclusion

In this paper, we study the applicability of the *classical regeneration* to the estimation of the extreme behaviour of the steady-state performance measures in some queueing systems. The obtained theoretical results are then applied to compare the waiting time extremal indexes in two $M/G/1$ systems with stochastically ordered Pareto service times. The regenerative estimates of the extremal indexes of waiting times depending on the level of exceedance (percentile) are presented. Simulation confirms theoretical results concerning the ordering of extremal indexes. In particular, this ordering is confirmed for three $M/G/1$ systems with exponential, Pareto and Exponential-Pareto mixture service times, respectively.

References

1. Asmussen, S.: Applied Probability and Queues: Stochastic Modelling and Applied Probability, 2nd edn. Springer-Verlag, New York (2003). https://doi.org/10.1007/b97236
2. Asmussen, S., Kluppelberg, C., Sigman, K.: Sampling at subexponential times with queueing applications. Report TUM M9804 (1998)
3. Bertail, P., Clémençon, S., Tressou, J. : Regenerative block-bootstrap confidence intervals for the extremal index of Markov chains. In: Proceedings of the International Workshop in Applied Probability (2008)
4. Embrechts, P., Klüppelberg, C., Mikosch, T.: Modelling Extremal Events. AM, vol. 33, p. 660. Springer, Heidelberg (1997). https://doi.org/10.1007/978-3-642-33483-2
5. berg de Haan, L., Ferreira, A. : Extreme Value Theory: An Introduction. Springer Series in Operations Research and Financial Engineering, p. 491. Springer, Heidel (2006). https://doi.org/10.1007/0-387-34471-3
6. Hooghiemstra, G., Meester, L.E.: Computing the extremal index of special Markov chains and queues. Stochastic Process. Appl. **65**(2), 171–185 (1996). https://doi.org/10.1016/S0304-4149(96)00111-1
7. Iglehart, D.L.: Extreme values in GI/G/1 queue. Ann. Math. Stat. **43**(2), 627–635 (1972). https://doi.org/10.1214/aoms/1177692642
8. Leadbetter, M.R., Lindgren, G., Rootzin, H. : Extremes and Related Properties of Random Sequences and Processes. Springer, New York (1983). https://doi.org/10.1007/978-1-4612-5449-2
9. Morozov, E.: An extended regenerative structure and queueing network simulation. Preprint No 1995–08/ISSN 0347–2809. Department of Mathmatics, Chalmers University, Gothenburg, Sweden (1995)
10. Peshkova, I., Morozov, E., Maltseva, M.: On comparison of multiserver systems with exponential-pareto mixture distribution. In: Gaj, P., Gumiński, W., Kwiecień, A. (eds.) CN 2020. CCIS, vol. 1231, pp. 141–152. Springer, Cham (2020). https://doi.org/10.1007/978-3-030-50719-0_11
11. Peshkova, I., Morozov, E., Maltseva, M.: On comparison of multiserver systems with two-component mixture distributions. In: Vishnevskiy, V.M., Samouylov, K.E., Kozyrev, D.V. (eds.) DCCN 2020. CCIS, vol. 1337, pp. 340–352. Springer, Cham (2020). https://doi.org/10.1007/978-3-030-66242-4_27

12. Resnick, S.I.: Extreme Values, Regular Variation and Point Processes. SSORFE, p. 334. Springer, New York (1987). https://doi.org/10.1007/978-0-387-75953-1

13. Rootzen, H.: Maxima and exceedances of stationary Markov chains. Adv. Appl. Prob. **20**(2), 371–390 (1988). https://doi.org/10.2307/1427395

14. Smith, R.L.: The extremal index for a Markov chain. J. Appl. Prob. **29**(1), 37–45 (1992). https://doi.org/10.2307/3214789

15. Smith, L., Weissman, I.: Estimating the extremal index. J. Royal Stat. Soc. Ser. B Methodol **56**(3), 515–528 (1994). https://doi.org/10.1111/J.2517-6161.1994.TB01997.X

16. Whitt, W.: Comparing counting processes and queues. Adv. Appl. Prob. **13**, 207–220 (1981)

Evaluation of the Performance Parameters of a Closed Queuing Network Using Artificial Neural Networks

A. V. Gorbunova[(✉)] [iD] and Vladimir Vishnevsky [iD]

V.A. Trapeznikov Institute of Control Sciences of Russian Academy of Sciences,
65, Profsoyuznaya Street, Moscow 117997, Russia
avgorbunova@list.ru, vishn@inbox.ru

Abstract. The article presents a novel approach to solving complex problems of queuing theory in general and to the analysis of closed queuing networks, in particular. The main idea is to combine a simulation on a limited set of input parameters with data mining methods. As an example of applying this approach, the paper evaluates the main performance parameters of a closed exponential queuing network, for which, as is known, methods of analytical and numerical analysis have been developed with some restrictions on its structure. The specified network was chosen in order to qualitatively test the proposed approach, as well as to demonstrate its effectiveness on a very large amount of input data, which is an essential distinguishing feature of this work. The maximum approximation error of the estimates of the mean sojourn times of customers obtained in the study, as well as their average number in the network nodes, does not exceed 5%. The new approach is universal in nature, and its field of application is not limited to studies of only exponential networks, therefore, it can be used to study more complex queuing networks.

Keywords: Closed queuing network · Average response time · Multilayer perceptron · Artificial neural networks · Machine learning methods

1 Introduction

Interest in the study of both open and closed queuing networks (QN) lies in the possibility of their use for modeling real physical systems. One of the most demanded areas of research, and one of the most common examples is the use of QNs in various configuration options as models of computer networks, which, in turn, have already become an integral part of modern life. Thus, computer networks provide their users with a fairly wide range of services, which include e-mail, and all kinds of news services, and work with remote databases and much more. In addition, distance learning, telemedicine, teleconferences are implemented on the basis of computer networks, which, in the light of the well-known

© Springer Nature Switzerland AG 2021
V. M. Vishnevskiy et al. (Eds.): DCCN 2021, LNCS 13144, pp. 265–278, 2021.
https://doi.org/10.1007/978-3-030-92507-9_22

events of 2020, clearly demonstrated their relevance, although the demand for improving network technologies has existed and will always exist.

Thus, a rather rapid and continuous growth in the number of computer networks and their users leads to the need to develop the theoretical foundations of computer network design and, as a consequence, to develop new approaches to the analysis of the main performance characteristics of the constructed analytical network models [1].

Since the present work is devoted specifically to closed QNs, we will dwell in more detail on the description of the existing methods of their study, the main advantages and disadvantages of these methods. It should be noted that there have not been any significant progress in the direction of developing new approaches to studying closed networks, and there are not so many methods of analysis themselves, all of which depend on the network belonging to a certain type.

All approaches to the analysis of closed networks can be divided into exact and approximate. First, let's turn to the exact research methods. In this regard, it is necessary to mention an analogue of Jackson's theorem in the case of open QNs, but already for closed networks. This is the so-called Gordon-Newell theorem, which extends its action to homogeneous closed networks with exponential service times and buffers of unlimited capacity [2]. It follows from this theorem that the stationary distribution of the number of customers at the nodes of a closed network has a multiplicative form.

Later, the theorem was generalized by the BCMP theorem, named after the authors of the article (Baskett, Chandi, Muntz and Palacios), in which a network of the same type was first described, extending the multiplicative form of stationary probabilities to a wider range of networks, but, nevertheless, with the constraint on the form of the service time distribution function, which must be either exponential or simply have a rational Laplace-Stieltjes representation, and in the latter case the number of servers in the network nodes must be equal to one or to the total number of customers circulating in the network [3].

Knowing the stationary probabilities, it is possible to determine the main probabilistic and temporal characteristics of the entire network. However, the key problem that arises when calculating the state probabilities is to calculate the normalizing constant present in the expression defining them. Its direct computation requires significant resources and time consuming even with a relatively small network dimension. This circumstance has led to the emergence of special recurrent methods for calculating the normalizing constant. One of the first methods, which later became the basis for many other later developed algorithms, is the Buzen method or, as it is also called, the convolution algorithm [4]. Although there are some difficulties here, namely, there is a high probability of obtaining a machine zero, i.e. to get the results to be zeroed, or their overflow in the computer memory, when, for example the number of nodes or the number of customers simultaneously in the network increase.

There is also another recurrent method, which is comparatively more computationally efficient, since it does not require knowledge of stationary probabilities and, accordingly, preliminary calculation of the normalizing constant. It refers to the so-called mean value analysis (MVA) and is based on the equality of the intensities of flows of customers entering and leaving the same node, as well as on Little's formula [5]. As a result of applying this iterative method, it is possible to calculate such important performance characteristics for any network as the average sojourn times of customers in the network nodes and the average number of customers in each node; in addition, other indicators can be calculated, for example, the marginal distribution of the queue length in a particular network's node. However, despite the absence of a problem with zeroing and/or overflow, with an increase in the number of nodes and the total number of customers, the computational complexity of the algorithm also grows, and with it the required amount of computational resources and time spent on performing calculations. Moreover, it is conceivable that the required costs may exceed the capabilities of even modern computers, then the only solution is the use of approximate methods of analysis.

If we consider closed QNs that do not satisfy the BCMP theorem, then exact methods, as a rule, are developed only for certain types of networks consisting of no more than two nodes, and with rare exceptions with more than one server in each of the subsystems [6–9], as well as a limited set of types of distributions for servicing on the devices themselves, for example, exponential or phase distribution, Cox distribution [10]. At the same time, it is worth noting that there are separate works in which queues to network nodes can be finite, which, of course, leads to the appearance of blocking before or after service. For example, [11,12] proposes an exact solution for nonlinear topology networks with finite storage capacity in single-line nodes and, accordingly, various blocking options, but with exponential service times on servers.

We now turn to a discussion of approximate methods. Despite the fact that exact solution methods are known in the case of networks with exponential nodes and buffers of unlimited capacity, a considerable number of research works have been devoted to the development of approximate calculation procedures in order to analyze their performance. Most of the approximation algorithms are based on the MVA, and their advantage lies in the greater efficiency and, accordingly, the execution speed in comparison with the exact methods. Approximate methods have also been developed for exponential networks of a similar architecture, but already with limited buffer capacity. Most of the procedures are also based on the MVA or on the decomposition approach, the application of which is not limited to exponential service times, therefore, we will further consider this method in the context of general distributions of service times.

The main idea of the decomposition method is to divide the network into subsystems, analyze them separately, and then combine the results to obtain the characteristics of the network as a whole. The role of subsystems can be either loosely coupled subnets or the network nodes themselves. In the latter case, it is already assumed that various approaches to the analysis of queuing systems (QS) are applied. In this case, an additional difficulty will be in determining the main parameters of the QS functioning, namely, the parameters of the input and output flows for each of the nodes. Therefore, the decomposition approach is quite often combined with diffusion approximation, which, in turn, is used to analyze more general models of nodes like $GI|G|1$. Moreover, this method turned out to be effective, with some restrictions, in the case of open queuing networks with a general distribution function of the service time [13–15]. However, its accuracy strongly depends on the load level of network fragments, i.e., the higher the load level, the better the approximation, as well as the value of the variation coefficient and the specific type of service time distribution, at that the method error can sometimes be unacceptable.

Decomposition is sometimes also understood as a transition to the consideration of equivalent networks. Thus, a fairly efficient iterative procedure was proposed by R.A. Marie, which is called by the name of the author [16,17]. It is based on Norton's theorem and refers to methods that allow to analyze networks with nodes like $\cdot|G|m$. In order to assess the performance of an individual node, the rest of the network is transformed into a single node with an exponential distribution of the service time, depending on the number of customers located there. Then the service rates in the composition node are calculated, which is not quite simple. After that, a system of equations is compiled for an isolated node with general distribution function of the service time and the intensity of the input flow, depending on the load, which has a solution in a closed form. As a result, it is possible to obtain the marginal probability distribution of the number of customers in the analyzed node. Each of the network nodes is examined in a similar way. However, in this case, it is still assumed that general distribution function of the service time in the nodal fragments of the network has a rational Laplace representation, i.e., it should be approximated by a generalized Cox distribution with different parameters depending on the value of the coefficient of variation of the original function.

In the case of networks with a finite buffer capacity and general distribution functions of service times at the nodes, there are not so many methods for their approximate analysis. So, for example, in [18], the analysis of such a network actually boils down to its approximation by an equivalent network with unlimited buffer capacity at the nodes and then applying the Marie's method. Alternatively, one can proceed from the analysis of a closed network to the study of an open network with similar characteristics; other decomposition options are also possible, but, as a rule, in combination with known methods for analyzing networks without blocking.

As mentioned above, there are no major variations towards the emergence of new methods for studying QNs. Nevertheless, among the latest works, one can note, for example, [19], in which the author adapted the method of mean value analysis for exponential networks with blocking and single-server nodes. In [20] authors evaluated the parameters of exponential closed network with blocking for multi-server nodes using a combination of the decomposition method, which actually consists in splitting each network node into two QSs, and MVA. In [21] a decomposition method was proposed for networks with blocking and general distribution of service time.

2   Analysis of Closed QNs Using ANN

In the previous section, a brief description of the main methods for studying closed queuing networks was given. Summarizing the above, we can conclude that for networks not from the BCMP category, the known methods under certain conditions can lead either to unacceptable errors in the estimated characteristics, or the calculations themselves can take a significant amount of time comparable to the time spent on simulation. Moreover, large time costs and serious computational difficulties can arise for networks that have the property of a multiplicative form, provided, for example, a large network dimension and a large number of customers circulating in it, which, in general, is typical for real computer networks. Therefore, the authors of the current work propose a new approach to obtaining the characteristics of closed QNs. The idea is to combine simulation with various data mining techniques and the use of artificial neural networks (ANN) in particular. As already mentioned, simulation can be time-consuming, but if using simulation within a given interval for individual values of the input parameters to obtain a set of corresponding values of the characteristics of interest, then it is possible to train a neural network on the data obtained, which will give an estimate for any intermediate values of the input parameters that affect the operation of the network.

As a result, it will only be necessary to spend time on simulation, but for a limited set of input parameters, as well as on training a neural network or some other intelligent model. The forecasting process itself does not actually require time. Thus, it will be possible to obtain estimates of the main performance indicators of the required accuracy for a general QN without any restrictions on the service time distribution function, the number of servers in the network nodes, or on the buffer capacity.

If an effective computational algorithm has been developed for assessing the probabilistic-temporal characteristics of a certain network or queuing system, but it requires too much time and computational costs, then it is possible, similarly using this algorithm, to first obtain estimates of characteristics for a limited set of input parameters, and then train the neural network and solve the forecasting problem, as was done, for example, in [22] for a non-Markov QS with a "warm-up". Nevertheless, there are not so many efficient approximate methods for analyzing queuing networks, and their accuracy strongly depends on the properties of the network itself, therefore, simulation in this context is more advantageous and has a more universal character.

As for the construction of a simulation model, there are several options. It is possible to use specialized software applications developed for these purposes. Judging by one of the recent reviews of [23], there are three most popular applications: GPSS World, AnyLogic and Arena. All of these software environments are commercial products, although some of them have free versions with limited functionality. As an alternative to implemented options, you can develop your own simulation model, for example, in an absolutely free Python software environment with a fairly large set of functions, feature and libraries, including those for training ANN.

The good perspectives of this approach in relation to the analysis of an open queuing network with nodes of the type $\cdot|G|1$, as well as to the analysis of a fork-join system, was demonstrated in the papers [24] and [25], accordingly, and, as expected, is not limited only to open QNs or various kinds of QSs. Therefore, in order to extend the scope of this approach in the context of solving problems of queuing theory and at the same time to verify its effectiveness, we will use the described method to analyze a closed queuing network. Since the main task is to demonstrate the effectiveness of the new approach on a large amount of input data, as an illustrative example, consider an exponential closed network, which will be discussed in the next section.

3   Closed QN Model and Numerical Experiment

Consider a closed QN consisting of M nodes, each of which is a single-server QS with a buffer of unlimited capacity. The order in which service requests are selected at each node is determined by the first-in-first-out (FCFS) discipline. The distribution function of the service time on the server of the i-th node is exponential with the parameter μ_i. In a closed network, customers do not come from outside and do not leave the network, therefore their number is constant. We denote their number by K. Customers move from one network node to another in accordance with the route matrix $\Theta = (\theta_{ij})$, where θ_{ij} is the probability of an instant transition to the j-th node after service on the server of the i-th node. Since customers cannot leave the network, the following condition must be satisfied for the routing matrix

$$\sum_{j=1}^{M} \theta_{ij} = 1, \quad i = \overline{1, M}.$$

We denote by λ_i the intensity of the output flow from the node i. The scheme of the described network is shown in Fig. 1, and $\theta_j = \sum_{i=1}^{M}$ is the total probability of transition into the j-th queuing subsystem. The characteristics of the presented network can be obtained in an analytical form, however, due to the fact that we want to analyze the effectiveness of the new approach, we will follow it.

The first step is to develop a simulation model of the network. For this purpose, programming language Python was chosen instead of a canned software. Now let's set specific values for the network parameters. Let the matrix of transition probabilities between the network nodes be equal to

$$\Theta = \begin{pmatrix} 0.1\ 0.3\ 0.4\ 0.1\ 0.1 \\ 0.2\ 0.1\ 0.2\ 0.3\ 0.2 \\ 0.3\ 0.1\ 0.1\ 0.3\ 0.2 \\ 0.2\ 0.2\ 0.1\ 0.3\ 0.2 \\ 0.1\ 0.1\ 0.2\ 0.3\ 0.3 \end{pmatrix},$$

the number of nodes is $M = 5$, and the number of customers is $K = 100$. As the estimated performance indicators of the network, we will choose the average

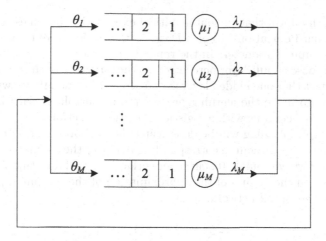

Fig. 1. Scheme of a closed queuing network.

number of customers and the average response time of customers at each of the nodes, which we denote by N_i and v_i, respectively, $i = \overline{1,5}$. The input data, which is subject to change, is the service rate of each of the servers μ_i. All parameters μ_i step from 3.0 to 4.0 in 0.25 increments (Table 1). As a result, a total of 3125 datasets would need to be obtained using the simulation model.

Table 1. Input data for training artificial neural networks.

No.	μ_1	μ_2	μ_3	μ_4	μ_5
1	3.00	3.00	3.00	3.00	3.00
2	3.00	3.00	3.00	3.00	3.25
3	3.00	3.00	3.00	3.25	3.00
4	3.00	3.00	3.00	3.25	3.25
5	3.00	3.00	3.25	3.00	3.00
...	...	...	...	...	...
3125	4.00	4.00	4.00	4.00	4.00

Next, we will train the neural network also in the Python software environment. To do this, we divide the resulting sample into a training and a test one in the ratio of 80% and 20%, respectively. On the training set, we will actually train the ANN, and to assess the training we will use the remaining 20% that did not participate in the training, but will allow us to assess how well the trained neural network will solve the forecasting problem in reality for new input data.

To improve the quality of the forecast using the ANN, the data of both the training and test samples were standardized and normalized. By standardization we mean bringing the set of input data for each of the parameters affecting the system to such a form that its mathematical expectation becomes equal to zero, and the standard deviation—to one. Data normalization here means scaling the values of the input parameters in the range from 0 to 1.

We will choose a multilayer perceptron as the structure of the neural network. Despite the fact that one hidden layer would already be enough, we will still focus on two layers to make the learning process more controlled. Each hidden layer will consist of 10 neurons with a logistic activation function $\varphi(x) = 1/(1 + e^{-x})$ on each of them. Learning will be done using the backpropagation method. The output will be only one neuron corresponding to one of the estimated parameters, that is, in fact, we will build 10 neural networks (Fig. 2). Thus, this should have a positive effect on the quality of the approximation of the estimated performance indicators of the closed network.

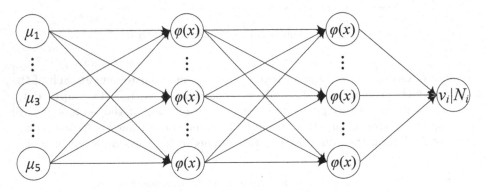

Fig. 2. Diagram of a three-layer perceptron used to evaluate the performance characteristics of a closed QN.

To assess the quality of the forecast produced by the neural network, we will use the mean square error (MSE), mean absolute error (MAE), and mean absolute percentage error (MAPE), which are determined by the following expressions

$$MSE = \frac{1}{n}\sum_{j=1}^{n}(y_j - \widehat{y}_j)^2, \quad MAE = \frac{1}{n}\sum_{j=1}^{n}|y_j - \widehat{y}_j|,$$

$$MAPE = \frac{1}{n}\sum_{j=1}^{n}\left|\frac{(y_j - \widehat{y}_j)}{y_j}\right| \cdot 100\%,$$

where $\widehat{y}_j$ is the estimate of the investigated characteristic, obtained using ANN on a test or some other sample for the j-th input dataset. In our case, we mean the average number of customers in one of the nodes of the QN N_i or the response time by customers in a specific node of the network v_i, $i = \overline{1,5}$. The value y_j is the real value of the estimated characteristic obtained as a result of closed network simulation, $j = \overline{1,n}$, and n is the number of datasets in the sample.

Of course, one could have spent much more time on the process of building the optimal architecture of a neural network (the number of hidden layers and neurons in them), the choice of activation functions on neurons, as well as a specific training method, which are not so few, and other things, but all the listed manipulations allow to get already quite good result, as can be seen from the Table 3.

Table 2. Input data for evaluating the performance characteristics of the closed queuing network using ANN.

No.	μ_1	μ_2	μ_3	μ_4	μ_5
1	3.10	3.10	3.10	3.10	3.10
2	3.10	3.10	3.10	3.10	3.20
3	3.10	3.10	3.10	3.20	3.10
4	3.10	3.10	3.10	3.20	3.20
5	3.10	3.10	3.20	3.10	3.10
...	...	...	...	...	...
59049	3.90	3.90	3.90	3.90	3.90

Table 3. Approximation errors when calculating the estimates of the performance characteristics of a closed queuing network using ANN on a test dataset.

Evaluated characteristic	Type of error		
	MSE	MAE	MAPE, %
v_1	0.000012	0.002821	0.306338
v_2	0.000006	0.001542	0.185111
v_3	0.000042	0.003707	0.301359
v_4	0.001378	0.027710	0.128102
v_5	0.000449	0.013940	0.893533
N_1	0.000019	0.003138	0.115259
N_2	0.000043	0.004066	0.227245
N_3	0.000691	0.018952	0.646385
N_4	0.011222	0.076211	0.101165
N_5	0.002921	0.032369	0.845084

Now, in order to finally verify the performance of the constructed neural network and the quality of the estimates obtained with its help, we will make a forecast for each of the above performance characteristics of the closed QN for the values of the service time intensities varying in the same range as before [3.0, 4.0], but with a step of 0.1, for example (Table 2). Thus, the number of different sets of input data will already be 59049 units. This is about 19 times more than the number of datasets that were required to conduct simulation to train the neural network in this case. Recall that, unlike simulation modeling, ANN forecast is made almost instantly, so the benefits from their use are obvious.

We have the opportunity to check the operation of the ANN on such a significant amount of data due to the choice of the exponential closed QN as the object of research, since the performance characteristics of this network can be calculated analytically and, accordingly, the necessary comparative analysis can be carried out. To calculate the normalizing constants present in expressions for stationary probabilities, we have programmed in Python Buzen's algorithm, which can be found in more detail in the original [4] or, for example, in [1]. After that, in the same software environment, the average characteristics of the nodes of the closed network were calculated.

Table 4. Approximation errors when calculating the estimates of the performance characteristics of a closed queuing network using ANN for the input dataset from Table 2.

Evaluated characteristic	Type of error		
	MSE	MAE	MAPE, %
v_1	0.000014	0.002876	0.284568
v_2	0.000003	0.001275	0.161997
v_3	0.000120	0.003230	0.235173
v_4	0.004063	0.055876	0.227230
v_5	0.000239	0.009265	0.639649
N_1	0.000904	0.010727	0.287468
N_2	0.000024	0.003369	0.203218
N_3	0.001016	0.018032	0.551451
N_4	0.015708	0.087965	0.104054
N_5	0.010458	0.042261	0.765629

The average values of errors on the test sample containing 20% randomly selected data from the general set are presented in Table 3, and the average values of the same types of errors, but for the input parameters from Table 2 are presented in Table 4. As can be seen from the table, the average error values differ, but this is not surprising, since the number of data sets in the test sample is many times smaller, although the average relative error of approximations for all the estimated characteristics does not exceed 1%, which is a good result.

In order to understand what is hidden behind the averaged results, let us analyze in more detail the structure of the relative errors included in the MAPE. Since it will obviously be difficult to draw any conclusions on the basis of a graph containing almost 60 thousand points, we will construct histograms of the frequencies of falling into separate intervals of relative approximation errors with a reflection of the cumulative effect in percentage terms for clarity (Fig. 3, 4, 5, 6 and 7).

As can be seen from the graphs, the approximation error generally does not exceed 5%. Moreover, in the best case, the maximum value of the relative approximation error does not exceed 1.55%, which is true for the values v_1, v_2, v_4 and N_2 (Fig. 3, 4b, 6a), for the indicators N_1 and N_4 the maximum in the first case does not exceed, and in the second case it is equal to 3% (Fig. 5b, 7a), for v_5—does not exceed 3.7% (Fig. 5a). In the conditionally worst case, the relative approximation errors fall in the range from 4% to 5% for the values v_3, N_3 and N_5 (Fig. 4a, 6b, 7b), however, the number of such errors in total does not exceed half a percent of their total number, that is, 59049, which indicates a good quality of the ANN-based estimates of the main performance indicators of the closed network.

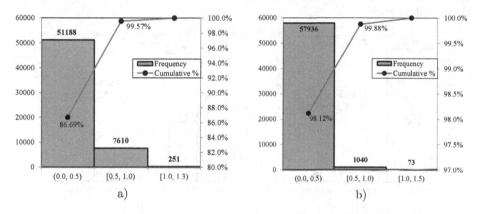

Fig. 3. Histogram of the frequency of values of relative approximation errors (%) for estimates: *a)*—average time v_1, *b)*—average time v_2, obtained using a neural network.

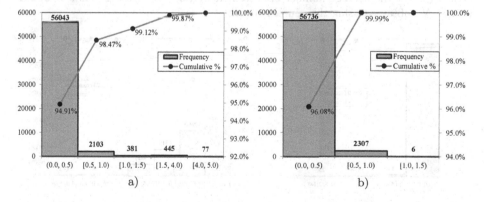

Fig. 4. Histogram of the frequency of values of relative approximation errors (%) for estimates: *a)*—average time v_3, *b)*—average time v_4, obtained using a neural network.

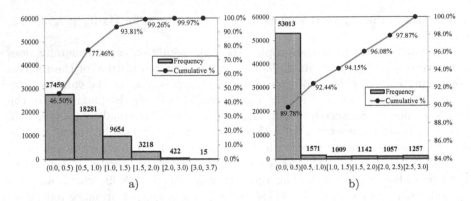

Fig. 5. Histogram of the frequency of values of relative approximation errors (%) for estimates: *a)*—average time v_5, *b)*—average number of customerss N_1, obtained using the neural network.

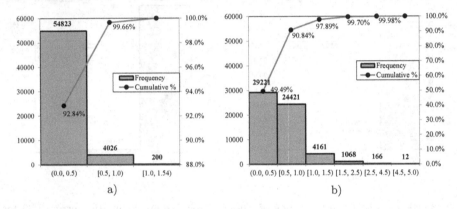

Fig. 6. Histogram of the frequency of values of relative approximation errors (%) for estimates: *a)*—the average number of customers N_2, *b)*—the average number of customers N_3, obtained using the neural network.

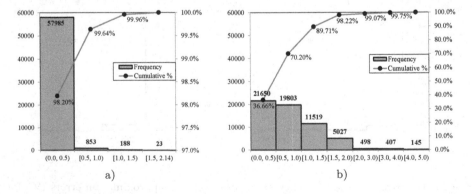

Fig. 7. Histogram of the frequency of values of relative approximation errors (%) for estimates: *a)*—the average number of customers N_4, *b)*—the average number of customers N_5, obtained using the neural network.

4 Conclusion

The article presents a new approach to studying the performance characteristics of closed queuing networks, and also provides a detailed analysis of the effectiveness of this approach. The novelty of the approach resides in the use of ANNs to predict the main characteristics of network performance in combination with preliminary simulation modeling on a limited dataset. Moreover, as shown by a numerical experiment, the accuracy of prediction by neural networks is quite high even in the case of choosing, perhaps, not the most optimal ANN architecture. However, if necessary, the forecast accuracy can be expected to be increased in the case of large time expenditures on the selection of the main parameters of the neural network and the choice of the algorithm for its training.

In this paper, as an illustration of the application of the new approach, as well as its effectiveness, a closed QN was considered, which, as is known, lends itself to analytical analysis with some restrictions. So, for example, to carry out efficient computations according to the Buzen's algorithm, a not too large dimension of the state space of the network is desirable. However, most of the QNs, which serve as models of real physical systems, cannot be investigated analytically. Therefore, the described approach, in the opinion of the authors, has good perspectives for solving complex problems in queuing theory and the analysis of closed QNs with general distribution functions of service times and limited buffers at the nodes, in particular, which is planned to be confirmed by new studies in the future.

References

1. Dudin, A.N., Klimenok, V.I., Vishnevsky, V.M.: The Theory of Queuing Systems with Correlated Flows, 1st edn. Springer, Cham (2020). https://doi.org/10.1007/978-3-030-32072-0
2. Gordon, W.J., Newell, G.F.: Closed queuing systems with exponential servers. Oper. Res. **15**, 254–265 (1967)
3. Baskett, F., Chandy, K.M., Muntz, R.R., Palacios, F.G.: Open, closed and mixed networks of queues with different classes of customers. J. ACM **22**, 248–260 (1975)
4. Buzen, J.P.: Computational algorithms for closed queueing networks with exponential servers. Commun. ACM **16**(9), 527–531 (1973)
5. Reiser, M., Lavenberg, S.S.: Mean-value analysis of closed multichain queueing networks. J. Assoc. Comput. Mach. **27**(2), 313–322 (1980)
6. Boxma, O.J., Donk, P.: On response time and cycle time distribution in a two-stage cyclic queue. Perf. Eval. **2**(3), 181–194 (1982)
7. Daduna, H.: Two-stage cyclic queues with nonexponential servers: steady-state and cyclic time. Oper. Res. **34**(3), 455–459 (1986)
8. Balsamo, S., Donatiello, L.: On the cycle time distribution in a two-stage cyclic network with blocking. IEEE Trans. Softw. Eng. **15**(11), 1206–1216 (1989)
9. Akyildiz, I.F., von Brand, H.: Exact solutions for networks of queues with blocking-after-service. Theor. Comput. Sci. **125**(1), 111–130 (1994)

10. Carbini, S., Donatiello, L., Iazeolla, G.: An efficient algorithm for the cycle time distribution in two-stage cyclic queues with a non-exponential server. In: Proceedings of the International Seminar in Teletraffic Analysis and Computer Performance Evaluation, pp. 99–115. North Holland Publishing Co., Amsterdam (1986)

11. Balsamo, S., Clo, M.: A convolution algorithm for product-form queueing networks with blocking. Ann. Oper. Res. **79**, 97–117 (1998)

12. Clo, M.C.: MVA for product-form cyclic queueing networks with blocking. Ann. Oper. Res. **79**, 83–96 (1998)

13. Baynat, B., Dallery, Y.: A unified view of product-form approximation techniques for general closed queueing networks. Perf. Eval. **18**(3), 205–224 (1993)

14. Kuhn, P.J.: Approximate analysis of general queuing networks by decomposition. IEEE Trans. Commun. **27**(1), 113–126 (1979)

15. Whitt, W.: The queueing network analyzer. Bell Syst. Tech. J. **62**(9), 2779–2815 (1983)

16. Marie, R.A.: An approximate analytical method for general queueing networks. Trans. Softw. Eng **5**(5), 530–538 (1979)

17. Marie, R.A.: Calculating equilibrium probabilities for $\lambda(n)/c_k/1/n$ queues. In: Proceedings of the 1980 International Symposium on Computer Performance Modelling, Measurement and Evaluation, Toronto, pp. 117–125 (1980)

18. Akyildiz, I.F.: General closed queueing networks with blocking. In: Proceedings of the 12th IFIP WG 7.3 International Symposium on Computer Performance Modelling, Measurement and Evaluation, Brussels, pp. 283–303 (1988)

19. Yuzukirmizi, M.: Closed finite queueing networks with multiple servers and multiple customer types. Ph.D. Dissertation. Department of Industrial Engineering and Operations Research at the University of Massachusetts. Amherst Campus (2005)

20. Smith, J.M.G., Barnes, R.: Optimal server allocation in closed finite queueing networks. Flex. Serv. Manuf. J. **27**(1), 58–85 (2014). https://doi.org/10.1007/s10696-014-9202-2

21. Dayar, T., Meri, A.: Kronecker representation and decompositional analysis of closed queueing networks with phase-type service distributions and arbitrary buffer sizes. Ann. Oper. Res. **164**(1), 193–210 (2008)

22. Khomonenko, A.D., Yakovlev, E.L.: Nejrosetevaya approksimaciya harakteristik mnogokanal'nyh nemarkovskih sistem massovogo obsluzhivaniya [Neural network approximation of characteristics of multichannel non-Markovian queuing systems]. Trudy SPIIRAN [SPIIRAS Proceedings] **4**(4), 81–93 (2015). (in Russian) (2020)

23. Dias, L.M.S., Vieira, A.A.C., Pereira, G.A.B., Oliveira, J.A.: Discrete simulation software banking – a top list of the worldwide most popular and used tools. In: Proceedings of the 2016 Winter Simulation Conference (WSC), pp. 1060–1071. IEEE, Washington (2016)

24. Gorbunova, A.V., Vishnevsky, V.M., Larionov, A.A.: Evaluation of the end-to-end delay of a multiphase queuing system using artificial neural networks. In: Vishnevskiy, V.M., Samouylov, K.E., Kozyrev, D.V. (eds.) DCCN 2020. LNCS, vol. 12563, pp. 631–642. Springer, Cham (2020). https://doi.org/10.1007/978-3-030-66471-8_48

25. Gorbunova, A.V., Vishnevsky, V.M.: Estimating the response time of a cloud computing system with the help of neural networks. Adv. Syst. Sci. Appl. **20**(3), 105–112 (2020)

Example of Degrading Network Slicing System in Two-Service Retrial Queueing System

Faina Moskaleva[1](✉) [ID], Ekaterina Lisovskaya[1] [ID], Lyubov Lapshenkova[1],
Sergey Shorgin[2] [ID], and Yuliya Gaidamaka[1,2] [ID]

[1] Peoples' Friendship University of Russia (RUDN University),
6 Miklukho-Maklaya Street, Moscow 117198, Russian Federation
{moskaleva-fa,lisovskaya-eyu,1032172790,gaydamaka-yuv}@rudn.ru
[2] Federal Research Center "Computer Science and Control" of the Russian
Academy of Sciences (FRC CSC RAS), 44-2 Vavilov Street,
Moscow 119333, Russian Federation
sshorgin@ipiran.ru

Abstract. Network slicing is defined as one of the main components of
fifth-generation mobile communications that can solve the problem of
colossal growth in data volume traffic in cellular networks. This paper
reveals the concept of the probability of slice degradation as the example of a model built using a retrial queueing system. First, we build a
mathematical model, for defining of the degradation probability, then we
analyze it using the example of two tenant case. The purpose of this study
is to develop a framework that will allow the infrastructure provider to
calculate such bandwidth thresholds for each slice which do not violate
the requirements for the probability of degradation and maximizing the
revenue from the tenant.

Keywords: Access control · Radio resource slicing · Performance
measure · Retrial queueing system · Isolation · Degradation

1 Introduction

Network slicing is a key technology that allows network operators to provide
their physical infrastructure to support various services with different requirements [13]. The key feature of network slicing for ensuring performance and high
quality of service is isolation, which limits the influence of slices on each other.
Isolation is a fundamental property of network slicing that provides performance

The results of Sects. 1, 2, 3 were obtained within RSF grant (project No. 21-79-00142,
recipient F.M.). The results of Sects. 4, 7 were obtained within the RUDN University
Strategic Academic Leadership Program (recipient E.L.). The results of Sects. 5, 6 were
obtained by Y.G. The results of Sect. 8 were obtained within the RUDN University
Strategic Academic Leadership Program (recipient F.M.).

© Springer Nature Switzerland AG 2021
V. M. Vishnevskiy et al. (Eds.): DCCN 2021, LNCS 13144, pp. 279–293, 2021.
https://doi.org/10.1007/978-3-030-92507-9_23

and security guarantees for each client when different clients use network segments for services with conflicting performance requirements [24]. Using isolation and resource sharing strategies on the radio interface is a rather entangled process [19]. A network slice can span multiple networks, such as a radio access network, a message transport network, and a core network [23]. However, the infrastructure provider does not guarantee the availability of the required number of slices to meet the needs of a particular service provider [18].

Network slicing was firstly introduced in 3GPP in Release 14 [1] in the context of the "vehicular-to-everything" (V2X) – communication between vehicles and other communication devices. Later, in versions 15 and 16, some clarifications were added, especially virtual operators and service packages with similar Quality of Service (QoS) requirements were introduced. According to 3GPP, a network slice is an end-to-end entity supporting prescribed traffic service requirements throughout, including the core and access network of a Physical Mobile Network Operator (PMNO) [6]. The use cases are flexible and include the provision of services to third parties such as Mobile Virtual Network Operators (MVNOs), providers, etc.

It is expected that the slicing concept introduced by 3GPP in Release 14 [1], as part of one of the core functions of 5G cellular systems, will greatly simplify entry into the MVNO market, as well as provide differentiated quality for various network services [16,20]. In the coming years its functionality is projected to become the main trend in the world of mobile communications [8,15]. It will provide a layered network structure, allowing resource sharing with logical isolation between several tenants or network operators and/or services in a multi-domain context [14]. Based on a set of requirements and associated mechanisms that provide resource allocation both over the air interface and over the core network, the technology promises to virtualize PMNO resources in a secure manner, providing a high degree of isolation between MVNO [2].

In network slicing technology the first and most important question is the strategy for allocating the available bandwidth among the slice tenants. Thus, the authors in [21] consider three levels of access control and, using the Markov chain apparatus, investigate various strategies for radio resource management in scenarios with several tenants and several 5G services, including services with both guaranteed and non-guaranteed data transfer rate. In [5] it is proposed to distribute the resources of the system based on the criteria of fairness. The authors in [11] propose a variety of resources sharing alternatives using a set of spectrum planning strategies that provide a degree of flexibility in resource allocation for each tenant.

The slice isolation is an important requirement controlling each slice of the network operation without interference from others, that provides performance, management, security and privacy. This characteristic means that an explosion of traffic in one slice of the network should not negatively affect the QoS of other slices [17]. The authors of [3,9] propose to provide slice isolation by providing the tenant with a predetermined amount of bandwidth, since this approach makes the provided resources independent from each other. However, this can lead to inefficient use of resources, especially in large service areas with various traffic requirements [11].

The isolation is closely related to slice degradation, which occurs as a result of negative impact on QoS. The scientific community has proposed different approaches to isolating slices that make it possible to efficiently use the available resources. In [10] the isolation of two slices with the help of thresholds for the guaranteed number of users that the infrastructure provider can accept for service is considered, and the fairness of resource allocation is achieved by solving the corresponding optimization problem. The authors in [24] propose an algorithm for network slicing, implemented at the base station, aimed at efficient use of resources [7], fair distribution of resources between users and isolation of slices by QoS. The concept of degradation is not sufficiently formalized by the engineering and scientific communities. For example, in [22], the authors define degradation states of the system as those in which one or more users accepted for service are not provided with the required amount of resources to ensure the appropriate data transfer rate.

The overview of existing work shows that there is still uniform definition of network slicing, and issues such as isolation and degradation of slices are the most relevant at the moment. Thus, the purpose of this work is to give one of definition of degradation and the probability of degradation for base station resource slicing using loss system with retrials. We also define the optimization problem that allows the infrastructure provider to calculate the threshold values of the throughput for each slice, without violating the requirements for the degradation probability, while obtaining the maximum revenue from the tenant.

According to 3GPP TS 23.501, a network slice is a logical network that provides a prescribed set of network capabilities [4]. In the context of a 5G system, a slice instance is an end-to-end aggregation of virtual resources in which a set of virtual network functions are created and connected through the virtual network. More specifically, following the GSMA [13], a 5G slice is a set of 5G network functions and associated device functions set up within the 5G system that is tailored to support the communication service to a particular type of user or service. In network slicing technology the first and most important question is the strategy for distributing the available bandwidth among the slice tenants. Bandwidth is a metric that indicates the ratio of a limited number of information units that pass through a channel, node, or network per unit of time. Slice isolation is an important requirement which controls that each slice of the network operates without interference from others and that provides performance, management, security, and privacy. This property means that traffic explosion in one slice of the network should negatively affect the QoS of other slices [17].

This work is organized as follows. In the first section, a system model is built for the application of network slicing technology, which makes it possible to calculate such an efficiency indicator as the probability of slice degradation. In the next section, a mathematical model in terms of a retrial queueing system is presented, which allows you to investigate the key indicators of the system's efficiency, including the degradation probability. Next, an optimization problem is formulated to find threshold values for data transmission rates in a slice, which will allow an infrastructure provider to guarantee a certain level of service. In the last section, an example of numerical analysis is carried out and the corresponding conclusions are drawn.

2   System Model

We model the operation of the Slicing Manager (SliM), which distributes the total bandwidth C [Gbps] of the base station between the tenants of K slices. We assume that each slice is for one type of service (e.g., video streaming, video conferencing, gaming, file transfer, web browsing), and therefore the traffic in each slice is homogeneous in terms of characteristics and QoS requirements [24]. Let the bandwidth required by each session in one slice be constant and the same for all sessions in the slice. By session we mean a period of the user's activity from initial service request to the completion or break of service.

The bandwidth threshold value is defined for each slice, stated in the SLA concluded between an infrastructure provider and a tenant. According to SLA, the infrastructure provider guarantees the provision of bandwidth on demand to the tenant with some probability. Slice degradation refers to those states of the network when the corresponding tenant was not provided with the required bandwidth.

We note that the considered SliM model assumes dynamic division of base station radio resources for their efficient use, i.e. the bandwidth of each slice is not strictly determined, but can be used by other tenants during their peak loads.

3   Mathematical Model

Let the resource queueing system receive K arrival flows (Fig. 1), corresponding to requests for data transmission from users K of slices. The processes are Poisson with arrival rates $\lambda_1, \ldots, \lambda_K$. Each customer of the k-th arrival flow requires b_k units of resource to serve. The system has C resource units. If there is enough free resource in the system, then the customer gets up for service, the duration of which is exponentially distributed with the parameter μ_k, while occupying one server and b_k resource units. If the free resource in the system turns out to be insufficient for servicing, then the customer goes to the orbit, where it carries out an exponentially distributed random delay with the parameter σ_k, after which it makes the next attempt to get up for service. The orbits for each of the arrival processes have unlimited capacity.

We define a stochastic process $\{\mathbf{X}(t), t \leq 0\}$, where

$$\mathbf{X}(t) = (\mathbf{N}(t), \mathbf{I}(t)),$$

$$\mathbf{N}(t) = (N_1(t), \ldots, N_K(t)),$$

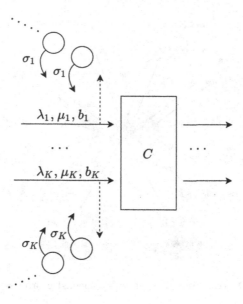

Fig. 1. Scheme of the resource retrial queueing system

$N_k(t)$ is the customers number of the k-process in the service at the time t, $t > 0$,

$$\mathbf{I}(t) = (I_1(t), \dots, I_K(t)),$$

$I_k(t)$ is the customers number of the k-process in the orbit at the time t, $t > 0$, $k = 1, 2$.

Then the states space of the process $\mathbb{X}(t)$ has the form:

$$\mathbb{X} = \left\{ (\mathbf{n,i}) : \sum_{k=1}^{K} b_k n_k \leq C,\ i_k \geq 0,\ k = 1, \dots, K \right\},$$

where $\mathbf{n} = (n_1, \dots, n_K)$, n_k is current state of the process $N_k(t)$ and $\mathbf{i} = (i_1, \dots, i_K)$, i_k is current state of the process $I_k(t)$.

4   Probability Distribution Approximation

For illustration of the definition of the degradation probability proposed in this paper, let us consider the case of two arrival flows. The condition for steady-state regime in the two-service system is defined as follows:

$$\frac{\lambda_1}{\mu_1} b_1 + \frac{\lambda_2}{\mu_2} b_2 < C.$$

The state transition diagram is shown in Fig. 2.

Let us proceed to obtaining the steady-state distribution for further finding the probabilistic characteristics of interest for us. Let $P(n_1, n_2, i_1, i_2)$,

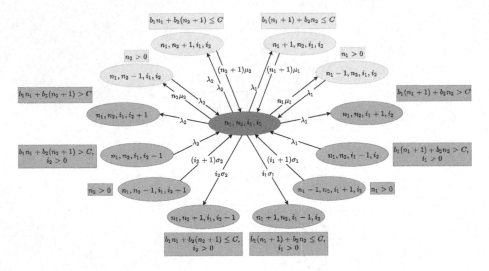

Fig. 2. Transition diagram of a stochastic process $\mathbf{X}(t)$

$(n_1, n_2, i_1, i_2) \in \mathbb{X}$, be the stationary probability distribution of states of the process $\mathbf{X}(t)$:

$$P(n_1, n_2, i_1, i_2) = \lim_{t \to \infty} P\{N_1(t) = n_1, N_2(t) = n_2, I_1(t) = i_1, I_2(t) = i_2\}.$$

Then system of equilibrium equations has the form:

$$[\lambda_1 + \lambda_2 + n_1\mu_1 + n_2\mu_2 + i_1\sigma_1 \cdot I(b_1(n_1 + 1) + b_2 n_2 \le C, i_1 > 0)$$
$$+ i_2\sigma_2 \cdot I(b_1 n_1 + b_2(n_2 + 1) \le C, i_2 > 0)] \cdot P(n_1, n_2, i_1, i_2)$$
$$= \lambda_1 \cdot I(n_1 > 0) \cdot P(n_1 - 1, n_2, i_1, i_2) + \lambda_2 \cdot I(n_2 > 0) \cdot P(n_1, n_2 - 1, i_1, i_2)$$
$$+ \lambda_1 \cdot I(b_1(n_1 + 1) + b_2 n_2 > C, i_1 > 0) \cdot P(n_1, n_2, i_1 - 1, i_2)$$
$$+ \lambda_2 \cdot I(b_1 n_1 + b_2(n_2 + 1) > C, i_2 > 0) \cdot P(n_1, n_2, i_1, i_2 - 1)$$
$$+ (n_1 + 1)\mu_1 \cdot P(n_1 + 1, n_2, i_1, i_2) \cdot I(b_1(n_1 + 1) + b_2 n_2 \le C)$$
$$+ (n_2 + 1)\mu_2 \cdot P(n_1, n_2 + 1, i_1, i_2) \cdot I(b_1 n_1 + b_2(n_2 + 1) \le C)$$
$$+ (i_1 + 1)\sigma_1 \cdot I(n_1 > 0) \cdot P(n_1 - 1, n_2, i_1 + 1, i_2)$$
$$+ (i_2 + 1)\sigma_2 \cdot I(n_2 > 0) \cdot P(n_1, n_2 - 1, i_1, i_2 + 1).$$

It is impossible to obtain an exact solution to a system with an infinite number of equations. Therefore, to obtain an approximation of the probability distribution of the considered four-dimensional stochastic process we will use the method proposed and detailed in [12]. As a result of applying the method, we formulate the following statement.

Approximation of the stationary probability distribution of a four-dimensional stochastic process $(N_1(t), N_2(t), I_1(t), I_2(t))$ under the condition of proportionally growing delay in orbits ($\sigma_k = \gamma_k \cdot \sigma, \sigma \to 0$) is the product of:

- $\mathbf{R}$ is a matrix of elements $R(n_1, n_2)$ – a stationary joint probability distribution of the two-dimensional process $(N_1(t), N_2(t))$, which is a solution to the system of equations

$$[(\mathbf{A} + \lambda_1\mathbf{B_1} + \lambda_2\mathbf{B_2}) - x_1\gamma_1(\mathbf{C_1} - \mathbf{D_1}) - x_2\gamma_2(\mathbf{C_2} - \mathbf{D_2})]\mathbf{R} = 0,$$

$$\mathbf{E}\left[\lambda_1\mathbf{B_1} - x_1\gamma_1\mathbf{D_1}\right]\mathbf{R} = 0, \quad \mathbf{E}\left[\lambda_2\mathbf{B_2} - x_2\gamma_2\mathbf{D_2}\right]\mathbf{R} = 0, \quad \mathbf{ER} = 1.$$

where $\mathbf{A}, \mathbf{B_1}, \mathbf{B_2}, \mathbf{C_1}, \mathbf{C_2}, \mathbf{D_1}, \mathbf{D_2}$ are operators, which are specified in the following form:

$$\mathbf{AR} = \begin{cases} -(\lambda_1 + \lambda_2)R(n_1, n_2) + \mu_1 R(n_1 + 1, n_2) + \mu_2 R(n_1, n_2 + 1), & (1) \\ -(\lambda_1 + \lambda_2 + n_1\mu_1 + n_2\mu_2)R(n_1, n_2) + \lambda_1 R(n_1 - 1, n_2) \\ \quad + \lambda_2 R(n_1, n_2 - 1) + (n_1 + 1)\mu_1 R(n_1 + 1, n_2) \\ \quad + (n_2 + 1)\mu_2 R(n_1, n_2 + 1), & (2) \\ -(\lambda_1 + \lambda_2 + n_1\mu_1 + n_2\mu_2)R(n_1, n_2) + \lambda_1 R(n_1 - 1, n_2) \\ \quad + \lambda_2 R(n_1, n_2 - 1), & (3) \\ -(\lambda_1 + \lambda_2 + n_1\mu_1 + n_2\mu_2)R(n_1, n_2) + \lambda_1 R(n_1 - 1, n_2) \\ \quad + \lambda_2 R(n_1, n_2 - 1) + (n_2 + 1)\mu_2 R(n_1, n_2 + 1), & (4) \\ -(\lambda_1 + \lambda_2 + n_1\mu_1 + n_2\mu_2)R(n_1, n_2) + \lambda_1 R(n_1 - 1, n_2) \\ \quad + \lambda_2 R(n_1, n_2 - 1) + (n_1 + 1)\mu_1 R(n_1 + 1, n_2), & (5) \end{cases}$$

$$\mathbf{B_1R} = \begin{cases} 0, & (1), (2), (5) \\ R(n_1, n_2), & (3), (4) \end{cases}$$

$$\mathbf{B_2R} = \begin{cases} 0, & (1), (2), (4) \\ R(n_1, n_2), & (3), (5) \end{cases}$$

$$\mathbf{C_1R} = \begin{cases} R(n_1, n_2), & (1), (2), (5) \\ 0, & (3), (4) \end{cases}$$

$$\mathbf{C_2R} = \begin{cases} R(n_1, n_2), & (1), (2), (4) \\ 0, & (3), (5) \end{cases}$$

$$\mathbf{D_1R} = \begin{cases} 0, & (1) \\ R(n_1 - 1, n_2), & (2)-(5) \end{cases}$$

$$\mathbf{D_2R} = \begin{cases} 0, & (1) \\ R(n_1, n_2 - 1), & (2)-(5) \end{cases}$$

where $\mathbf{E}$ is an operator that sums functions overall available values of n_1, n_2, and x_1, x_2 are the normalized customers numbers means in orbits, and conditions $(1), (2), (3), (4), (5)$ are defined as follows
- $(1) : b_1 n_1 + b_2 n_2 = 0$,
- $(2) : (b_1(n_1 + 1) + b_2 n_2 \leq C) \cap (b_1 n_1 + b_2(n_2 + 1) \leq C)$,
- $(3) : (b_1(n_1 + 1) + b_2 n_2 > C) \cap (b_1 n_1 + b_2(n_2 + 1) > C)$,
- $(4) : (b_1(n_1 + 1) + b_2 n_2 > C) \cap (b_1 n_1 + b_2(n_2 + 1) \leq C)$,
- $(5) : (b_1(n_1 + 1) + b_2 n_2 \leq C) \cap (b_1 n_1 + b_2(n_2 + 1) > C)$.

- approximation of the two-dimensional distribution of the customers numbers in orbits, which coincides with the two-dimensional Gaussian probability distribution with a vector of means

$$\frac{1}{\sigma}[x_1, x_2]$$

and the covariance matrix

$$\frac{1}{\sigma}\begin{bmatrix} K_{11} & K_{12} \\ K_{12} & K_{22} \end{bmatrix},$$

whose elements are normalized variances and covariance, are the solution to the system of equations

$$(\lambda_1 \mathbf{B_1} - \gamma_1 x_1 \mathbf{D_1} - \gamma_1 K_{11}(\mathbf{C_1} - \mathbf{D_1}) - \gamma_2 K_{12}(\mathbf{C_2} - \mathbf{D_2}))\mathbf{R}$$

$$+ (\mathbf{A} + \lambda_1 \mathbf{B_1} + \lambda_2 \mathbf{B_2} - \gamma_1 x_1(\mathbf{C_1} - \mathbf{D_1}) - \gamma_2 x_2(\mathbf{C_2} - \mathbf{D_2}))\mathbf{g_1} = 0,$$

$$(\lambda_2 \mathbf{B_2} - \gamma_2 x_2 \mathbf{D_2} - \gamma_1 K_{12}(\mathbf{C_1} - \mathbf{D_1}) - \gamma_2 K_{22}(\mathbf{C_2} - \mathbf{D_2}))\mathbf{R}$$

$$+ (\mathbf{A} + \lambda_1 \mathbf{B_1} + \lambda_2 \mathbf{B_2} - \gamma_1 x_1(\mathbf{C_1} - \mathbf{D_1}) - \gamma_2 x_2(\mathbf{C_2} - \mathbf{D_2}))\mathbf{g_2} = 0,$$

$$\mathbf{E}\mathbf{g_1} = 0, \quad \mathbf{E}\mathbf{g_2} = 0,$$

$$\mathbf{E}[(\lambda_1 \mathbf{B_1} + \gamma_1(x_1 - 2K_{11})\mathbf{D_1})\mathbf{R} + 2(\lambda_1 \mathbf{B_1} - \gamma_1 x_1 \mathbf{D_1})\mathbf{g_1}] = 0,$$

$$\mathbf{E}[(\lambda_2 \mathbf{B_2} + \gamma_2(x_2 - 2K_{22})\mathbf{D_2})\mathbf{R} + 2(\lambda_2 \mathbf{B_2} - \gamma_2 x_2 \mathbf{D_2})\mathbf{g_2}] = 0,$$

$$\mathbf{E}[-K_{12}(\gamma_1 \mathbf{D_1} + \gamma_2 \mathbf{D_2})\mathbf{R} + (\lambda_2 \mathbf{B_2} - \gamma_2 x_2 \mathbf{D_2})\mathbf{g_1} + (\lambda_1 \mathbf{B_1} - \gamma_1 x_1 \mathbf{D_1})\mathbf{g_2}] = 0.$$

5 Degradation States Space

In the proposed mathematical model, by the degradation states of the k-th arrival process (k-th slice) we mean the states of the resource queueing system when the total amount of the occupied resource by the customers of the k-th process is less than the threshold value C_k^*, and the customers number in the corresponding orbit is greater than zero. An illustration of the options for sharing the total bandwidth between two slices and degradation is shown in Fig. 3, where the red indicates the total amount of the occupied resource by the customers of the first slow, and blue – the second flow.

Let us define the space of degradation states as

$$\mathbb{D}_k(C_k^*) = \{(\mathbf{n}, \mathbf{i}) \in \mathbb{X} : n_k b_k < C_k^*, i_k > 0\}, k = 1, \ldots, K.$$

Consider two examples with $C = 5, b_1 = 1, b_2 = 2$.

Example 1. Case $C_1^* + C_2^* \leq C$. Let $C_1^* = 1, C_2^* = 3$ (Fig. 4). Then

$$\mathbb{D}_1(C_1^*) = \{(n_1, n_2, i_1, i_2) : n_1 = 0, \forall n_2, i_1 > 0, \forall i_2\},$$

$$\mathbb{D}_2(C_2^*) = \{(n_1, n_2, i_1, i_2) : \forall n_1, n_2 \in [0, 1], \forall i_1, i_2 > 0\}.$$

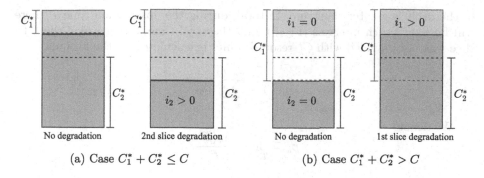

(a) Case $C_1^* + C_2^* \le C$ (b) Case $C_1^* + C_2^* > C$

Fig. 3. Illustration of sharing resource's between two slices

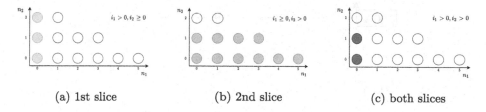

(a) 1st slice   (b) 2nd slice   (c) both slices

Fig. 4. Illustration of degradation states ($C_1^* = 1, C_2^* = 3, C = 5$)

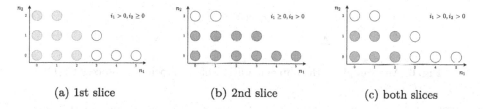

(a) 1st slice   (b) 2nd slice   (c) both slices

Fig. 5. Illustration of degradation states ($C_1^* = C_2^* = 3, C = 5$)

Example 2. Case $C_1^* + C_2^* > C$. Let $C_1^* = C_2^* = 3$ (Fig. 5). Then

$$\mathbb{D}_1(C_1^*) = \{(n_1, n_2, i_1, i_2) : n_1 \in [0, 1, 2], \forall n_2, i_1 > 0, \forall i_2\},$$

$$\mathbb{D}_2(C_2^*) = \{(n_1, n_2, i_1, i_2) : \forall n_1, n_2 \in [0, 1], \forall i_1, i_2 > 0\}.$$

6 Degradation Probability Definition

Figure 6 shows an example of the implementation of the stochastic process $C_k(t) = b_k N_k(t), k = 1, ..., K$, changes in the total resource amounts occupied by the customers of the k-th arrival flow during time T. Time intervals $\Delta_1, \Delta_2, ...$ correspond the fall of the current volume of the occupied resource $C_k(t)$ below the threshold value C_k^* resource units, provided that the number of customers in the orbit is greater than zero. In total, during the observation time T there will be $n_k(T)$ such intervals (Fig. 6), and the total time during which the k-th slice was not provided with C_k^* resource units is calculated by the formula

$$\Delta_k(T) = \sum_{n=1}^{n_k(T)} \Delta_n.$$

We denote

$$\beta_k \overset{\Delta}{=} \lim_{T \to \infty} \frac{\Delta_k(T)}{T}. \tag{1}$$

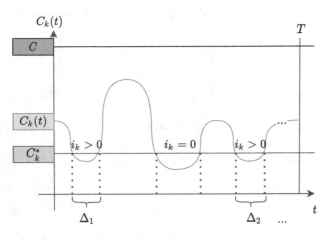

Fig. 6. An example of the implementation of the stochastic process $C_k(t)$

The degradation probability corresponds to the probability of the staying time for a stochastic process $\mathbf{X}(t)$ in the set $\mathbb{D}_k(C_k^*)$, which depends on the threshold value C_k^* and coincides with β_k calculated by formula (1). The degradation probability can also be calculated as

$$\beta_k(C_k^*) = \sum_{\mathbb{D}_k(C_k^*)} p(\mathbf{n}, \mathbf{i}).$$

7   Optimization Problem

For a given vector of threshold values of the degradation probabilities $\beta^* = (\beta_1^*, \dots, \beta_K^*)$, it is necessary to find the throughput vector $\mathbf{C}^* = (C_1^*, \dots, C_K^*)$, such that

$$(C_1^*, \dots, C_K^*) = \arg\max(w_1 C_1 + \dots + w_K C_K) \tag{2}$$

subject to

$$\sum_{k=1}^{K} w_k = 1,$$

$$\beta_k(C_k^*) < \beta_k^*, \ k = 1, \ldots, K.$$

Here, the weights w_k are assigned according to the priority of the tenant, for example, the higher priority is assigned to the tenant bringing the higher revenue to the infrastructure provider. We will solve the optimization problem (2) by the uniform search method. The essence of the method is as follows:

- the range of possible values of each $C_k, k = 1, \ldots, K$, is divided into $(n_k + 1)$ uniform parts by division points

$$C_k^{(i)} = C + i \frac{C}{(n_k + 1)}, i = 1, \ldots, n_k;$$

- calculating the values of the function $F(C_1, \ldots, C_K) = w_1 C_1 + \cdots + w_K C_K$ at the points $C_k^{(i)}, k = 1, \ldots, K, i = 1, \ldots, n_k$, by successive comparison we find the point $(C_1^*, \ldots, C_K^*)$ such that $F(C_1^*, \ldots, C_K^*) = \max F(C_1, \ldots, C_K)$;
- the solution to the optimization problem will be a set of values $C_1^*, \ldots, C_K^*$.

We note that the problem of determining the applicability range of the probability approximation of states distribution of the resource queueing system is solved by the same method.

8 Numerical Examples

Solving the Optimization Problem. Let us calculate the values of the probability of degradation (Fig. 7) for all possible pairs of values C_1^*, C_2^* and the following parameters of the system

$$C = 5, b_1 = 1, b_2 = 2, \lambda_1 = \lambda_2 = 2, \mu_1 = \mu_2 = 2, \sigma_1 = \sigma_2 = 3 \cdot \sigma, \sigma = 0.01.$$

Let the threshold values of the degradation probability be $\beta_1^* = \beta_2^* = 0.8$ and it is necessary that the degradation probability β_k calculated by definition satisfies the condition $\beta_k \le \beta_k^*$. Then the set of solutions for a 1st slice will be the values $(C_1^*, C_2^*) \in \mathbb{S}_1$:

$$\mathbb{S}_1 = \{(1,1), (1,2), (1,3), (1,4), (1,5), (2,1), (2,2), (2,3), (2,4), (2,5)\},$$

and for a 2nd slice $(C_1^*, C_2^*) \in \mathbb{S}_2$:

$$\mathbb{S}_2 = \left\{ \begin{array}{l} (1,1), (1,2), (1,3), (1,4), (2,1), (2,2), (2,3), (2,4), (3,1), (3,2), \\ (3,3), (3,4), (4,1), (4,2), (4,3), (4,4), (5,1), (5,2), (5,3), (5,4) \end{array} \right\}.$$

The intersection of these sets will give the set of all possible values (C_1^*, C_2^*) for which the condition $\beta_k \le \beta_k^*, k = 1, 2$ will be satisfied:

$$\mathbb{S} = \{(1,1), (1,2), (1,3), (1,4), (2,1), (2,2), (2,3), (2,4)\}.$$

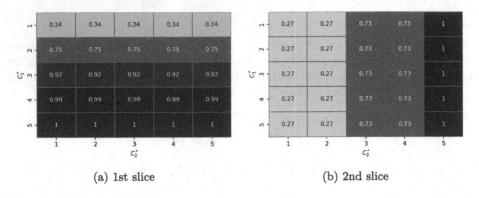

(a) 1st slice   (b) 2nd slice

Fig. 7. Degradation probability values

Then we choose from $\mathbb{S}$ such a pair (C_1^*, C_2^*) so that the sum $w_1 C_1^* + w_2 C_2^*$ is maximal. Let, for example, the weights w_1, w_2 have the meaning of the revenue share from each tenant from a unit of guaranteed bandwidth. When $w_1 = w_2 = 0.5$, the infrastructure provider receives the same revenue from each tenant. Then solution $(2, 4)$ will give the maximum revenue to the tenant 3 units. When $w_1 = 0.1, w_2 = 0.9$, 3.8 units of the tenant's revenue. As we can see, the values of the weights do not affect the solution of the optimization problem, but only affect the value of the revenue function, since the function is linear.

Thus, the solution to the optimization problem will be the throughput values $C_1^* = 2$ for the 1st slice and $C_2^* = 4$ for the 2nd slice, which the infrastructure provider can offer tenants to conclude an SLA, while guaranteeing not to provide the specified bandwidth with probability $\beta_k^* = 0.8, k = 1, 2$.

Identifying the Area of Applicability of the Approximation. Let us pose the problem of determining σ, for which the approximation can be applied from the point of view of the threshold value of the relative error of the average customers numbers in orbits and their variances. For the following initial system parameters

$$C = 5, b_1 = 1, b_2 = 2, \lambda_1 = \lambda_2 = 2, \mu_1 = \mu_2 = 2, \sigma_1 = \sigma_2 = 3 \cdot \sigma, \sigma = 0.01.$$

Figure 8 shows the change in the relative error between the mean and variance obtained using the approximation and the simulation modeling, as a function of the approximation parameter σ. Thus, for example, if the infrastructure provider considers the relative error $\Delta \leq 0.1$ acceptable for its calculations, then according to the graphs in Fig. 8, we see that with $\sigma \leq 0.06$ the approximation can be used to calculate the probabilistic and numerical characteristics.

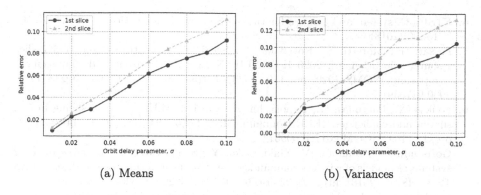

(a) Means   (b) Variances

Fig. 8. Relative error for orbit delay parameter

9 Conclusion

The paper proposes a definition of slice degradation for network slicing technology in 5G networks on the example of a SliM functioning model at one base station, built using a retrial resource queueing system. A system model of the base station was built, in which it is supposed to serve several slices by one infrastructure supplier, as well as a mathematical model of the proposed system. The definitions of degradation and the degradation probability were given for the system built. An optimization problem for the degradation probability was formulated, the solution of which is the bandwidth thresholds that the infrastructure provider guarantees for each slice. For two-service model (two slices), analytical results were obtained. The numerical example is presented to illustrate the proposed framework. The purpose of further research is to formalize the definition of degradation in general terms.

Acknowledgements. The authors are grateful to Prof. Konstantin Samouylov for the approach to determining the degradation probability.

References

1. 3GPP: Service requirements for V2X services (Release 14). 3GPP TS 22.185 V 14.3.0 (2017)
2. Ahmad, I., Kumar, T., Liyanage, M., Okwuibe, J., Ylianttila, M., Gurtov, A.: Overview of 5G security challenges and solutions. IEEE Commun. Stand. Mag. **2**(1), 36–43 (2018). https://doi.org/10.1109/MCOMSTD.2018.1700063
3. Banchs, A., de Veciana, G., Sciancalepore, V., Costa-Perez, X.: Resource allocation for network slicing in mobile networks. IEEE Access **8**, 214696–214706 (2020). https://doi.org/10.1109/ACCESS.2020.3040949

4. ETSI: 5G: system architecture for the 5G System (Release 15). 3GPP TS 23.501 V 15.2.0 (2018)
5. Fossati, F., Moretti, S., Rovedakis, S., Secci, S.: Decentralization of 5G slice resource allocation. In: 2020 IEEE/IFIP Network Operations and Management Symposium, NOMS 2020, pp. 1–9 (2020). https://doi.org/10.1109/NOMS47738. 2020.9110391
6. Foukas, X., Patounas, G., Elmokashfi, A., Marina, M.K.: Network slicing in 5G: survey and challenges. IEEE Commun. Mag. **55**(5), 94–100 (2017). https://doi. org/10.1109/MCOM.2017.1600951
7. Gorbunova, A., Naumov, V., Gaidamaka, Y., Samouylov, K.: Resource queuing systems as models of wireless communication systems. Informatika i ee Primeneniya **12**(3), 48–55 (2018). https://doi.org/10.14357/19922264180307
8. Gudkova, I., et al.: Service failure and interruption probability analysis for licensed shared access regulatory framework. In: 2015 7th International Congress on Ultra Modern Telecommunications and Control Systems and Workshops (ICUMT), pp. 123–131 (2015). https://doi.org/10.1109/ICUMT.2015.7382416
9. Li, J., Liu, J., Huang, T., Liu, Y.: DRA-IG: the balance of performance isolation and resource utilization efficiency in network slicing. In: 2020 IEEE International Conference on Communications (ICC), ICC 2020, pp. 1–6 (2020). https://doi.org/ 10.1109/ICC40277.2020.9149324
10. Moskaleva, F., Lisovskaya, E., Gaidamaka, Y.: Resource queueing system for analysis of network slicing performance with QoS-based isolation. In: Dudin, A., Nazarov, A., Moiseev, A. (eds.) Information Technologies and Mathematical Modelling. Queueing Theory and Applications, pp. 198–211. Springer, Cham (2021). https://doi.org/10.1007/978-3-030-72247-015
11. Munoz, P., Adamuz-Hinojosa, N., Navarro-Ortiz, J., Sallent, O., Pérez-Romero, J.: Radio access network slicing strategies at spectrum planning level in 5G and beyond. IEEE Access **8**, 79604–79618 (2020). https://doi.org/10.1109/ACCESS. 2020.2990802
12. Nazarov, A., Phung-Duc, T., Izmailova, Y.: Multidimensional central limit theorem of the multiclass M/M/1/1 retrial queue. Lect. Notes Comput. Sci. **12563**(4), 298–310 (2020). https://doi.org/10.1007/978-3-030-66471-8_23
13. NGMN: 5G White Paper. Technical report (2015)
14. Oladejo, S.O., Falowo, O.E.: 5G network slicing: a multi-tenancy scenario. In: 2017 Global Wireless Summit (GWS), pp. 88–92 (2017). https://doi.org/10.1109/GWS. 2017.8300476
15. Ometov, A., Kozyrev, D., Rykov, V., Andreev, S., Gaidamaka, Y., Koucheryavy, Y.: Reliability-centric analysis of offloaded computation in cooperative wearable applications. Wirel. Commun. Mob. Comput. **2017**, 1–15 (2017). https://doi.org/ 10.1155/2017/9625687
16. Ordonez-Lucena, J., Ameigeiras, P., Lopez, D., Ramos-Munoz, J.J., Lorca, J., Folgueira, J.: Network slicing for 5G with SDN/NFV: concepts, architectures, and challenges. IEEE Commun. Mag. **55**(5), 80–87 (2017). https://doi.org/10.1109/ MCOM.2017.1600935
17. Pérez-Romero, J., Sallent, O.: Optimization of multitenant radio admission control through a semi-Markov decision process. IEEE Trans. Veh. Technol. **69**(1), 862–875 (2020). https://doi.org/10.1109/TVT.2019.2951322
18. Qin, Y., et al.: Enabling multicast slices in edge networks. IEEE Internet Things J. **7**(9), 8485–8501 (2020). https://doi.org/10.1109/JIOT.2020.2991107

19. Richart, M., Baliosian, J., Serrat, J., Gorricho, J.: Resource slicing in virtual wireless networks: a survey. IEEE Trans. Netw. Serv. Manage. **13**(3), 462–476 (2016). https://doi.org/10.1109/TNSM.2016.2597295
20. Samdanis, K., Costa-Perez, X., Sciancalepore, V.: From network sharing to multi-tenancy: the 5G network slice broker. IEEE Commun. Mag. **54**(7), 32–39 (2016). https://doi.org/10.1109/MCOM.2016.7514161
21. Vilà, I., Pérez-Romero, J., Sallent, O., Umbert, A.: Characterization of radio access network slicing scenarios with 5G QoS provisioning. IEEE Access **8**, 51414–51430 (2020). https://doi.org/10.1109/ACCESS.2020.2980685
22. Vilà, I., Sallent, O., Umbert, A., Pérez-Romero, J.: Guaranteed bit rate traffic prioritisation and isolation in multi-tenant radio access networks. In: 2018 IEEE 23rd International Workshop on Computer Aided Modeling and Design of Communication Links and Networks (CAMAD), pp. 1–6 (2018). https://doi.org/10.1109/CAMAD.2018.8515006
23. Wei, F., Feng, G., Sun, Y., Wang, Y., Qin, S., Liang, Y.C.: Network slice reconfiguration by exploiting deep reinforcement learning with large action space. IEEE Trans. Netw. Serv. Manage. **17**(4), 2197–2211 (2020). https://doi.org/10.1109/TNSM.2020.3019248
24. Yarkina, N., Gaidamaka, Y., Correia, L.M., Samouylov, K.: An analytical model for 5G network resource sharing with flexible SLA-oriented slice isolation. Mathematics **8**(7) (2020). https://doi.org/10.3390/math8071177

Durability Evaluation of a Distributed Communication Network of Weather Stations

Evgeny Golovinov[1], Dmitry Aminev[1,2], Dmitry Kozyrev[2,3](✉) (iD),
and Vladimir Kulygin[4]

[1] All-Russian Research Institute for Hydraulic Engineering and Land Reclamation
(VNIIGiM), Moscow, Russia
evgeny@golovinov.info, aminev.d.a@ya.ru
[2] V. A. Trapeznikov Institute of Control Sciences of Russian Academy of Sciences,
65 Profsoyuznaya Street, Moscow 117997, Russia
[3] Peoples' Friendship University of Russia (RUDN University),
6 Miklukho-Maklaya St, Moscow 117198, Russian Federation
kozyrev-dv@rudn.ru
[4] HSE University, 20 Myasnitskaya Street, Moscow 101000, Russia

Abstract. We consider the problem of ensuring the durability indicators of a distributed communication automatic weather stations network (AWSN), consisting of access points and weather stations (WS), remote from each other at distances of up to several tens of kilometers and connected to the nearest access points and to each other by communication channels. An expert assessment of the durability indicators was carried out and the modes of operation of the distributed AWSN were determined. A model of the complex load factor and resource of the distributed AWSN has been elaborated. A method for assessing and a methodology for determining the durability indicators of AWSN has been developed, which, based on the initial data for AWSN elements, the operating model of the elements and the calculated values of the complex load factor for each of the modes, calculates the gamma-percentage resource before the decommissioning of the AWSN element, based on the criterion of the limit state of the AWSN. The modeling and calculation of durability indicators of the distributed AWSN have been carried out according to the proposed method.

Keywords: Weather station · Distributed network · Mobile communications · Access point · Meteorological measurements · Agrometeorological parameters · WiFi · Durability · Resource

1   Introduction

In the the present-day realities of the development of the digital economy, including agriculture, one of the tasks is to equip the space with means of automatic

This paper has been supported by the RUDN University Strategic Academic Leadership Program and funded by RFBR according to the research project number 19-29-06043.

monitoring of agrometeorological parameters, including agricultural fields. Meteorological stations are technical means for registering such agrometeorological parameters as temperature, humidity of the surface layer of the atmosphere, precipitation, temperature and soil moisture at different depths. These data are used to determine the irrigation rate, quality, and uniformity of crops in the field, yield and efficiency of the farm as a whole. The above factors determine stringent requirements for the reliability and durability of weather stations [1–3].

Located on agricultural lands, the AWSN, which has a multilevel star network topology, consists of equipment of automatic weather stations (AWS), communication channels (CS) of access points to the global Internet via mobile communication or WiFi, which can be far from the AWS at distances up to several tens of kilometers [4] (Fig. 1).

Fig. 1. Distributed communication network of automatic weather stations

The hardware equipment of meteorological stations and the features of its functioning were considered in [6–8] and others. An automatic weather station (AWS) is defined as a means that automatically transmits or records observations obtained via measuring instruments. In AWS, measurements of meteorological elements are converted into electrical signals using sensors. The signals are then processed and converted into meteorological data. The resulting information is finally transmitted via cable or wireless facilities, or automatically stored on a data carrier. In [7] a micro-meteorological station was presented, which can measure temperature, relative humidity, pressure, and wind speed with high accuracy. An AWS (Automatic weather station) consists of a micro-controller, an anemometer, a measurement system, an indication system, and a power management system. In [6,8], an IoT weather station was presented as a tool or device that provides information about the weather in an adjacent environment. For example, the AWS can provide detailed information about ambient temperature, atmospheric pressure, humidity, etc. Therefore, this device mainly measures temperature, pressure, humidity, light intensity, rain intensity. The prototype contains various types of sensors with which all of the above parameters can be measured.

Automatic weather station networks (AWSNs) are increasingly used to collect weather data for agricultural and other bioclimatic purposes. The use of AWS networks grew rapidly during the 1980s due to improvements in self-powered data logging systems and computer communications.

The reliability model of the AWSN configuration was studied in [3] with the help of the multidimensional alternating stochastic processes. Also, an assessment of sets of spare parts, tools and accessories (SPTA) for AWSN was carried out [4]. A suitable complement to probabilistic reliability analysis of complex systems is structural sensitivity analysis [5].To complete the definition of reliability as an indicator of quality, it is necessary to create a model of the complex load factor and resource of the AWSN in order to assess its durability.

Despite the research carried out in this area [9–11], the problem of determining the durability of a distributed network of weather stations has not been sufficiently developed. Therefore, the problem of finding and researching existing and developing new concepts, methods, models, algorithms and methods for determining the durability indicators of a distributed AWSN is relevant.

2   Problem Setting

The initial data for determining the durability of the AWSN are its topology, which is determined by the coverage area, the structure of the AWS, the type of CS, data on the load factors and the resource of elements and communication lines, as well as the criteria for the system operability, the time schedule and operating conditions. The performance criteria of the AWSN are determined by the requirements for the coverage area, that is, the area where the AWSN can monitor agrometeorological parameters. The AWSN coverage area is conventionally divided into areas with respect to access points (Fig. 2). In general, AWSs are located at each access point (AWS in the diagram, AWS * - intermediate AWS with WiFi repeaters involved). At that, their number for each access point may be different.

The durability of the AWSN should be at least 10 years with round-the-clock operation. Cellular communication lines and WiFi access points can be used as communication channels. The restoration of the AWSN's operability should be ensured by cold backup (by replacing the failed components from the spare parts kit), and the recovery time τ_R should not exceed 24 h.

The generalized structure of an AWS and its interaction with access points and the external environment is shown in Fig. 2 (left).

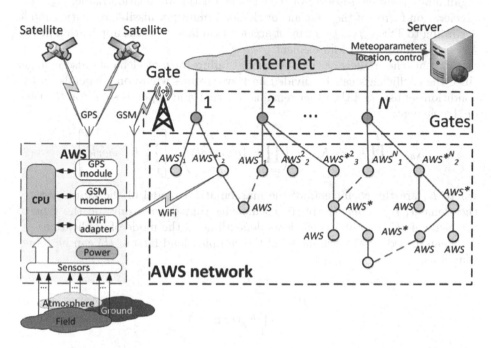

Fig. 2. Block diagram of the interaction of the AWS with each other and access points

The core of an AWS is a microcontroller (MC) that receives and processes data from weather parameters sensors and a GPS receiver, controls data transmission via a GSM modem and a WiFi module. Telemetry and geolocation data are transmitted by a GSM modem via the Internet to a monitoring server for further processing, analysis and presentation to the operator. One can monitor the location of the AWS and read the meteorological parameters at any time. The Power supply in the diagram is a typical battery pack.

3   Mathematical Models of the Complex Load Factor and Resource of Elements and a Method for Assessing the Durability Indicators of the AWSN

Since during life tests the acceleration factor is calculated similarly to the factor for the operational failure rate, to assess the lifetime of the elements, depending on the mode of use, the base model must be modified:

$$T_{p\gamma}^* = \frac{T_{p\gamma}}{K_1} \Rightarrow \frac{T_{p\gamma}\,TS}{T_{p\gamma_{work}}} \sim \frac{K_{ac\,work} \cdot \lambda_b}{K_{ac}\,TS \cdot \lambda_b} \tag{1}$$

where: $T_{p\gamma\,TS}$ – gamma-percentage resource of the element for the maximum permissible operating mode according to the technical specifications (TS); $T_{p\gamma\,work}$ – gamma-percentage resource of the element for the application mode; $K_{ac\,TS}$ – acceleration factor of the element for the maximum permissible operating mode according to TS; $K_{ac\,work}$ – element acceleration factor for the application mode; λ_b –basic failure rate of an element.

At that, the models of the operational failure rate of elements also change, since the coefficients can be divided into two groups where one depends on the conditions of use, and the other remains constant and depends solely on the type of the element:

$$K_{ac} = \prod_{i=1}^{I} K_i \Rightarrow K_{ac} = \left(\prod_{i=1}^{I} K_i(general) \right) \cdot \left[\prod_{j=1}^{J} K_j(work) \right], \qquad (2)$$

where K_i are the coefficients of the mathematical model of the failure rate of the element, $K_j(general)$ are the coefficients that depend only on the type of the element; $K_j(work)$ – coefficients depending on the mode of application.

Then, based on (1), the model of the complex load factor (Π) can be represented as:

$$\Pi = \frac{\prod\limits_{j=1}^{J} K_j(work)}{\prod\limits_{j=1}^{J} K_j(max_{TS})} \qquad (3)$$

where $K_j(work)$ is the value of the j-th coefficient of the mathematical model of the failure rate of the element in the operating mode; $K_j(max_{TS})$ is the value of the j-th coefficient of the mathematical model of the failure rate of an element in the maximum permissible mode according to TS.

To assess the durability indicators of the AWSN, the method shown in Fig. 3 was developed.

In accordance with the method, the initial data are first formed on the elements from the AWSN: the type of the element, the modes of its use (operating temperature of the element, power, etc.) and the operating conditions (group of equipment). On the basis of these data, the values of the complex coefficient are calculated according to the mathematical model (6) for each of the segments of the model of the operation of all elements. Further, based on the operating model of the elements and the calculated values of the complex load factor for each of the modes, the values of the gamma-percentage resource of the elements are calculated in a cyclic mode according to the mathematical model (4). Based on the criterion of the limiting state of the AWSN, indicated in the terms of reference for the AWSN, and the obtained data on the gamma-percentage resources of the elements, the value of the gamma-percentage resource of the AWSN is determined, and on its basis other necessary durability indicators of the AWSN are calculated.

The AWS operation model can contain L operating modes, while its elements can contain either L operating modes, or N operating modes and M standby

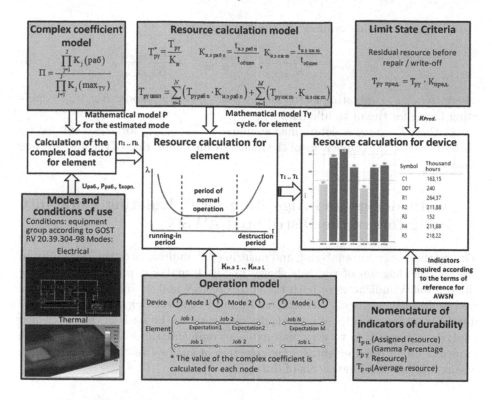

Fig. 3. Method for calculating the resource of the distributed AWSN

modes, in this case, the mathematical model of the gamma-percentage resource of the element, taking into account the operating model, will be as follows:

$$T_{p\gamma\,cycle} = \sum_{n=1}^{N} \left(T_{p\gamma\,work\,n} \cdot K_{op.rate\,work\,n}\right) + \sum_{n=1}^{M} \left(T_{p\gamma\,idle\,m} \cdot K_{op.rate\,idle\,m}\right) \quad (4)$$

where: $T_{p\gamma\,work\,n}$ – gamma-percentage resource of the element in the n-th mode of operation; $T_{p\gamma\,idle\,m}$ – gamma-percentage resource of the element in the m-th idle mode; $K_{op.rate\,work\,m}$ – coefficient of the operation rate of the element of the n-th mode of operation; $K_{op.rate\,idle\,m}$ is the coefficient of the operation rate of the element of the m-th idle mode.

The operation rate factor for the n-th operating mode and the m-th idle mode according to (5) is determined by the expressions:

$$K_{op.rate\,work\,n} = \frac{t_{op.rate\,work\,n}}{t_{total}}, \quad K_{op.rate\,idle\,n} = \frac{t_{op.rate\,idle\,n}}{t_{total}} \quad (5)$$

where: $t_{op.rate\,work\,n}$ – operating time in the n-th mode; $t_{op.rate\,idle\,n}$ – waiting time in the m-th mode; t_{total} - total operating time of the AWS.

To take into account the operating conditions in model (5), the values of the gamma-percentage resource of the element of the n-th operating mode and the

m-th idle mode are determined by the expressions:

$$T^*_{p\gamma\,work\,n} = \frac{T_{p\gamma\,work\,n}}{K_{oper.\,n}}, \quad T^*_{p\gamma\,idle\,n} = \frac{T_{p\gamma\,idle\,n}}{K_{oper.\,m}}, \tag{6}$$

where: $K_{oper.\,n}$ – operating factor for the n-th operating mode; $K_{oper.\,m}$ – operating factor for the m-th idle mode.

At that, $T_{p\gamma\,idle\,m}$ is determined similarly to $T_{p\gamma\,work\,m}$ for Π_i calculated by Formula (3) for the coefficients of the model of the failure rate of the idle/storage mode.

4 Methodology and Algorithm for Calculating Durability Indicators of the Distributed AWSN

The methodology for analyzing and ensuring the indicators of the durability of AWSN, the diagram of which is shown in Fig. 4, makes it possible to evaluate the durability indicators of both the entire AWSN and its individual nodes, as well as identify "critical" elements and analyze the coefficients that affect the durability characteristics of a particular element. On the basis of the performed analysis it is possible to give reasonable recommendations for changes in the modes of use or requirements for the selection of analogs for this element, which ensures the required indicators of durability.

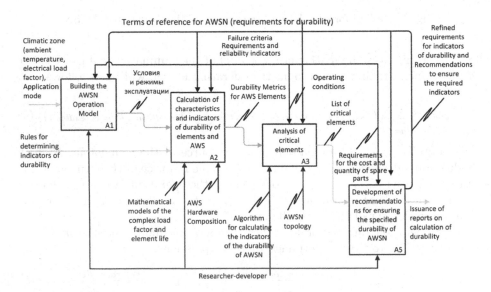

Fig. 4. Methodology for calculating the durability of the distributed AWSN

The initial data for the application of the proposed method are data on the climatic zone (temperature gradient of the region), and the periods of active operation of the distributed AWSN (dates of sowing and harvesting of the proposed crops). In block A1, a model for the operation of the AWSN is formed, on its basis a list of modes and operating conditions of the elements is formed. In addition, to calculate the durability indicators of AWS elements, in block A2, mathematical models of the complex load factor and resource of the element, data on the composition of the AWS equipment, as well as requirements for the durability of AWSN are required.

In practice, with the exception of sensors for which handbooks [12,13] do not show the dependence on ambient temperatures, the critical elements of a weather station in high environmental conditions are integrated circuits. As a rule, the temperature rise for integrated circuits (RAM and processors) operating in sparing modes is 20–30° C relative to the ambient temperature. Batteries are critical elements in cold conditions. For microcircuits, the model of the operational failure rate λ_O in the operating mode, given in the handbook [12], has the form:

$$\lambda_O = \lambda_b \cdot K_{c.t.} \cdot K_{case} \cdot K_v \cdot K_{op.mode} \cdot K_{AC}, \tag{7}$$

where λ_b – basic failure rate;
$K_{c.t.}$ – coefficient depending on the complexity of microcircuits and temperature;
K_{case} – coefficient depending on the case type;
K_v – coefficient depending on the supply voltage;
$K_{op.mode}$ – coefficient depending on the operating mode;
K_{AC} – acceptance coefficient.

Dividing the coefficients of the model (7) into groups in accordance with (2), we obtain a particular form of the mathematical model of the complex load factor (3) for microcircuits:

$$\Pi = \frac{K_{c.t.}(work)}{K_{c.t.}(max_{TS})} \tag{8}$$

For the battery in operation mode, according to the mathematical models given in the handbook [12], provided that the maximum permissible temperature of the battery (for lithium-ion batteries used at the weather station it is 50° C) is not exceeded, the load factor does not depend on temperature.

The obtained indicators of the AWS durability are analyzed in block A3, taking into account the operating conditions and topology of the entire complex of AWSN nodes, using the algorithm for calculating the durability indicators. As a result of the analysis, a list of critical elements is formed, on the basis of which recommendations are developed in block A5 to ensure the specified durability of the AWSN. In case of successful completion of all operations at the output, a report is issued on the durability evaluation and recommendations for the rotation of AWSN elements are made.

According to the algorithm for calculating the durability indicators (Fig. 5), first, the terms of reference for the AWSN are analyzed, the minimum required service lifetime is specified, and the service lifetime required under the terms of reference is introduced. Based on the analysis of the operating conditions (climatic features of the region where the AWSN should be installed, the time of planting and harvesting of the proposed agricultural crops), an operating model is formed and introduced. Next, a sequential analysis of each of the AWSN nodes is carried out in the operating mode and the storage (idle) mode, the current mode is determined.

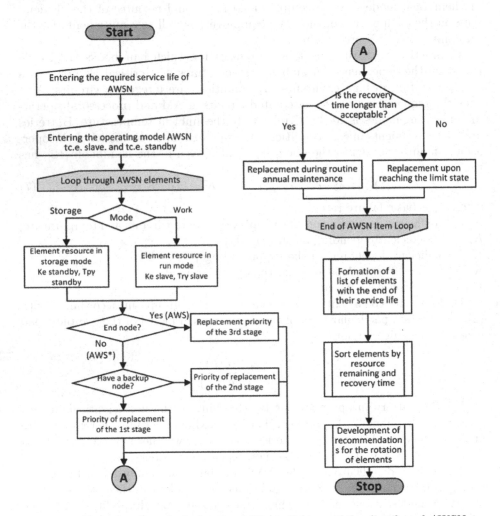

Fig. 5. Algorithm for calculating durability indicators of the distributed AWSN

The resource evaluation of each of the components of the AWSN nodes in operation/storage mode is carried out, taking into account the model of operation and conditions of application using the above mathematical models.

In case a node is the end node (AWS) the elements, the resource of which is lower than the required for AWSN are added to the list for "replacement of the 3rd order". If a node is not the end one (AWS*), but after its failure the link with the group of the end nodes connected to it can be carried out through another (backup) node - the elements whose resource is lower than the required for AWSN are added to the list for "replacement of the 2nd order". In other cases (AWS*) the elements, which resource is lower than the required for AWSN, are added to the list for "replacement of the 1st order".

The recovery time of the elements included in the "replacement priorities" lists are assessed in accordance with the territorial location of the AWSN node and the replacement time data. If the recovery time is less than the permissible time, and the element is in the list of "replacement priorities" of the 2nd and 3rd order - it is added to the list "replace upon reaching a limit state". Elements from the lists of "replacement priorities" of the 1st order and elements, which restoration time is more than permissible, are added into the list of replacement "during the planned annual AWSN service".

Based on the lists formed in previous paragraphs, a list of elements with an ending resource is formed by year of operation of the AWSN. Within each year of operation, lists of the most vulnerable elements are formed, the restoration of which is planned in the event of reaching a limit state. Based on the lists of the most vulnerable elements for each year, recommendations are developed for the rotation of elements from AWS * nodes of high responsibility to the AWS nodes, which failure will not lead to the disconnection of the group of nodes connected to it.

5   Durability Evaluation of the Distributed AWSN

From the point of view of agriculture, 65% of the territory of the Russian Federation is in the zone of risky agriculture, in addition to having a rather complex relief, requiring fragmentation and territorial separation of cultivated land plots over a large area, often remote from each other. Therefore, to control such territories, the use of geographically distributed weather stations is especially relevant.

For example, for the Irkutsk region, which produces 10% of all products of the Siberian Federal District, which has a sharply continental climate with large temperature differences [14], the temperature gradient is shown in Fig. 6.

In the summer months, the average daytime temperature is $+24\,^{\circ}\mathrm{C}$, the average night temperature is $+10\,^{\circ}\mathrm{C}$. In spring and autumn, the average daytime temperature is $+7\,^{\circ}\mathrm{C}$, the average night temperature is $0\,^{\circ}\mathrm{C}$. In the winter months, the average daytime temperature is $-15\,^{\circ}\mathrm{C}$, the average nighttime temperature is $-22\,^{\circ}\mathrm{C}$. Moreover, in the summer months, on sunny days (34% of days), the temperature can reach $35\,^{\circ}\mathrm{C}$ and higher. In the winter months, drop to $-45\,^{\circ}\mathrm{C}$. Based on the above data, it follows that the weather station is in operation mode for 7 months and 5 months is in storage (idle) mode. In this case, the values of the ambient temperature are given in Table 1.

Показатель	Янв.	Фев.	Март	Апр.	Май	Июнь	Июль	Авг.	Сен.	Окт.	Нояб.	Дек.	Год
Абсолютный максимум, °C		10,2	20,0	29,2	34,5	35,6	37,2	34,7	29,7	24,5	14,4	5,3	37,2
Средний максимум, °C	-12,8	-7,8	-3,4	9,4	18,1	22,7	24,8	22,2	15,7	7,7		-10,6	7,3
Средняя температура, °C	-17,8	-14,4		2,5	10,2	15,4	18,3	15,9	9,2		-7,6	-15,4	
Средний минимум, °C	-22	-19,6	-12,2		3,6	9,3	13,0	10,9	4,5		-11,8	-19,1	
Абсолютный минимум, °C	-49,7	-44,7	-37,3		-14,3	-6			-11,9		-49,4	-46,3	-49,7
Норма осадков, мм	13	8	12	18	37	76	114	91	52	21	20	16	480

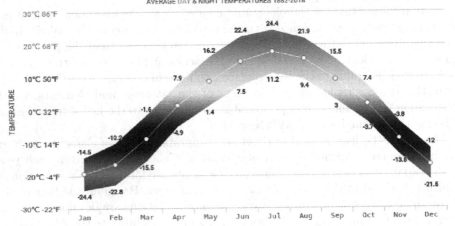

Irkutsk Russia Average Monthly Temperatures

AVERAGE DAY & NIGHT TEMPERATURES 1882-2018

Fig. 6. Climate of the northeastern part of the Russian Federation

Table 1. Temperatures for the northeastern part of the Russian Federation

Season	Duration	Temperature (approx.)
Summer day	5 months	+25...+35 °C
Summer night	3 months * 1/2 days	0...+15 °C
Spring, Autumn	4 months	0...+15 °C
Winter	5 months	−45...0 °C

To estimate the resource of microcircuits and a battery in cyclic mode, a mathematical model (5) is used using coefficients (6) and (7). A special case for the selected operating model can be represented as a model:

$$T_{p\gamma\,cycle} = K_{op.rate\,work\,1} \cdot \frac{T_{p\gamma\,work\,1}}{K_{op.mode\,work}} + K_{op.rate\,work\,2} \cdot \frac{T_{p\gamma\,work\,2}}{K_{op.mode\,work}} +$$

$$+K_{op.rate\,idle} \cdot \frac{T_{p\gamma\,idle}}{K_{op.mode\,idle}} \tag{9}$$

where $K_{op.rate\,work\,1}$, $K_{op.rate\,work\,2}$ and $K_{op.rate\,idle}$ are the operating rate factors in the operating and storage (idle) modes (in accordance with Table 2);

$T_{p\gamma\,work\,1}$, $T_{p\gamma\,work\,2}$ and $T_{p\gamma\,idle}$ – gamma-percentage resource in operating and storage modes (obtained as a result of calculation using Formula (6) using complex load factors in accordance with Table 2);

$K_{op.mode\,work}$ $K_{op.mode\,idle}$ – coefficients depending on operating modes in operation and idle modes (in accordance with Table 2);

The operating rate factors in accordance with the operation model $K_{op.rate}$, the complex load factors in accordance with the operation model, and the factors depending on the operating modes are disclosed in Table 2.

Table 2. Operation intensity factors

Coefficient	Value	
$K_{op.rate\,work\,1}$	0,13	
$K_{op.rate\,work\,2}$	0,45	
$K_{op.rate\,idle}$	0,42	
	For microcircuits	For battery
$\Pi_{work\,1}$	0,85	1
$\Pi_{work\,2}$	0,5	1
$K_{op.mode\,work}$	1,5[a]	1,5[a]
$K_{op.mode\,idle}$	1,4[b]	2,86[b]

[a] for operation group 1.10 (work in temporary structures or in the open air)
[b] during the storage period they vary in the range from 0 °C to −20 °C, which is possible according to the storage conditions according to the classification of the handbooks [12, 13].

In accordance with the above mathematical models and the values of the coefficients, we determine the hardware service life time, based on the service life time of the integrated circuits (IC):

$$T_{service}(IC) = 100500\,hours \approx 11\,years \tag{10}$$

$$T_{service}(battery) = 5\,years \cdot (7/12\,months) + (5\,years/2.86) \cdot (5/12\,months) =$$
$$= 3.65\,years. \tag{11}$$

The most critical element of the system in these conditions is the battery. The storage period of the battery in accordance with the conditions is reduced by 2.86 times compared to keeping it in a heated storage. That, in total, will reduce the battery life (in accordance with (10)) by 16 months.

Hence, it is advisable to give a recommendation to dismantle the weather station with rechargeable batteries, with their subsequent transfer to a heated room, for a period of inactivity. Thus, the service life time of the weather station is provided for 10 years, given the one-time replacement of batteries after 5 years of operation and given the replacement of sensors every 2 years.

6   Conclusions

The proposed models of the complex load coefficient and the models for assessing the AWS resource taking into account the light operating modes, together with the method for assessing the AWS resource with a complex operating model, can significantly increase the accuracy of calculating the AWS durability indicators at the design stage of the equipment.

The methodology and algorithm for calculating the durability indicators of the AWSN determines the sequence of calling procedures during the calculation, allows one to take into account the AWSN operation model, modes and conditions for the use of elements. The result of the execution of the algorithm is the calculated durability indicators of the AWSN and the formed histogram of resources, allowing to analyze, identify and determine the replacement time of potentially unreliable elements.

The algorithm provides for a rotary method of replacing WS components with installation of newer nodes either on the most distant nodes or on key (routing) nodes. This leads to a decrease in the costs of operating weather stations for periods of more than 10 years, taking into account the following factors:

- the geographical distribution and routing of data from long-range meteorological sites through the nodes closely located to the base station;

- the admissibility of data absence from a single node within 24 h;

- the organization of self-testing systems with informing an operator in the event of faults;
- the constant availability of SPTA sets (optimized according to the minimum stock criterion).

The algorithm also takes into account the physics of the process that determines the increase in the failure probability upon the resource depletion, but does not entail an immediate failure at the end of the service life. So, for example, an immediate failure of batteries is rare, but it is characterized by a decrease in capacity, the level of which is easily controlled by the microprocessor of the weather station, and the sensors need to be replaced every 2 years.

References

1. Lisnianski, A., Frenkel, I. (ed.): Recent advances in system reliability: signatures, multi-state systems and statistical inference (Springer series in reliability engineering) 2012, p. 323. https://doi.org/10.1007/978-1-4471-2207-4_1
2. Gertsbakh, I., Shpungin, Y., Vaisman, R. Ternary Networks, SpringerBriefs in Electrical 61 and Computer Engineering, Ternary networks: Reliability and Monte Carlo, p. 62 (2014). https://doi.org/10.1007/978-3-319-06440-6

3. Aminev, D., Golovinov, E., Kozyrev, D., Larionov, A., Sokolov, A.: Reliability evaluation of a distributed communication network of weather stations. In: Vishnevskiy V., Samouylov K., Kozyrev D. (eds.) Distributed Computer and Communication Networks. DCCN 2019. Lecture Notes in Computer Science, vol. 11965, pp. 591–606, 2019. Springer, Cham. https://doi.org/10.1007/978-3-030-36614-8_45

4. Golovinov, E., Aminev, D., Tatunov, S., Polesskiy, S., Kozyrev, D. Optimization of SPTA acquisition for a distributed communication network of weather stations. In: Vishnevskiy V.M., Samouylov K.E., Kozyrev D.V. (eds.) Distributed Computer and Communication Networks. DCCN 2020. Lecture Notes in Computer Science, vol. 12563, pp. 666–679, 2020. Springer, Cham. https://doi.org/10.1007/978-3-030-66471-8_51

5. Kala, Z.: Estimating probability of fatigue failure of steel structures // Acta et Commentationes Universitatis Tartuensis de Mathematica 23(2), 245–254 (2019). https://doi.org/10.12697/ACUTM.2019.23.21

6. Suciu, G., Ijaz, H., Zatreanu, I., Drăgulinescu, A.M.: Real time analysis of weather parameters and smart agriculture using IoT. In: Poulkov V. (eds.) Future Access Enablers for Ubiquitous and Intelligent Infrastructures. FABULOUS 2019. Lecture Notes of the Institute for Computer Sciences, Social Informatics and Telecommunications Engineering, vol. 283, pp. 181–194, 2019. Springer, Cham. https://doi.org/10.1007/978-3-030-23976-3_18

7. Sarkar, I., Pal, B., Datta, A., Roy, S.: Wi-Fi-Based Portable Weather Station for Monitoring Temperature, Relative Humidity, Pressure, Precipitation, Wind Speed, and Direction. Advances in Intelligent Systems and Computing, pp. 399-404 (2019). https://doi.org/10.1007/978-981-13-7166-0_39

8. Ahmad, L., Kanth, R.H., Parvaze, S., Mahdi, S.S.: Automatic weather station. Exper. Agrometeorol. A Pract. Manual 83-87 (2017). https://doi.org/10.1007/978-3-319-69185-5_12

9. Zhadnov, V.V., Kulygin, V.N., Zotov, A.N. Method for predicting the durability of electronic equipment. / Radio Electronics, Computer Science, Control. 2019, Issue 2, p 34–43. 10p

10. Aleksandrovich, I.I., Sergeevich, K.P., Nikolaevich, P.S., Vladimirovich, Z.V.: Estimation of durability indices of integrated microcircuit communication network. In: International Siberian Conference on Control and Communications (SIBCON) 2016, pp. 1–4 (2016). https://doi.org/10.1109/SIBCON.2016.7491837

11. MIL-STD-810 - Military standards for equipment durability MIL-STD

12. Handbook "Nadezhnost ERY" [Reliability of electrical radio products], Moscow, Russia, p. 641 (2006)

13. FIDES guide 2009 Edition A - Reliability Methodology for Electronic Systems, p. 465 (2010)

14. http://www.pogodaiklimat.ru/climate/30710.htm

Unreliable Retrial Queueing System with a Backup Server

Valentina Klimenok[1], Alexander Dudin[1], and Olga Semenova[2(✉)]

[1] Department of Applied Mathematics and Computer Science,
Belarusian State University, Minsk 220030, Belarus
{klimenok,dudin}@bsu.by
[2] Institute of Control Sciences of Russian Academy of Sciences, Moscow, Russia

Abstract. In this paper we consider an unreliable retrial queueing system consisting of the main unreliable server, the reliable backup server and the orbit of infinite size. Customers and breakdowns arrive to the main server according to Markovian arrival processes (MAPs). If the main server is busy at a customer arrival moment, then the customer goes to the orbit. If at the moment of a primary or an orbital customer arrival the main server is under repair and the backup server is idle, then the customer goes to the backup server. If during the service by the backup server the main server restores, the customer goes to this server and is serviced as a new. If at the time of the arrival of a primary or an orbital customer the main server is repaired and the backup one is busy, then the customer goes to orbit. If a breakdown arrives at the busy main server, then the under-served customer is transferred to the backup server and is serviced as a new. Service times and repair times have phase type (PH) distributions. The system under consideration can be used when modeling the hybrid communication system presented by the main optic channel and the reserve radio channel. We investigate the stable system operation, calculate the stationary distribution and a number of performance measures. We also present illustrative numerical examples to show the behavior of the performance measures of the system.

Keywords: Unreliable queueing system · Repeated attempts · Backup server · Markovian arrival process · Phase type distribution · Stationary distribution · Performance measures

1 Introduction

The rapid development of technology has given rise to many problems that require the construction of mathematical models and further research, which requires a skillful and creative approach to the problem with the involvement of a string of scientific fields. Gigantic amounts of data exchanged by millions of companies, as well as ordinary Internet users, together represent a cyclopean traffic flow, which requires the provision of proper bandwidth, which inevitably grows every year. Due to this fact, recently the development of ultra-high-speed

© Springer Nature Switzerland AG 2021
V. M. Vishnevskiy et al. (Eds.): DCCN 2021, LNCS 13144, pp. 308–322, 2021.
https://doi.org/10.1007/978-3-030-92507-9_25

and reliable communication means is underway. One of the promising directions is related to the development of hybrid communication systems combining laser and radio technologies.

The technology of laser atmospheric optical communication lines (FSO – free-space optics) has become recently widespread. This technology allows transmitting data through the atmosphere by using the near infra-red emission and its sequential detection by an optical photo-detector. The atmospheric optical communication channels have the following compelling advantages:

- High bandwidth.
- Protection from unauthorized access and secrecy.
- The system is tolerant to interference and does not causes it itself.
- The speed and simplicity of an FSO network deploying.

Besides the main advantages of the FSO networks, they suffer from serious shortcoming, namely, the dependence of the communication channel availability on the weather conditions and the necessity to provide an optical path between the emitter and the receiver. Heavy weather conditions like snow or fog can significantly reduce the range of effectiveness of an atmospheric laser communication. Therefore, to ensure operator values, one has to resort to using hybrid solutions based on the backup radio channels in conjunction with an optical channel.

Hybrid communication systems with different architectures are widely discussed in the literature, see, e.g., [1–8]. Due to the high practical need for hybrid communication systems, a significant number of studies of this class of systems by means of mathematical modeling have recently appeared, see, for example, the monograph [10] and references therein.

In this paper we consider a queueing system that differs substantially from previous mathematical models of hybrid communication systems by the presence of repeated calls. The system under consideration can be applied to describe a hybrid system where a backup reliable IEEE 802.11n wireless radio channel is considered completely reliable and replaces an optical channel when the latter is interrupted due to adverse weather conditions.

We describe the operation of the system by a multidimensional Markov chain, calculate its stationary distribution and a number of the system performance characteristics. The behavior of the main performance characteristics is investigated numerically.

2   Mathematical Model

We consider a retrial queuing system presented by two servers. One of servers (the main one) is unreliable, and the second one (backup server) is reliable. Interpretation is as follows: the main server is the laser channel (FSO), and the backup server is the IEEE 802.11n wireless channel. Under the influence of weather conditions the FSO channel may fail and immediately begin to recover. During recovery, data is transmitted through the backup radio channel. As the

FSO channel is completely recovered, the backup one is turned off until the next breakdown of the main server.

Customers enter the system in the MAP described by an irreducible continuous time Markov chain $\nu_t, t \geq 0$, called the underlying process of the MAP, and $(W+1) \times (W+1)$ matrices D_0 and D_1. The Markov chain has a finite state space $\{0, 1, \ldots, W\}$. Customers in the MAP can be generated only when the underlying process jumps. The rates of transitions of the process $\nu_t, t \geq 0$, which are accompanied by the generation of a customer, are specified by the matrix D_1, and the rates of "idle" transitions are specified by the off-diagonal elements of matrix D_0. Matrix $D = D_0 + D_1$ presents an infinitesimal generator of $\nu_t, t \geq 0$. The fundamental rate of arrivals is determined as $\lambda = \theta D_1 e$ where θ is a row vector representing the stationary distribution of the Markov chain $\nu_t, t \geq 0$. Here vector θ is calculated as the unique solution to the system $\theta D = 0$, $\theta e = 1$. Here and in the sequel e is a column vector consisting of ones, and 0 is a row vector consisting of zeroes. The variation of intervals between successive arrivals is defined by $c_{var}^2 = 2\lambda\theta(-D_0)^{-1}e - 1$. The coefficient of correlation of the successive intervals between arrivals is given by $c_{cor} = (\lambda\theta(-D_0)^{-1}(D_1)(-D_0)^{-1}e - 1)/c_{var}^2$. For detail about a MAP, its history and properties see, e.g., [11].

If at the time of a primary customer arrival the main server is busy, then the customer is sent to the orbit (a virtual place for such customers) from where he/she attempts to get the service at random times distributed exponentially with the parameter $\alpha > 0$. It is assumed that the volume of the orbit is unlimited.

If at the time of a primary or an orbital customer arrival the main server is under repair, and the backup server is free, then the customer is taken over by the backup server and starts being serviced. If during the service by the backup server the main server restores, the customer goes to the restored server and is serviced as a new. If at the time of a primary or an orbital customer arrival the main server is under repair and the backup one is busy, then the customer goes to orbit.

Breakdowns arrive at the main server in the MAP and are governed by the Markov chain $\eta_t, t \geq 0$, with the phase space $\{0, 1, \ldots, V\}$. The transition rates of the MAP are described by the matrices H_0 and H_1. The breakdown fundamental rate is defined as $h = \gamma H_1 e$, where γ is a row vector consisting of the steady state probabilities of the underlying process $\eta_t, t \geq 0$.

We assume that the service time of a customer at the server j is of the phase type (PH) of order $M^{(j)}$ with irreducible representation $(\beta^{(j)}, S^{(j)})$. The service rate at the jth server is calculated by the formula $\mu^{(j)} = -[\beta^{(j)}(S^{(j)})^{-1}e]^{-1}$, $j = 1, 2$. For more information about a PH distribution see, e.g., [9].

The repair time of the main server has the PH distribution of order R with an irreducible representation (τ, T). The repair rate is given by $\tau = -(\tau T^{-1}e)^{-1}$.

The structure of the system is shown in Fig. 1.

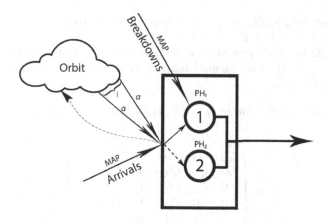

Fig. 1. The structure of the system

3   Process Describing the system States

Here we describe a system state at time t and let

- i_t be the orbit size (the number of customers there), $i_t \geq 0$;
- $n_t = 0$, if the main server is available for service and idle; $n_t = 1$, if the main server is available for service and busy; $n_t = 2$, if the main server is repaired, and the backup server is idle; $n_t = 3$, if the main server is under repair, and the backup one is busy;
- $m_t^{(j)}$ be the service phase at server j, $m_t^{(j)} = \overline{1, M^{(j)}}, j = 1, 2$;
- ϑ_t be the repair phase, $\vartheta_t = \overline{1, R}$;
- ν_t and η_t be the states of the processes $\nu_t, t \geq 0$ and $\eta_t, t \geq 0$ describing arrivals of customers and breakdowns, respectively, $\nu_t = \overline{0, W}, \eta_t = \overline{0, V}$.

The system behavior is governed by a regular irreducible continuous-time Markov chain $\xi_t, t \geq 0$, with the following state space

$$\Omega = \{(i, n, \nu, \eta), i \geq 0, n = 0, \nu = \overline{0, W}, \eta = \overline{0, V}\} \bigcup$$

$$\{(i, n, \nu, \eta, m^{(1)}), i \geq 0, n = 1, \nu = \overline{0, W}, \eta = \overline{0, V}, m^{(1)} = \overline{1, M^{(1)}}\} \bigcup$$

$$\{(i, n, \nu, \eta, \vartheta), i \geq 0, n = 2, \nu = \overline{0, W}, \eta = \overline{0, V}, \vartheta = \overline{1, R}\} \bigcup$$

$$\{(i, n, \nu, \eta, m^{(2)}, \vartheta), i \geq 0, n = 3, \ \nu = \overline{0, W}, \eta = \overline{0, V}, \vartheta = \overline{1, R}, m^{(2)} = \overline{1, M^{(2)}}\}.$$

In what follows, we enumerate the states of the process $\xi_t, t \geq 0$, in the lexicographic order. The subset of states having value i of the countable component i_t is called the level i. Let us denote by $Q_{i,j}$ the matrix of the intensities of the process $\xi_t, t \geq 0$, transitions from level i to level j. We also use the following notation:

- I and O are an identity and a zero matrix, respectively;
- $\bar{W} = W + 1$, $\bar{V} = V + 1$, $a = \bar{W}\bar{V}$;
- $\otimes$, $\oplus$ are Kronecker's product and sum of matrices, respectively;
- $diag\{Z_i, i = \overline{1,k}\}$ is a diagonal matrix with diagonal entries presented by blocks Z_i, $i = \overline{1,k}$.

Lemma 1. *The Markov chain* $\xi_t, t \geq 0$, *is defined by the following infinitesimal generator:*

$$Q = \begin{pmatrix} Q_{0,0} & Q_{0,1} & 0 & 0 & 0 & \cdots \\ Q_{1,0} & Q_{1,1} & Q_{1,2} & 0 & 0 & \cdots \\ 0 & Q_{2,1} & Q_{2,2} & Q_{2,3} & 0 & \cdots \\ 0 & 0 & Q_{3,2} & Q_{3,3} & Q_{3,4} & \cdots \\ \vdots & \vdots & \vdots & \vdots & \vdots & \ddots \end{pmatrix},$$

where

$$Q_{i,i-1} = i\alpha \begin{pmatrix} O_a & I_a \otimes \beta^{(1)} & O & O \\ O & O_{aM^{(1)}} & O & O \\ O & O & O_{aR} & I_a \otimes \beta^{(2)} \otimes I_R \\ O & O & O & O_{aM^{(2)}R} \end{pmatrix}, i \geq 1,$$

$$Q_{i,i} =$$

$$\begin{pmatrix} D_0 \oplus H_0 - i\alpha I_a & D_1 \otimes I_{\bar{V}} \otimes \beta^{(1)} & I_W \otimes H_1 \otimes \tau & O \\ I_a \otimes S_0^{(1)} & D_0 \oplus H_0 \oplus S^{(1)} & O & I_W \otimes H_1 \otimes e_{M^{(1)}} \otimes \beta^{(2)} \otimes \tau \\ I_a \otimes T_0 & O & D_0 \oplus H \oplus T - i\alpha I_{aR} & D_1 \otimes I_{\bar{V}} \otimes \beta^{(2)} \otimes I_R \\ O & I_a \otimes \beta^{(1)} \otimes e_{M^{(2)}} \otimes T_0 & I_a \otimes S_0^{(2)} \otimes I_R & D_0 \oplus H \oplus S^{(2)} \oplus T \end{pmatrix},$$

$$i \geq 0,$$

$$Q_{i,i+1} = \begin{pmatrix} O_a & O & O & O \\ O & D_1 \otimes I_{\bar{V}M^{(1)}} & O & O \\ O & O & O_{aR} & O \\ O & O & O & D_1 \otimes I_{\bar{V}M^{(2)}R} \end{pmatrix}, i \geq 0,$$

where $H = H_0 + H_1$, $S_0^{(j)} = -S^{(j)}e$.

Corollary 1. *The Markov chain* $\xi_t, t \geq 0$, *belongs to the class of asymptotically quasi-Toeplitz Markov chains (AQTMCs).*

Proof. Let us denote by $A^{(i)}$ the matrix with diagonal entries coinciding with the moduli of the diagonal entries of $Q_{i,i}$, $i \geq 0$. It follows from [12] that the chain $\xi_t, t \geq 0$, is an $AQTMC$, if there are limits

$$Y_k = \lim_{i \to \infty} (A^{(i)})^{-1} Q_{i,i+k-1}, k \in \{0, 2\}, \tag{1}$$

$$Y_1 = \lim_{i \to \infty} (A^{(i)})^{-1} Q_{i,i} + I \tag{2}$$

and the matrix $Y_0 + Y_1 + Y_2$ is stochastic.

When finding the matrix Y_k note that each of the matrices $A^{(i)}, i > 0$, can be represented as a block diagonal matrix $diag\{A_0^{(i)}, A_1, A_2^{(i)}, A_3\}$ where the size of the block with subscript n is equal to the size of the corresponding diagonal block of the matrix $Q_{i,i}$. Taking into account these notation, we get the matrices Y_k in the form

$$Y_0 = \begin{pmatrix} O_a\, I_a \otimes \beta^{(1)} & O & O \\ O & O_{aM^{(1)}} & O & O \\ O & O & O_{aR}\, I_a \otimes \beta^{(2)} \otimes I_R & O \\ O & O & O & O_{aM^{(2)}R} \end{pmatrix},$$

$$Y_1 = \left(Y_1^{(1)} \mid Y_1^{(2)} \right),$$

where

$$Y_1^{(1)} = \begin{pmatrix} O & O \\ A_1^{-1}(I_a \otimes S_0^{(1)}) & A_1^{-1}(D_0 \oplus H_0 \oplus S^{(1)}) + I \\ O & O \\ O & A_3^{-1}(I_a \otimes \beta^{(1)} \otimes \mathbf{e}_{M^{(2)}} \otimes T_0) \end{pmatrix},$$

$$Y_1^{(2)} = \begin{pmatrix} O & O \\ O & A_1^{-1}(I_{\bar{W}} \otimes H_1 \otimes \mathbf{e}_{M^{(1)}} \otimes \beta^{(2)} \otimes \tau) \\ O & O \\ A_3^{-1}(I_a \otimes S_0^{(2)} \otimes I_R) & A_3^{-1}(D_0 \oplus H \oplus S^{(2)} \oplus T) + I \end{pmatrix},$$

$$Y_2 = \begin{pmatrix} O_a & O & O & O \\ O & A_1^{-1}(D_1 \otimes I_{\bar{V}M^{(1)}}) & O & O \\ O & O & O_{aR} & O \\ O & O & O & A_3^{-1}(D_1 \otimes I_{\bar{V}M^{(2)}R}) \end{pmatrix}.$$

Thus, we see that limits (1)-(2) exist and, as it easy verified, $Y_0 + Y_1 + Y_2$ is a stochastic matrix. This completes the proof of the corollary.

4 Steady-State Analysis

4.1 Stability Condition

Theorem 1. (i) *The Markov chain $\xi_t, t \geq 0$, is stable if the inequality holds:*

$$\lambda < \pi_1 S_0^{(2)} + \pi_1 S_0^{(1)}, \tag{3}$$

where the vectors π_1, π_2 are calculated as

$$\pi_1 = \mathbf{y}_1 (\mathbf{e}_{\bar{V}} \otimes I_{M^{(2)}} \otimes \mathbf{e}_R), \quad \pi_2 = \mathbf{y}_2 (\mathbf{e}_{\bar{V}} \otimes I_{M^{(1)}}),$$

and the vector $\mathbf{y} = (\mathbf{y}_1, \mathbf{y}_1)$ is the unique solution to the system of linear algebraic equations (SLAE)

$$\mathbf{y} \begin{pmatrix} H \oplus (S^{(2)} + S_0^{(2)}\beta^{(2)}) \oplus T\, I_{\bar{V}} \otimes \beta^{(1)} \otimes \mathbf{e}_{M^{(2)}} \otimes T_0 \\ H_1 \otimes \mathbf{e}_{M^{(1)}} \otimes \beta^{(2)} \otimes \tau & H_0 \oplus (S^{(1)} + S_0^{(1)}\beta^{(1)}) \end{pmatrix} = \mathbf{0}, \quad \mathbf{y}\mathbf{e} = 1. \tag{4}$$

(ii) *If inequality (3) has the opposite sign, then the Markov chain $\xi_t, t \geq 0$, is not ergodic.*

Proof. Let $Y(z)$ denote the generating function of the matrices $Y_k, k = 0, 1, 2$, i.e., $Y(z) = Y_0 + Y_1 z + Y_2 z^2, |z| \leq 1$. Then, according to [12], the *AQTMC* $\xi_t, t \geq 0$, is ergodic if it holds

$$\frac{d}{dz}(zI - Y(z))|_{z=1} > 0 \tag{5}$$

and the process $\xi_t, t \geq 0$, is non-ergodic if inequality (3) has the opposite sign.

Further, we exploit the fact that for the block matrix, the permutation of its block rows and the corresponding block columns does not change its determinant. Calculating the matrix $Y(z)$ and permuting the first and the fourth block columns and rows, we get the matrix, say $\tilde{Y}(z)$. Using the above property of a determinant, we come to the conclusion that inequality (5) is equivalent to the following inequality:

$$\frac{d}{dz} det(zI - \tilde{Y}(z))|_{z=1} > 0. \tag{6}$$

The determinant in (6) can be represent in the form

$$det(zI - \tilde{Y}(z)) = det \begin{pmatrix} A(z) & B(z) \\ C & D(z) \end{pmatrix}, \tag{7}$$

where

$$A(z) = -zA \begin{pmatrix} C_2 + z(D_1 \otimes I_{\bar{V}M^{(2)}R}) & I_a \otimes \beta^{(1)} \otimes e_{M^{(2)}} \otimes T_0 \\ I_{\bar{W}} \otimes H_1 \otimes e_{M^{(1)}} \otimes \beta^{(2)} \otimes \tau & C_1 + z(D_1 \otimes I_{\bar{V}M^{(1)}}) \end{pmatrix}, \tag{8}$$

$$B(z) = -zA \begin{pmatrix} I_a \otimes S_0^{(2)} \otimes I_R & O \\ O & I_a \otimes S_0^{(1)} \end{pmatrix}, \tag{9}$$

$$C = - \begin{pmatrix} I_a \otimes \beta^{(2)} \otimes I_R & O \\ O & I_a \otimes \beta^{(1)} \end{pmatrix}, \quad D(z) = zI, \quad A = diag\{A_3^{-1}, A_1^{-1}\}. \tag{10}$$

Using the property of a determinant of a block matrix, we can write

$$det(zI - \tilde{Y}(z)) = det[A(z) - B(z)D^{-1}(z)C] det D(z). \tag{11}$$

Then ergodicity condition (6) can be rewritten as

$$\frac{d}{dz}\{det[A(z) - B(z)D^{-1}(z)C] det D(z)\}|_{z=1} > 0. \tag{12}$$

Substituting the expressions (8) and (9) for $A(z), B(z), D(z), C$ in (12) and the further simplification result in

$$\frac{d}{dz}\left\{det\left[-z\begin{pmatrix} C_2 + z(D_1 \otimes I_{\bar{V}M^{(2)}R}) & I_a \otimes \beta^{(1)} \otimes e_{M^{(2)}} \otimes T_0 \\ I_{\bar{W}} \otimes H_1 \otimes e_{M^{(1)}} \otimes \beta^{(2)} \otimes \tau & C_1 + z(D_1 \otimes I_{\bar{V}M^{(1)}}) \end{pmatrix}\right.\right.$$

$$-\begin{pmatrix} I_a \otimes S_0^{(2)} \beta^{(2)} \otimes I_R & O \\ O & I_a \otimes S_0^{(1)} \beta^{(1)} \end{pmatrix}\Bigg)\Bigg]\Bigg\}_{z=1} > 0, \tag{13}$$

where

$$C_1 = D_0 \oplus H_0 \oplus S^{(1)}, \ C_2 = D_0 \oplus H \oplus S^{(2)} \oplus T.$$

Following the proof of Theorem 2 in [12] we can show that inequality (13) is equivalent to

$$\frac{d}{dz} \mathbf{x}\Bigg[z \begin{pmatrix} C_2 + z(D_1 \otimes I_{\bar{V}M^{(2)}R}) & I_a \otimes \beta^{(1)} \otimes \mathbf{e}_{M^{(2)}} \otimes T_0 \\ I_{\bar{W}} \otimes H_1 \otimes \mathbf{e}_{M^{(1)}} \otimes \beta^{(2)} \otimes \tau & C_1 + z(D_1 \otimes I_{\bar{V}M^{(1)}}) \end{pmatrix} \Bigg]_{z=1} \mathbf{e} < 0, \tag{14}$$

where the vector $\mathbf{x}$ is determined as the $SLAE$

$$\mathbf{x}\begin{pmatrix} C_2 + D_1 \otimes I_{\bar{V}M^{(2)}R} + I_a \otimes S_0^{(2)} \beta^{(2)} & I_a \otimes \beta^{(1)} \otimes \mathbf{e}_{M^{(2)}} \otimes T_0 \\ I_{\bar{W}} \otimes H_1 \otimes \mathbf{e}_{M^{(1)}} \otimes \beta^{(2)} \otimes \tau & C_1 + D_1 \otimes I_{\bar{V}M^{(1)}} + I_a \otimes S_0^{(1)} \beta^{(1)} \end{pmatrix} = \mathbf{0},$$

$$\mathbf{xe} = 1. \tag{15}$$

Substituting in (14)-(15) expressions for the matrices C_1, C_2 and differentiating in (14), we obtain the inequality

$$\mathbf{x}\begin{pmatrix} D_1 \otimes I_{\bar{V}M^{(2)}R} + I_a \otimes S^{(2)} \otimes I_R \\ D_1 \otimes I_{\bar{V}M^{(1)}} + I_a \otimes S^{(1)} \end{pmatrix} \mathbf{e} < 0, \tag{16}$$

where vector $\mathbf{x}$ is determined as the unique solution to the following $SLAE$:

$$\mathbf{x}\begin{pmatrix} D \oplus H \oplus (S^{(2)} + S_0^{(2)} \beta^{(2)}) \oplus T & I_a \otimes \beta^{(1)} \otimes \mathbf{e}_{M^{(2)}} \otimes T_0 \\ I_{\bar{W}} \otimes H_1 \otimes \mathbf{e}_{M^{(1)}} \otimes \beta^{(2)} \otimes \tau & D \oplus H_0 \oplus (S^{(1)} + S_0^{(1)} \beta^{(1)}) \end{pmatrix} = \mathbf{0}, \ \mathbf{xe} = 1. \tag{17}$$

Let vector $\mathbf{x}$ has the form

$$\mathbf{x} = (\boldsymbol{\theta} \otimes \mathbf{y}_1, \ \boldsymbol{\theta} \otimes \mathbf{y}_2), \tag{18}$$

where $\mathbf{y} = (\mathbf{y}_1, \mathbf{y}_2)$ is a stochastic vector. Taking into account that $\boldsymbol{\theta}D = \mathbf{0}$, $\mathbf{ye} = 1$ and $\boldsymbol{\theta}D_1\mathbf{e} = \lambda$, we reduce inequality (16) and system (17) to the following form

$$\lambda + \mathbf{y}_1(\mathbf{e}_{\bar{V}} \otimes S^{(2)} \mathbf{e} \otimes \mathbf{e}_R) + \mathbf{y}_2(\mathbf{e}_{\bar{V}} \otimes S^{(1)})\mathbf{e} < 0. \tag{19}$$

$$\mathbf{y}\begin{pmatrix} H \oplus (S^{(2)} + S_0^{(2)} \beta^{(2)} \oplus T) I_{\bar{V}} \otimes \beta^{(1)} \otimes \mathbf{e}_{M^{(2)}} \otimes T_0 \\ H_1 \otimes \mathbf{e}_{M^{(1)}} \otimes \beta^{(2)} \otimes \tau & H_0 \oplus (S^{(1)} + S_0^{(1)} \beta^{(1)}) \end{pmatrix} = \mathbf{0}, \ \mathbf{ye} = 1. \tag{20}$$

System (20) coincides with system (4) in the statement of the theorem.

Now we will prove that inequality (19) is equivalent to inequality (3). Using the property of the Kronecker product, we can write inequality (19) as

$$\lambda < \mathbf{y}_1(\mathbf{e}_{\bar{V}} \otimes I_{M^{(2)}} \otimes \mathbf{e}_R)S_0^{(2)} + \mathbf{y}_2(\mathbf{e}_{\bar{V}} \otimes I_{M^{(1)}})S_0^{(1)} < 0. \tag{21}$$

Inequality (21) is equivalent to inequality (3) in the statement of the theorem. This proves the statement (i) of the theorem.

The statement (ii) of the theorem follows from the theory of $AQTMC$ according to which the asymptotically quasi-Toeplitz Markov chain is non-ergodic if the sufficient ergodicity condition has the opposite sign.

Remark 1. Inequality (3) has the following physical meaning: the vector π_k specifies the distribution of the phases of the PH service process by server k in heavy traffic conditions, $k = 1, 2$. Then the right-hand side of inequality (3) denotes the rate of the output flow, while the left-hand side of this inequality presents the input flow rate. It is intuitively clear that the system is stable (the Markov chain ξ_t is ergodic) if the input flow rate is less than the output flow rate.

4.2   Steady-State Probability Vectors. Performance Measures

Let stability condition (3) hold below. It means that the stationary state probability distribution of the Markov chain ξ_t exists. Suppose the states of the Markov chain are enumerated in the lexicographic order and introduce the row vectors $\mathbf{p}_i$ formed by the stationary state probabilities corresponding to level i of the first (denumerable) component i_t of the chain. In order to calculate the vectors $\mathbf{p}_i, i \geq 0$, we will apply an algorithm developed in [12] for a general $AQTMC$. In our case we adapted this algorithm to the case of a block tridiagonal generator.

ALGORITHM

1) Compute the matrix G as the minimal non-negative solution of the following matrix equation:
$$G = Y_0 + Y_1 G + Y_2 G^2.$$

2) Compute the matrices $G_i, i \geq 0$, from the reverse recursion equation
$$G_i = \left(-Q_{i+1,i+1} - Q_{i+1,i+2} G_{i+1}\right)^{-1} Q_{i+1,i}, \tag{22}$$

When implementing this step, we use the existence of the limit $\lim\limits_{i \to \infty} G_i = G$ (this fact follows from the asymptotic properties of the $AQTMC$ under consideration) to find the initial condition for the backward recursion equation. To do this, we choose some integer i_0, set $G_{i_0+1} = G$, compute G_{i_0} by the equation (22) and check the condition $\|G_{i_0} - G\| < \epsilon$. If the condition is satisfied, then we assume all matrices G_i for $i \geq i_0$ equal to G. The remaining matrices G_i are calculated from the backward recursion equation (22). If the condition is not met for this i_0, then by some algorithm we select a new (larger) value i_0.

3) Compute the matrices $\bar{Q}_{i,i}, \bar{Q}_{i,i+1}$ by the formulas
$$\bar{Q}_{i,i} = Q_{i,i} + Q_{i,i+1} G_i, \ i \geq 0, \quad \bar{Q}_{i,i+1} = Q_{i,i+1}, \ i \geq 0.$$

4) Compute the matrices F_i from the recurrent relations:

$$F_0 = I, F_i = F_{i-1}\bar{Q}_{i-1,i}\left(-\bar{Q}_{i,i}\right)^{-1}, i \geq 1.$$

5) Compute the vector $\mathbf{p}_0$ as the unique solution of the $SLAE$

$$\mathbf{p}_0(-\bar{Q}_{0,0}) = \mathbf{0}, \quad \mathbf{p}_0 \sum_{i=0}^{\infty} F_i \mathbf{e} = 1.$$

6) Compute the vectors $\mathbf{p}_i$ by the formulas $\mathbf{p}_i = \mathbf{p}_0 F_i, i \geq 0$.

As soon as the steady state vectors $\mathbf{p}_i, i \geq 0$, have been calculated, we are able to derive formulas for important performance characteristics of the system under consideration. Below we give some of them.

- Probability that the current orbit size is i: $p_i = \mathbf{p}_i \mathbf{e}$.
- The average orbit size $L = \sum_{i=1}^{\infty} i p_i$.
- Variance of the orbit size $V = \sum_{i=1}^{\infty} i^2 p_i - L^2$.
- Probability that the main server is under repair

$$P_{repair} = \sum_{i=0}^{\infty} \mathbf{p}_i \begin{pmatrix} \mathbf{0}^T_{a(1+M^{(1)})} \\ \mathbf{e}_{aR(1+M^{(2)})} \end{pmatrix}.$$

- Probability that the main server is fault-free $P_{fault-free} = 1 - P_{repair}$.
- Probability that the main server is fault-free and free

$$P_{idle}^{(1)} = \sum_{i=0}^{\infty} \mathbf{p}_i \begin{pmatrix} \mathbf{e}_a \\ \mathbf{0}^T_{a(M^{(1)}+R+RM^{(2)})} \end{pmatrix}.$$

- Probability that the main server is being repaired and the reserve server is idle

$$P_{idle}^{(2)} = \sum_{i=0}^{\infty} \mathbf{p}_i \begin{pmatrix} \mathbf{0}^T_{a(1+M^{(1)})} \\ \mathbf{e}_{aR} \\ \mathbf{0}^T_{aRM^{(2)}} \end{pmatrix}.$$

- Probability that an arriving primary customer immediately goes for service by the main server (without visiting the orbit)

$$P_{imm} = \frac{1}{\lambda} \sum_{i=0}^{\infty} \mathbf{p}_i \begin{pmatrix} I_{\bar{W}} \otimes \mathbf{e}_{\bar{V}} \\ I_{\bar{W}} \otimes \mathbf{0}^T_{\bar{V}(M^{(1)}+R+RM^{(2)})} \end{pmatrix} D_1 \mathbf{e}.$$

5   Numerical Results

In this section, we present some numerical results to demonstrate feasibility of the developed algorithms and to bring out the qualitative nature of the queue under study. To obtain these results we performed two numerical experiments to investigate the behavior of the system performance measures depending on the parameters of the system.

In the **first experiment** we analyse the dependence of the mean number of orbital customers, L, on the input rate λ for varying breakdown rates. In this experiment the following input data are used.

The MAP of customers is defined by the matrices

$$D_0 = \begin{pmatrix} -1.349076 & 1.09082 \times 10^{-6} \\ 1.09082 \times 10^{-6} & -0.043891 \end{pmatrix}, \quad D_1 = \begin{pmatrix} 1.340137 & 0.008939 \\ 0.0244854 & 0.0194046 \end{pmatrix}.$$

This MAP has $c_{var}^2 = 9.621426$, $c_{cor} = 0.407152$, $\lambda = 1$.

MAP of breakdowns is characterised by the matrices

$$H_0 = \begin{pmatrix} -8.6 & 0.001 \\ 0.002 & -0.276 \end{pmatrix}, \quad H_1 = \begin{pmatrix} 8.5 & 0.099 \\ 0.02 & 0.254 \end{pmatrix}.$$

For this MAP $c_{var}^2 = 9.61425623$, $c_{cor} = 0,407152089$, $h = 1,77522951$.

To vary the values of λ and h in the course of the experiments, we will multiply the matrices D_0, D_1 and H_0, H_1 by the corresponding constants.

The PH distribution of the service time at the main server is defined as the Erlang distribution with the representation $(2, 20)$.

The PH distribution of the service time at the backup server is defined as the Erlang distribution with the representation $(2, 2)$.

The PH distribution of the repair time is defined as the hyper-exponential distribution with the representation (τ, T), where

$$\tau = (0.05, 0.95), \quad T = \begin{pmatrix} -1.86075 & 0 \\ 0 & -146.9994 \end{pmatrix}.$$

The repair time variation is $c_{var} = 5$.

The retrial rate is $\alpha = 1.5$.

Figures 2 and 3 depict the dependence of the mean number of orbital customers, L, on the input rate λ for varying rates of breakdowns: $h = 0.0001, 0.01, 0.1$. As it is seen from the figures, the curves increase with increasing the value of λ and the growth rate increases strongly in the range where the system load ρ is close to 1 (it corresponds to $\lambda = 9.96, 9.8, 9.22$ for $h = 0.0001, 0.01, 0.1$, respectively). The curves in Fig. 3 end at the points where the stability condition is violated. One more evident observation that the value of L increases with increasing the rate of breakdowns h.

In this experiment we have also analyzed the behavior of probability P_{imm} as the function of λ for varying values of breakdown rate h. As a result, we came to the conclusion that this probability decreases with λ and h increasing. It is

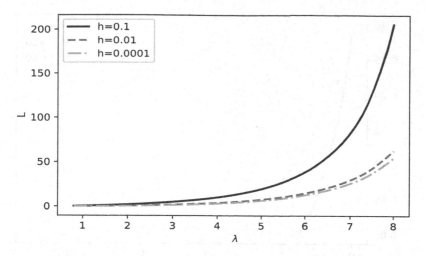

Fig. 2. L as a function of λ when $\lambda \in (0.8; 8]$ under various values of h

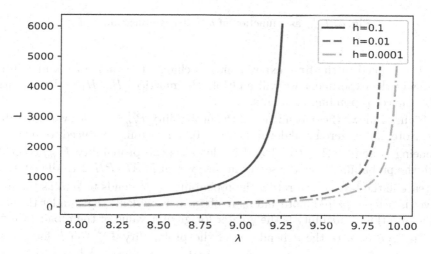

Fig. 3. L as a function of λ when $\lambda \in [8; 10)$ under various values of h

almost evident conclusion which follows from the fact that with increasing λ or (and) h the main server is being busy or repaired more and more often, therefore arriving primary calls are forced to go to the orbit.

In the **second experiment** we are interested in the dependence of the probability of fault-free and idle main server, $P_{idle}^{(1)}$, and the probability that the main server is being repaired and the backup server is idle, $P_{idle}^{(2)}$, on the breakdown rate h and repair rate τ.

In this experiment we consider MAP of customers from the **first** experiment and modify it in such a way that $\lambda = 7$. The retrial rate $\alpha = 1.5$. The shape of the of the MAP of the breakdowns and the distribution of the service time

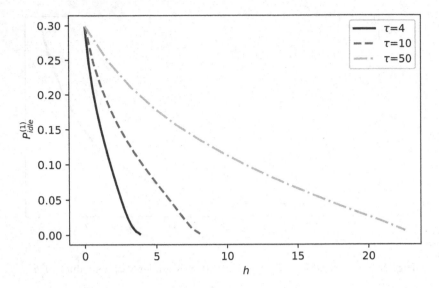

Fig. 4. $P_{idle}^{(1)}$ as a function of h under various values of τ

are also defined in the **first** experiment. To change the value of h and τ in the course of the experiment we will multiply the matrices H_0, H_1 and the matrix T by the corresponding constants.

Figure 4 shows the dependence of the probability $P_{idle}^{(1)}$ on h for varying values of τ. Note that at zero breakdown rate, $h = 0$, our system is reduced to the retrial queueing system $MAP/PH/1$ and in this case the probability $P_{idle}^{(1)}$ coincides with the probability that the server in the system $MAP/PH/1$ is idle. With an increase in the breakdown rate h the probability $P_{idle}^{(1)}$ tends to zero as the main server is being repaired more and more often. It is evident, that, under the same value of h this probability is the greater for greater value of the repair rate τ.

Figure 5 depicts the dependence of the probability $P_{idle}^{(2)}$ on h for varying values of τ. For any τ at the point $h = 0$ the probability $P_{idle}^{(2)}$ is equal to the probability that the backup server is idle. With an increase in the breakdown rate this probability first increases and then decreases to zero. Such a behavior is explained by the fact that with increasing h the joint probability $P_{idle}^{(2)}$ increases because the main server is being repaired more and more often. But, starting from a certain value of h, this probability begins to be more affected by the fact that the backup server has to serve customers more and more often and it is more and more busy. From this moment the probability $P_{idle}^{(2)}$ begins to decrease. The point of maximum of $P_{idle}^{(2)}$ shifts to the right with increasing τ. This is due to the fact that with an increase in the repair rate τ (with a decrease in the average repair time), the backup server can remain idle even with a greater breakdown rate h.

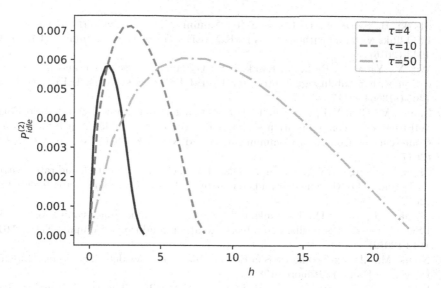

Fig. 5. $P_{idle}^{(2)}$ as a function of h under various values of τ

6   Conclusion

In this paper we analyzed the retrial queueing system with a single main unreliable server and an absolutely reliable backup server. The backup server replaces the main server when the latter is under repair. Customers arrive at the system in the MAP, breakdowns arrive at the main server in the another MAP. The service times at both servers and the repair time of the main server have PH distributions. We built the multi-dimensional Markov chain which describes the operation of the system, derive its ergodicity condition and calculate the stationary distribution. Using the steady-state probabilities, we got a number of performance measures of the system and investigated numerically their behavior. The results can be used in modeling the real hybrid communication system where a backup IEEE 802.11n wireless radio channel replaces the FSO channel in cases where the latter is interrupted due to adverse weather conditions.

References

1. Arnon, S., Barry, J., Karagiannidis, G., Schober, R., Uysal, M. (eds.): Advanced Optical Wireless Communication Systems. Cambridge University Press (2012)
2. Douik, A., Dahrouj, H., Al-Naffouri, T.Y., Alouini, M.S.: Hybrid radio/free-space optical design for next generation backhaul systems. IEEE Trans. Commun. **64**, 2563–2577 (2016)
3. Esmail, M.A., Fathallah, H., Alouini, M.S.: Outdoor FSO communications under FOG: attenuation modeling and performance evaluation. IEEE Photonics J. **8**, 1–22 (2016)

4. Makki, B., Svensson, T., Eriksson, T., Alouini, M.S.: On the performance of RF-FSO links with and without hybrid ARQ. IEEE Trans. Wirel. Commun. **15**, 4928–4943 (2016)
5. Wu, Y., Yang, Q., Park, D., Kwak, K.S.: Dynamic link selection and power allocation with reliability guarantees for hybrid FSO/RF systems. IEEE Access **5**, 13654–13664 (2017)
6. Zhang, V., Chu, Y.J., Nguyen, T.: Coverage algorithms for WiFO: a hybrid FSO-WiFi femtocell communication system. In: Proceedings of the 26th International Conference on Computer Communication and Networks (ICCCN), pp. 1–6. IEEE (2017)
7. Zhou, K., Gong, C., Wu, N., Xu, Z.: Distributed channel allocation and rate control for hybrid FSO/RF vehicular ad hoc networks. J. Opt. Commun. Netw. **9**, 669–681 (2017)
8. Shehaj, M., Nace, D., Kalesnikau, I., Pioro, M.: Link dimensioning of hybrid FSO/fiber networks resilient to adverse weather conditions. Comput. Netw. **161**, 1–13 (2019)
9. Neuts, M.: Matrix-Geometric Solutions in Stochastic Models. The Johns Hopkins University Press, Baltimore (1981)
10. Dudin, A.N., Klimenok, V.I., Vishnevsky, V.M.: The Theory of Queuing Systems with Correlated Flows. Springer, Cham (2020). https://doi.org/10.1007/978-3-030-32072-0
11. Lucantoni, D.: New results on the single server queue with a batch Markovian arrival process. Commun. Stat. Stoch. Models **7**, 1–46 (1991)
12. Klimenok, V.I., Dudin, A.N.: Multi-dimensional asymptotically quasi-Toeplitz Markov chains and their application in queueing theory. Queueing Syst. **54**, 245–259 (2006)

On k-out-of-n System Under Full Repair and Arbitrary Distributed Repair Time

Boyan Dimitrov[1] and Vladimir Rykov[2,3,4](✉)

[1] Department of Mathematics, Kettering University, Flint, MI, USA
[2] Peoples' Friendship University of Russia (RUDN University),
6 Miklukho-Maklaya St, Moscow 117198, Russian Federation
rykov-vv@rudn.ru
[3] Gubkin Oil and Gas Russian State University (Gubkin University),
65 Leninsky prospect, Moscow 119991, Russian Federation
[4] Institute for Transmission Information Problems (named after A.A. Kharkevich)
RAS, Bolshoy Karetny, 19, 4 Moscow, Russian Federation

Abstract. A k-out-of-n : F renewable system with exponentially distributed life-times, and arbitrary distributed repair times of its components under a partial repair regime has been investigated in the paper [1]. This paper deals with the same model under the full repair regime after the system failure. For this model we use Markovization method, based on adding of supplementary variables for the service process, describing system behavior in our analysis.

Keywords: k-out-of-n : F system · Arbitrary distributed repair times · Markovizaton method · Time-dependent and Stationary Process Probabilities

1 Introduction and Motivation

To provide a high level of systems reliability is widely used redundancy technique. One of form of redundancy is a k-out-of-n : F configuration, which is a repairable system that consists of n components in parallel redundancy, and fails when at least k of its components have failed. The failed components and the whole system are repaired by a single server. For a repairable k-out-of-n : F system there exist at least two possible scenarios for the system repairs when it fails:

- **Partial repair regime**, when after the system has failed its working components continue to function in previous regime. After the repair of some failed components the system enters state $k - 1$ with more than k working components;
- **Full repair regime**, when after the system has failed, a repair of the whole system begins. Having been repaired, the system starts working as a new one (it goes into state 0 with no failed components).

There are many real-world phenomena in different spheres of human activity, for example in telecommunication, transmission, transportation, electronics,

© Springer Nature Switzerland AG 2021
V. M. Vishnevskiy et al. (Eds.): DCCN 2021, LNCS 13144, pp. 323–335, 2021.
https://doi.org/10.1007/978-3-030-92507-9_26

manufacturing, service applications etc. that can be modeled by such type of models. Some applications of this model to the reliability of real world systems, such as high-altitude unmanned rotor-craft platforms [2], reliability analysis if oil and gas industry objects [3], systems for remote monitoring of underwater transport pipelines [4] are examples that come in mind.

Due to the wide practical applications, a lot of papers are devoted to the study of k-out-of-n systems. The bibliography on such studies is vast (see for example Trivedi [5], Chakravarthy et al. [6] and the bibliography therein). Most of these investigations deal with the systems under assumptions about components' Poisson failure flow and exponentially distributed repair times. In [7] M.S. Mustafa considered the k-out-of-n system with exponential life and general repair time distributions failed components with the help of embedded Markov chains method. The article [8] contains a detailed analysis of 2-out-of-n and 3-out-of-n systems with exponential life times and general repair time distributions.

In series of our previous works (bibliography can be found in Chap. 9 in [9]) for investigation of renewable systems with Poisson flow of components failures and arbitrarily distributed repair times has been used the so-called markovization method. Firstly, it has been introduced by D.R. Cox in 1955 [10] and consists in introduction of supplementary variables that allow describing the system behavior by a two-dimensional Markov process. This approach admits standing and proposing the way for solution of one of the principal problems for systems reliability analysis, namely: the problem of the sensitivity studies of the output characteristics with respect to the shape of input distributions. More detailed bibliography of the recent studies of a renewable k-out-of-n : F system one can find in [11], where the reliability function of k-out-of-n system has been calculated with the help of Markovization method. The same method has been used in the paper [12] for calculation of the system (and describing its behavior process) time-dependent state (t.d.s.p.'s) and the appropriate steady state probabilities (s.s.p.'s) for special case of 3-out-of-6 : F system under assumption about partial repair regime, when after the system failure its repair is prolonged in the same regime. In the paper [13] a general algorithm for the special class of hyperbolic system of partial differential equations as Kolmogorov forward system of differential equations for two-dimensional Markov process has been used[1].

In this paper we continue this study and apply this algorithm for finding closed form representations t.d.s.p.'s and respective s.s.p.'s for the renewable k-out-of-n : F system under assumption that after its failure the full system repair begins, and its duration has an arbitrary distribution, different from that for repair time of component. In a future project these results are planned to be used for the sensitivity analysis of the system characteristics as depending of the shape of components' repair time distributions. The paper is organized as follows. In the next section the problem setup and some notations are discussed. In the Sect. 3 we propose the extended Markov process for the system description and derive the Kolmogorov forward equation for its t.d.s.p.'s. The main results are presented

---

[1] Another approach for investigation of this kind of system is also possible and has been demonstrated in [14].

in Sect. 4. It contains an Algorithm for the solution of Kolmogorov equations. A representation of its s.s.p.'s in terms of Laplace transforms is obtained. Section 5 demonstrates the implementation of the proposed algorithm for an example. The paper ends with conclusion and proposed future studies.

2   Problem Set up Notations

Consider a repairable k-out-of-n system under the following assumptions. This is a repairable n-component system working in parallel. The system fails when k or more of its components are failed. Failed components are repaired by a single facility server. After the system fails, the entire system is supposed to go for repair. The following notations and additional assumptions are used:

- $\mathbb{P}\{\cdot\}$, $\mathbb{E}[\cdot]$ are symbols for the probability and expectation. Symbols $\mathbb{P}_i\{\cdot\}$, $\mathbb{E}_i[\cdot]$ are used for conditional probability and expectation, given that the initial state of the process is i. Here i stays for the number of failed components at the initial moment;
- The lifetimes of the systems' components are independent identically distributed (i.i.d.) random variables (r.v.'s), exponentially distributed with parameter α;
- Failed components are repaired by a single facility server. After repair each component starts working as a new one;
- Repair times are i.i.d. r.v.'s B_i $(i = 1, 2, \ldots, n)$ for each failed component. R.v.'s G_i $(i = 1, 2, \ldots, n)$ stays for full repair time of the system started with i failed components. The cumulative distribution functions (c.d.f.'s) are denoted by $B(t) = \mathbb{P}\{B_i \leq t\}$ and $G(t) = \mathbb{P}\{G_i \leq t\}$. The probability density functions (p.d.f.'s) have notations $b(x) = \dot{B}(x)$ and $g(x) = \dot{G}(x)$. The upper dot means the derivative with respect to time argument x. Note, that the random variables are independent.
- By $\beta(x)$ and $\gamma(x)$ are denoted the rates after x time spent on the repairs,

$$\beta(x) = \frac{b(x)}{1 - B(x)}, \quad \gamma(x) = \frac{g(x)}{1 - G(x)};$$

- The moment generation functions (m.g.f.'s) of the r.v.s B and G are denoted by

$$\tilde{b}(s) = \mathbb{E}[e^{-sB}] = \int_0^\infty e^{-st} dB(t), \quad \text{and} \quad \tilde{g}(s) = \mathbb{E}[e^{-sG}] = \int_0^\infty e^{-st} dG(t).$$

By b and g are symbolized their expectations,

$$b = \mathbb{E}[B] = \int_0^\infty (1 - B(t)) dt, \quad g = \mathbb{E}[G] = \int_0^\infty (1 - G(t)) dt.$$

To analyze the system process we introduce the following additional notations:

- Λ_i for the random time to fail of one of the system's component, when it starts in the state i;
- $\lambda_i = (n - i)\alpha$ for the intensity that one of $n - i$ working components will fail, when the system is in the state i;
- $E = \{0, 1, \ldots k\})$ for the set of system states, where state j means the number of failed components and k is the system failure state.
- Within this set of states we define the random process $J = \{J(t), \ t \geq 0\}$ by the relation

$$J(t) = j, \ \text{if at time} t \ \text{the system is in the state} \ j;$$

- System (and the process) state probabilities are defined as $\pi_j(t) = \mathbb{P}\{J(t) = j\}$;
- T is notation for the random time to the system failure when it starts as a new, $T = \inf\{t : \ J(t) = k\}$.

Suppose also that at the very beginning all system components are in working (UP) states. It means that the initial state of the process J is with zero failed components, $J(0) = 0$. We suppose that immediate components failures and repairs are impossible, and that the mean repair times are finite. This means that

$$B(0) = G(0) = 0, \quad b = \int\limits_0^\infty (1 - B(t))dt < \infty, \quad g = \int\limits_0^\infty (1 - G(t))dt < \infty.$$

Further we use the following notations for the system t.d.s.p.'s

$$\pi_j(t) = \mathbb{P}\{J(t) = j\},$$

and for the the s.s.p.'s

$$\pi_j = \lim_{t \to \infty} \pi_j(t).$$

We assume that they exist and can be calculated.

3   The Process Z

For the considered system behavior analysis we use so-called Markovization method, firstly proposed by D.R. Cox in 1955 [10]. It is based on introduction of complementary variables. Following this approach we introduce a two-dimensional stochastic process $Z = \{Z(t) = (J(t), X(t)), \ t \geq 0\}$, where the value $J(t)$ is the number of failed components at time t. As the complementary variable $X(t)$ we use the time spent by the component $J(t)$ in its state after its last enter in it. The state space of the process Z is

$$\mathcal{E} = \{0, (j, \mathbb{R}^+) : j = \overline{1, k}\}.$$

Due to the complementary variables, which are changed in the domain $0 \leq x \leq t < \infty$, the process Z is a Markov one. Denote its micro-state p.d.f.'s with respect to the complementary variable by

$$\pi_j(t; x) = \mathbf{P}\{J(t) = j, \ X(t) = x\} \ \ (j = \overline{1, k}).$$

The appropriate macro-state probabilities for $t \geq 0$ are

$$\pi_j(t) = \mathbf{P}\{J(t) = j\} = \int\limits_0^t \pi_j(t; x)dx.$$

Taking into account the probabilistic origin of these functions they are non-negative $\pi_j(t; x) \geq 0, \ \pi_j(t) \geq 0$.

To explain the behavior of the considered k-out-of-n system we present its transition graph in Fig. 1.

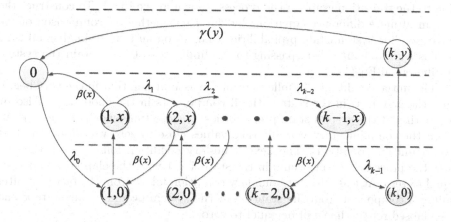

Fig. 1. Transition graph for the k-out-of-n system under full repair scenario.

To investigate the process Z behavior we consider the system of Kolmogorov forward partial differential equations for its t.d.s.p.'s in the corresponding range $0 \leq x \leq t < \infty$. Here for simplicity we use the notations D for operator $D = \left(\frac{\partial}{\partial t} + \frac{\partial}{\partial x}\right)$. In this notations the following theorem holds.

Theorem 1. *The system of Kolmogorov forward partial differential equations for the process Z t.d.s.p.'s has a form*

$$\dot{\pi}_0(t) = -\lambda_0\pi_0(t) + \int\limits_0^t \pi_1(t, x)\beta(x)dx + \int\limits_0^t \pi_k(t, x)\gamma(x)dx,$$

$$D\pi_1(t; x) = -(\lambda_1 + \beta(x))\pi_1(t; x),$$
$$D\pi_j(t; x) = -(\lambda_j + \beta(x))\pi_j(t; x) + \lambda_{j-1}\pi_{j-1}(t; x), \ \ (j = \overline{2, k-1}),$$
$$D\pi_k(t; x) = -\gamma(x))\pi_k(t; x). \tag{1}$$

It should be solved jointly under the initial

$$\pi_0(0) = 1, \quad \pi_i(0, x) = 0 \ \textit{for all} \ x \geq 0, \tag{2}$$

and boundary conditions

$$\pi_1(t, 0) = \lambda_0 \pi_0(t) + \int_0^t \pi_2(t; x)\beta(x)dx;$$

$$\pi_j(t, 0) = \int_0^t \pi_{j+1}(t; x)\beta(x)dx, \ (j = \overline{2, k-2});$$

$$\pi_{k-1}(t, 0) = 0;$$

$$\pi_k(t, 0) = \lambda_{k-1} \int_0^\infty \pi_{k-1}(t; x)dx. \tag{3}$$

Proof. To obtain this system of differential equations we firstly focus our attention to the transition graph of the process, shown in the Fig. 1. To construct the system of finite difference equations by the usual method of comparison of the corresponding system state probabilities changes on an infinitesimal small time epochs t and $t + \Delta t$. Then, passing to the limit $\Delta t \to 0$, we obtain the system of differential Eq. (1).

The initial conditions (2) follows from the assumption that in the very beginning the system is in the state with all components in the good state. Also by comparing of the system state probabilities at close time epochs t and $t + \Delta t$, when the complimentary variable take values close to zero we obtain the first two boundary conditions. The next boundary condition follows from the fact that the process Z never occurs in the state $k - 1$ with the elapsed repair time equal to zero. At least the last boundary condition follow from the fact that after failing a component from the state $k - 1$ the system occurs in the state k and the elapsed repair time will be equal to zero.

4   Main Results

The derived system (1) is a hyperbolic system of partial differential equations. For this type of systems the solution are obtained by the method of characteristics (see [15, 16]). This method consists in the construction of a family of curves with respect to some parameter u in the defined domain $0 \leq x \leq t < \infty$ of system (1). It is called *characteristics* that allow to transform each of the equation of the system (1) into usual differential equation along these curves.

In our case these curves satisfy to the system of ordinary differential equations:

$$\frac{dt}{du} = 1, \quad \frac{dx}{du} = 1, \tag{4}$$

in which, in the considered case, it is possible to choose the variable x as the parameter u. Then, by changing the coefficients before each partial differential equation of system (1) by their representation from (4) with respect to parameter

x, one can get the system of ordinary differential equations, which determine the functions π_j $(j = \overline{1, k})$ along the characteristics:

$$\frac{d}{dx}\pi_1(x) = -(\lambda_1 + \beta(x))\pi_1(x),$$

$$\frac{d}{dx}\pi_j(x) = -(\lambda_j + \beta(x))\pi_j(x) + \lambda_{j-1}\pi_{j-1}(x), \quad (j = \overline{2, k-1})$$

$$\frac{d}{dx}\pi_k(x) = -\gamma(x)\pi_k(x). \tag{5}$$

For continuation the solution of the given equations from the characteristics to the entire domain of the partial differential equations the initial and boundary conditions in the form (3) should be used. Based on this idea in the paper [13] an algorithm has been proposed for the solution of such systems. Applying this algorithm for our process the solutions are obtained.

The Algorithm

Step 1.
For the solution of the second Eq. (1) rewrite the equations for characteristics (4) and obtain the first of Eq. (5) in symmetric form as

$$dt = dx = -\frac{d\pi_1}{(\lambda_1 + \beta(x))\pi_1}. \tag{6}$$

The first integrals of system (6) are

$$dt = dx \quad \Rightarrow \quad t - x = C, \tag{7}$$

$$d\pi_1 = -(\lambda_1 + \beta(x))\pi_1 dx \quad \Rightarrow \quad \pi_1 = D_1 e^{-\lambda_1 x}(1 - B(x)). \tag{8}$$

The general solution of a partial differential equation is a continuous function of its first integrals [16]. Thus, following to this rule, the general solution of Eq. (6) along the characteristics $C = t - x$ is

$$D_1 = h_1(C) \quad \Rightarrow \quad \pi_1(t, x) = h_1(C)e^{-\lambda_1 x}(1 - B(x)), \tag{9}$$

where h_1 is an arbitrary continuous function, which should be found from the boundary conditions.

Step 2. The equations for the characteristics (4) and for the other functions π_j of (1) (for $j = \overline{2, k-1}$) in symmetric form are

$$dt = dx = -\frac{d\pi_j}{(\lambda_j + \beta(x))\pi_j + \lambda_{j-1}\pi_{j-1}}, \quad (j = \overline{2, k-1}). \tag{10}$$

Step 2.1. One of the first integrals of these equations for $j = 2$ is the same as before

$$dt = dx \quad \Rightarrow \quad t - x = C. \tag{11}$$

The other first integral of the Eq. (10) for $j = 2$, along the characteristic $t - x = C$ by using the expression for $\pi_1(t, x)$ from (8) can be found from the following equation:

$$\frac{d\pi_2}{dx} = -(\lambda_2 + \beta(x))\pi_2 + \lambda_1 h_1(C)e^{-\lambda_1 x}(1 - B(x)). \tag{12}$$

For its solution one should use the constant variation method. The general solution of the homogeneous part of this equation

$$\frac{d\check{\pi}_2}{dx} = -(\lambda_2 + \beta(x))\check{\pi}_2$$

has the form

$$\check{\pi}_2 = D_2 e^{-\lambda_2 x}(1 - B(x)).$$

Changing the constant D_2 by function $D_2(x)$, the constant variation method for unknown function $D_2(x)$ gives the following equation

$$D_2'(x) = \lambda_1 h_1(C)e^{-(\lambda_1 - \lambda_2)x}.$$

Its particular solution is

$$D_2(x) = -\frac{\lambda_1}{\lambda_1 - \lambda_2}h_1(C)e^{-(\lambda_1 - \lambda_2)x}.$$

Thus, the general solution of the heterogeneous Eq. (12) is

$$\pi_2 = \left(D_2 - \frac{\lambda_1}{\lambda_1 - \lambda_2}h_1(C)e^{-(\lambda_1 - \lambda_2)x}\right)e^{-\lambda_2 x}(1 - B(x)) =$$

$$= \left(D_2 e^{-\lambda_2 x} - \frac{\lambda_1}{\lambda_1 - \lambda_2}h_1(C)e^{-\lambda_1 x}\right)(1 - B(x)). \tag{13}$$

Putting $D_2 = h_2(C)$, where h_2 is an arbitrary smooth function, one can obtain the general solution of the corresponding partial differential equation along the characteristic in the form

$$\pi_2(t, x) = \left(h_2(t - x)e^{-\lambda_2 x} - h_1(t - x)\frac{\lambda_1}{\lambda_1 - \lambda_2}e^{-\lambda_1 x}\right)(1 - B(x)). \tag{14}$$

Step 2.2. (for cases when $k \geq 3$). The same approach can be used for the other equations. One of the first integrals for any of equations of (10) for $j = \overline{3, k - 1}$ is the same as before

$$dt = dx \quad \Rightarrow \quad t - x = C.$$

The construction of general solution for equations

$$\frac{d\pi_j}{dx} = -(\lambda_j + \beta(x))\pi_j + \lambda_{j-1}\pi_{j-1} \quad (j = \overline{3, k - 1})$$

gives the following result

$$\pi_j(t, x) = \Big(h_j(t - x)e^{-\lambda_j x} + H_{j-1}(t - x; x) \Big)(1 - B(x)), \qquad (15)$$

where $h_1, \dots h_j$ are arbitrary smooth functions, and the following notation is used

$$H_j(t - x; x) = \sum_{i=1}^{j} (-1)^i \frac{\lambda_{j-1} \cdots \lambda_{j-i}}{(\lambda_i - \lambda_{i+1}) \cdots (\lambda_i - \lambda_j)} h_i(t - x)e^{-\lambda_i x}.$$

Step 2.3. For the last equation with $j = k$ one of the first integrals is the same as before

$$dt = dx \quad \Rightarrow \quad t - x = C.$$

The construction of general solution for equations

$$\frac{d\pi_k(x)}{dx} = -\gamma(x))\pi_k(x)$$

gives the following result

$$\pi_k(t, x) = h_k(t - x)(1 - G(x)), \qquad (16)$$

where h_k is an arbitrary smooth function.

Step 3. Collecting these results with the help of the substitution

$$\hat{h}_i = (-1)^i \frac{\lambda_{j-1} \cdots \lambda_{j-i}}{(\lambda_i - \lambda_{i+1}) \cdots (\lambda_i - \lambda_j)} h_i, \quad i = \overline{1, j - 1}, \quad \hat{h}_j = h_j \qquad (17)$$

one can obtain the general solution of the considered forward Kolmogorov's system of Eq. (1) in the form

$$\pi_j(t; x) = \Big(\sum_{i=1}^{j} \hat{h}_i(t - x)e^{-\lambda_i x} \Big)(1 - B(x)) \quad (j = \overline{1, k - 1}),$$

$$\pi_k(t; x) = h_k(t - x)(1 - G(x)). \qquad (18)$$

Step 4. To complete the solution, it is necessary to calculate the functions h_i. We find them from the boundary conditions in terms of Laplace transforms.

Step 4.1. Using the substitution (17) one can find recursive equations for functions $\tilde{h}_i(s)$ $(i = \overline{1, k - 1})$ calculation. With the help of three first boundary conditions from (3) and the representation (17) it is possible to express the LT $\tilde{h}_j(s)$ of functions $h_j(t)$ $(j = \overline{1, k - 1})$ in terms of $\tilde{\pi}_0(s)^2$.

Step 4.2. From the last boundary conditions it follows

$$\pi_k(t; 0) = h_k(t) = \lambda_{k-1} \int_0^t \pi_{k-1}(t; x)dx.$$

---

[2] We omit here these cumbersome calculations, and demonstrate them with examples in the Sect. 5.

The substitution in expression (17) gives

$$\tilde{h}_k(s) = \lambda_{k-1} \sum_{1 \leq i \leq k-1} \left[\tilde{h}_i(s) \frac{1 - \tilde{b}(s + \lambda_i)}{s + \lambda_i} \right].$$

Step 5. Substitution of the expressions of functions $\tilde{h}_j(s)$ $(j = \overline{1, r})$ into the LT of the first equation of (1) allows to obtain an equation for $\tilde{\pi}_0(s)$.

Step 6. Substitution of the obtained solution into expressions for functions $\tilde{h}_j(s)$ and than into LT of functions $\pi_j(t)$ ends the solution of the system.

Stop

The possibilities of this algorithm are demonstrated in the next section by calculation of the t.d.s.p.'s and the s.s.p.'s for the special case of 2-out-of-n system.

5   The Example: 2-out-of-n : F Model

Consider a simplest example of the system 2-out-of-n : F. In this case the Kolmogorov forward system of equations take the form

$$\dot{\pi}_0(t) = -\lambda_0 \pi_0(t) + \int_0^t \pi_1(t, x)\beta(x)dx + \int_0^t \pi_2(t, x)\gamma(x)dx,$$

$$D\pi_1(t; x) = -(\lambda_1 + \beta(x))\pi_1(t; x),$$
$$D\pi_2(t; x) = -\gamma(x))\pi_2(t; x) \tag{19}$$

that should be solved jointly with initial

$$\pi_0(0) = 1, \quad \pi_1(0; x) = \pi_2(t; x) = 0 \quad \forall x \geq 0; \tag{20}$$

and boundary conditions

$$\pi_1(t, 0) = \lambda_0 \pi_0(t),$$

$$\pi_2(t, 0) = \lambda_1 \int_0^t \pi_1(t; x)dx, \tag{21}$$

Following to the algorithm we find the solution of the second from Eq. (19) in the form

$$\pi_1(t; x) = h_1(t - x)e^{-\lambda_1 x}(1 - B(x)) \tag{22}$$

where $h_1(x)$ is an arbitrary smooth function that should be find from boundary conditions as

$$\pi_1(t; 0) \equiv h_1(t) = \lambda_0 \pi_0(t)$$

and in terms of LT as

$$\tilde{h}_1(s) = \lambda_0 \tilde{\pi}_0(s). \tag{23}$$

Further the third of Eq. (19) has a solution

$$\pi_2(t; x) = h_2(t - x)(1 - G(x)) \tag{24}$$

with an arbitrary smooth function $h_2(x)$ for which from the boundary conditions one can find an expression

$$\pi_2(t; 0) \equiv h_2(t) = \lambda_1 \int\limits_0^t h_1(t - x)e^{-\lambda_1 x}(1 - B(x))$$

which in terms of LT gives

$$\tilde{h}_2(s) = \lambda_1 \tilde{h}_1(s)\frac{1 - \tilde{b}(s + \lambda_1)}{s + \lambda_1} = \lambda_0\lambda_1\frac{1 - \tilde{b}(s + \lambda_1)}{s + \lambda_1}\tilde{\pi}_0(s). \tag{25}$$

From the first of the Eq. (19) by substitution the expressions (22, 24) for functions $\pi_1(t; x)$ and $\pi_2(t; x)$ one can find an equation for $\pi_0(s)$

$$s\tilde{\pi}_0(s) - 1 = -\lambda_0\tilde{\pi}_0 + \tilde{h}_1(s)\tilde{b}(s + \lambda_1) + \tilde{h}_2(s)\tilde{g}(s).$$

Taking into account the expressions (23, 25) for functions $\tilde{h}_1(s)$ and $\tilde{h}_2(s)$ from here it follows

$$\tilde{\pi}_0(s) = \left[s + \lambda_0(1 - \tilde{b}(s + \lambda_1))\left(1 - \frac{\lambda_1\tilde{g}(s)}{s + \lambda_1}\right)\right]^{-1} =$$

$$= \left[s + \lambda_0(1 - \tilde{b}(s + \lambda_1))\frac{s + \lambda_1(1 - \tilde{g}(s))}{s + \lambda_1}\right]^{-1} =$$

$$= \frac{s + \lambda_1}{s(s + \lambda_1) + \lambda_0(1 - \tilde{b}(s + \lambda_1))(s + \lambda_1(1 - \tilde{g}(s)))}. \tag{26}$$

Further we calculate $\pi_1(s)$ and $\pi_2(s)$ from the expressions (22, 24) in terms of LT with the help of

$$\tilde{\pi}_i(s) = \int\limits_0^\infty e^{-st}\pi_i(t)dt = \int\limits_0^\infty e^{-st}\int\limits_0^t \pi_i(t; x)dx\,dt \quad (i = 1, 2).$$

After some algebra and substitution of $\pi_0(s)$ from (26) it gives

$$\tilde{\pi}_1(s) = \frac{\lambda_0(1 - \tilde{b}(s + \lambda_1))}{s(s + \lambda_1) + \lambda_0(1 - \tilde{b}(s + \lambda_1))(s + \lambda_1(1 - \tilde{g}(s)))},$$

$$\tilde{\pi}_2(s) = \frac{1}{s}\frac{\lambda_0\lambda_1(1 - \tilde{b}(s + \lambda_1))(1 - \tilde{g}(s))}{s(s + \lambda_1) + \lambda_0(1 - \tilde{b}(s + \lambda_1))(s + \lambda_1(1 - \tilde{g}(s)))}. \tag{27}$$

For the s.s.p.'s $\pi_i = \lim_{s \to 0} \tilde{\pi}_i(s)$ $(i = 0, 1, 2)$ it gives

$$\pi_0 = \frac{\lambda_1}{\lambda_1 + \lambda_0(1 - \tilde{b}(\lambda_1))(1 + \lambda_1 g)},$$

$$\tilde{\pi}_1 = \frac{\lambda_0(1 - \tilde{b}(\lambda_1))}{\lambda_1}\tilde{\pi}_0 = \frac{\lambda_0(1 - \tilde{b}(\lambda_1))}{\lambda_1 + \lambda_0(1 - \tilde{b}(\lambda_1))(1 + \lambda_1 g)},$$

$$\tilde{\pi}_2 = \lambda_0(1 - \tilde{b}(\lambda_1))g\tilde{\pi}_0 = \frac{\lambda_0\lambda_1(1 - \tilde{b}(\lambda_1))g}{\lambda_1 + \lambda_0(1 - \tilde{b}(\lambda_1))(1 + \lambda_1 g)}. \tag{28}$$

This result coincides with the one, obtained for the same model by the method of decomposable regenerative processes accepted for publication by the journal Mathematics [17]. Moreover, for the Markov model, where $B(x) = 1 - e^{-\frac{x}{b}}$ and $\tilde{b}(\lambda_1) = \frac{1}{1 + \lambda_1 b}$ the last formulas takes the form

$$\pi_0 = \lim_{s \to 0} s\pi_0(s) = \frac{\lambda_1(1 + \lambda_1 b)}{\lambda_1(1 + \lambda_1 b) + \lambda_0\lambda_1 b(1 + \lambda_1 g)} = \frac{1 + \lambda_1 b}{1 + b(\lambda_0 + \lambda_1 + \lambda_0\lambda_1 g)};$$

$$\pi_1 = \lim_{s \to 0} s\pi_1(s) = \frac{\lambda_0\lambda_1 b}{\lambda_1(1 + \lambda_1 b) + \lambda_0\lambda_1 b(1 + \lambda_1 g)} = \frac{\lambda_0 b}{1 + b(\lambda_0 + \lambda_1 + \lambda_0\lambda_1 g)};$$

$$\pi_2 = \lim_{s \to 0} s\pi_2(s) = \frac{\lambda_0\lambda_1^2 bg}{\lambda_1(1 + \lambda_1 b) + \lambda_0\lambda_1 b(1 + \lambda_1 g)} = \frac{\lambda_0\lambda_1 bg}{1 + b(\lambda_0 + \lambda_1 + \lambda_0\lambda_1 g)}.$$

The last results coincide with appropriate results calculated by solution of the balance equations obtained by the birth and death process.

Acknowledgements. The publication has been prepared with the support of the "RUDN University Strategic Academic Leadership Program" and funded by RFBR according to the research projects No.20-01-00575A (recipient V.V. Rykov).

References

1. Rykov, V., Ivanova, N.: Reliability and sensitivity analysis of a repairable k-out-of-$n : F$ system with general life- and repair times distributions. In: Baraldi, P., Di Maio, F., Zio, E. (eds.) Proceedings of the 30th European Safety and Reliability Conference and the 15th Probabilistic Safety Assessment and Management Conference. Copyright Oc ESREL2020-PSAM15 Organizers. Published by Research Publishing, Singapore (2020). ISBN/DOI 978-981-14-8593-0
2. Vishnevsky, V., Kozyrev, D., Rykov, V., Nguen, Z.: Reliability modeling of the rotary-wing flight module of a high-altitude telecommunications platform. Inf. Technol. Comput. Syst 4(20), 26–38 (2020). (in Russian)
3. Rykov, V.V., Sukharev, M.G., Itkin, V.Y.: Investigations of the potential application of k-out-of-n systems in oil and gas industry objects. J. Mar. Sci. Eng. **8**, 928 (2020). www.mdpi.com/journal/jmse, https://doi.org/10.3390/jmse8110928
4. Rykov, V., Kochueva, O., Farkhadov, M.: Preventive maintenance of a k-out-of-n system with applications in subsea pipeline monitoring. J. Mar. Sci. Eng. **9**, 85 (2021). https://doi.org/10.3390/jmse9010085

5. Trivedi, K.S.: Probability and Statistics with Reliability. Wiley, New York (2002)
6. Chakravarthy, S.R., Krishnamoorthy, A., Ushakumari, P.V.: A k-out-of-n reliability system with an unreliable server and Phase type repairs and services: the (N, T) policy. J. Appl. Math. Stoch. Anal. **14**(4), 361–380 (2001)
7. Moustafa, M.S.: Availability of K-out-of-N : G systems with exponential failure and general repairs. Econ. Qual. Control. **16**(1), 75–82 (2001)
8. Linton, D.G., Saw, J.G.: Reliability analysis of the k -out-of- n : f system. IEEE Trans. Reliab. **23**, 97–103 (1974)
9. Rykov, V.V.: On reliability of renewable systems. In: Vonta I., Ram M. (eds.) Reliability Engineering, Theory and Applications, pp. 173–196 (2018)
10. Cox, D.R.: The analysis of non-Markovian stochastic processes by the inclusion of supplementary variables. Math. Proc. Cambridge Phil. Soc. **51**(3), 433–441 (1955). https://doi.org/10.1017/S0305004100030437
11. Rykov, V., Kozyrev, D., Filimonov, A., Ivanova, N.: On reliability function of a k-out-of-n system with general repair time distribution. Prob. Eng. Inf. Sci. **35**, 1–18 (2020). https://doi.org/10.1017/S02699648200002
12. Rykov, V., Ivanova, N., Kozyrev, D.: Sensitivity analysis of a k-out-of-n: F system characteristics to shapes of input distribution. In: Lecture Notes in Computer Science (LNCS), vol. 12563, pp. 485–496. Springer, Heidelberg (2020). https://doi.org/10.1007/978-3-030-66471-837
13. Rykov, V., Filimonov, A.: Hyperbolic systems with multiple characteristics and some of their applications. Manag. Large Syst. **85**, 72–86 (2021)
14. Rykov, V.V.: Decomposable semi-regenerative processes: review of theory and applications to queueing and reliability systems. RT&A **16**(2(62)), 157–190 (2021)
15. Petrovsky, I.G.: Lectures on Theory of Ordinary Differential Equations. Nauka, Moscow (1984).(In Russian)
16. Rojdenstvensky, B.L., Tanenko, N.N.: Systems of Kvasi-Linear Equations, p. 688. Nauka, Moscow (1978). (In Russian)
17. Rykov, V., Ivanova, N., Kozyrev, D.: Application of decomposable semi-regenerative processes to the study of k-out-of-n systems. Mathematics **9**, 1933 (2021). https://doi.org/10.3390/math91611933

Using a Machine Learning Approach for Analysis of Polling Systems with Correlated Arrivals

Vladimir Vishnevsky[1], Olga Semenova[1]([✉]), and D. T. Bui[2]

[1] Institute of Control Sciences of Russian Academy of Sciences,
Profsoyuznaya Street 65, Moscow 117997, Russia
`vishn@inbox.ru`
[2] Moscow Institute of Physics and Technology, Institutskiy per. 9, Dolgoprudny,
Moscow Region 141701, Russia
`duytan@phystech.edu`

Abstract. The paper investigates stochastic polling systems using machine learning. $M/M/1$ and $MAP/M/1$-type polling systems with cyclic polling, as well as $M/M/1$-type polling systems with adaptive cyclic polling are considered. To train a machine model of a $M/M/1$-type polling system, we used the results of analytical calculations, and for other considered systems that do not allow exact analysis, we used the simulation results. Numerical examples are given, and it is shown that the results of machine learning are close enough to the results of analytical or simulation calculations.

Keywords: Machine learning · Polling systems · Correlated arrivals

1 Introduction

Polling systems are queuing systems with multiple queues and a common service device [1]. A server visits the queues and serves the accumulated customers there. The server visits the queues and serves the queued customers. Polling systems are widely used to evaluate the performance, design and optimizing the structure of telecommunication systems and networks, transport and traffic management systems, production systems and inventory management systems [2]. Despite a significant number of papers in the area, there remain a large number of unsolved problems, in particular, the polling systems with correlated arrivals or systems with limited queuing disciplines. In 2019 we developed the analytical and simulation software complex [3] covering a wide class of polling models used to study broadband wireless networks.

Here we propose to apply the machine learning method using artificial neural networks to analysis of polling systems. This area of research is new and, as the

The research is supported by the Russian Foundation for Basic Research, project no. 19-29-06043.

V. M. Vishnevskiy et al. (Eds.): DCCN 2021, LNCS 13144, pp. 336–345, 2021.
https://doi.org/10.1007/978-3-030-92507-9_27

results of calculations show, opens up new opportunities for studying queuing models that are not amenable or difficult to analyze within the framework of the theory of random processes. Note that only a few works in this direction are known in the literature [5,6]. Below we present the results of machine learning for $M/M/1$ and $MAP/M/1$ type polling systems with a cyclic polling and $M/M/1$ type polling systems with adaptive cyclic polling. To train a machine model of an $M/M/1$-type polling system, we used the analytical results and the results of simulation for other systems which have no exact solutions. The results of machine learning obtained are close enough to the results of analytical calculation or a simulation.

An artificial neural network can be defined as a computation model consisting of a networking structure presented by a set of neurons. A neural network structure is a set of parameters (weights). These parameters can be pre-defined or unknown and have to defined or improved to solve a specific problem. The structure of a neural network is presented by a set of parameters (weights) that need to be determined or improved on the basis of an existing set of weights to solve necessary problems. Artificial neural networks were first developed in the early 1940 s. ANNs are the forecasting tools used to build a mathematical model of the complex systems. Multilayer Perceptron (MLP) ANN [4] is the most famous ANN class. ANN MLPs usually have a feedforward architecture and are usually trained in backpropagation algorithms. MLP networks consist of one input layer and one output layer, with at least one additional hidden layer. For the task of selecting a function from the initial data, one hidden layer allows the neural network to approximate any function that is a continuous mapping from one finite space to another [7]. With two hidden layers, the network can represent an arbitrary decision boundary with any accuracy [8]. J. Heaton [9] shows that that the optimal number of neurons in the hidden layer is usually between the number of input parameters and the number of output parameters. The optimal number of neurons for the hidden layer is determined experimentally in order to minimize the error.

2 Neural Network Properties

Neural networks are widely used in many industries, such as retail, engineering, manufacturing, banking, insurance, healthcare, and others. Neural networks are used to detect relationships, recognize patterns, and predictions.

Most neural network models are of the following types:

- *Approximation* (or regression function). Approximation can be considered as the problem of choosing a function using the initial data. In this case, the neural network is trained on information represented by a data set consisting of a sample with input and target data, and the output data is the result of the neural network operation built after its training on a sample with target data.
- *Classification* (or pattern recognition). Classification can be defined as the process of assigning of an accepted template characterized by a set of features

to one of a prescribed number of classes. The input data includes a set of the sample characteristics. The targets define the class each template belongs to. The main goal of the classification problem is to simulate the posterior probabilities of class membership based on the input data.

The neural network is applied a teaching process to obtain the lowest possible errors by means of searching a set of parameters that match the neural network for the data set. The overall teaching process consists of two different concepts: error minimization and optimization algorithm.

The ways to find errors can be as follows:

- Mean squared error (MSE);
- Normalized squared error (NSE);
- Weighted squared error (WSE);
- Cross entropy error;
- Minkowski error (ME).

The following optimization algorithms are usually used to train a neural network:

- *Gradient descent* (GD). This is the simplest optimization algorithm. With this method, the parameters are updated each time in the direction of the negative gradient of the error index;
- *Conjugate gradient* (CG). The conjugate gradient algorithm searches along the conjugate directions which usually converges faster than the gradient descent directions;
- *Quasi-Newton method* (QNM). This method uses the Hessian of the error function which is a matrix of second derivatives to compute the direction of teaching. Because it uses high-order information, the teaching direction indicates the minimum of the error function with higher precision;
- *Levenberg-Marquardt algorithm* (LM). The algorithm is designed to approximate the second order learning rate without having to compute the Hessian.
- *Stochastic gradient descent* (SGD). The algorithm has a different nature than the algorithms above. Each time it updates the parameters multiple times based on the group data.
- *Adaptative linear momentum* (ADAM). The algorithm is similar to GD but implements a more complex method for calculating the teaching direction which usually provides faster convergence.

The optimization algorithm stops when the given stopping conditions hold. Some commonly used stopping conditions are:

- The parameter increment rate is less than the minimum value.
- The error is improved at one attempt less than the given number of attempts.
- Errors have been reduced to a goal.
- The rate of the error index gradient is set below the target.
- The maximum number of attempts has been reached.
- The maximum amount of computing time has been exceeded.
- The error in the selection subset increases over several attempts.

The process of training a neural network with a given number of hidden levels and neurons or a deep learning model goes through six stages:

- *Step 1.* Initialization: all neurons are given their initial weights;
- *Step 2.* Forward propagation: the input from the training set is passed through the neural network and the output data are calculated;
- *Step 3.* Error function: the error function captures the error between the known output data and the data that needs to be obtained using the constructed neural network model taking into account the weight of the current model (in other words, how far the model is from the known result);
- *Step 4.* Backpropagation: the purpose of backpropagation is to change the weight of the neurons to reduce the error;
- *Step 5.* Updating the weight: the weights are optimized according to the results of the backpropagation algorithm;
- *Step 6.* Iterating until convergence: since the weights are updated in a small steps it takes several iterations to train the network. The number of iterations required for convergence depends on the learning rate, network parameters, and the optimization method used.

Here, to study polling systems, we used the Levenberg-Marquardt algorithm (LM) since such an algorithm allows training a neural network more accurately and faster than other algorithms. In this case, the standard error (MSE) is taken as the error.

3   Machine Learning for an Asymmetric Cyclic Polling System with Gated Service Discipline

We consider a polling system with N ($N \geq 2$) queues with unlimited waiting space and a single server.

Queue i has a Poisson input of customers with parameter λ_i. Service times in queue i are i.i.d. random variables with distribution function $B_i(t)$ with the mean b_i and the second moment $b_i^{(2)}$. The server's switchover times between queues $(i-1)$ and i are i.i.d. random variables with distribution function $S_i(t)$ with the mean s_i and the second moment $s_i^{(2)}$, $i = \overline{1, N}$.

The server visits the queues in a cyclic order and serves the customers there accordingly to a gated service discipline, i.e. it serves only those customers which presented at the queue at a polling moment. Customers arrived during the queue service time should wait for the next server's visit to the queue.

The period of time the server spends visiting queues 1 through N is called a cycle. The mean cycle time is

$$C = \frac{\sum_{i=1}^{N} s_i}{1 - \sum_{i=1}^{N} \rho_i}$$

where $\rho_i = \lambda_i b_i$.

For this model, we train a neural network to get the mean sojourn times in queues. The exact formulas to calculate the mean sojourn times are obtained by U. Yechiali [10] and have the following form:

$$V_i = \frac{f_i(i,i)(1+\rho_i)}{2\lambda_i^2 C} + b_i, i = \overline{1,N} \tag{1}$$

where $f_i(j,k)$, $i,j,k = \overline{1,N}$ are the second moments of the queue lengths at polling moments and can be calculated as the solution of the linear system

$$f_{i+1}(j,k) = \lambda_j \lambda_k s_{i+1}^{(2)} + s_{i+1}\lambda_k f_i(j) + s_{i+1}\lambda_j f_i(k) + f_i(i)\lambda_j\lambda_k[2b_i s_{i+1} + b_i^{(2)}] +$$
$$+ f_i(i,i)\lambda_j\lambda_k b_i^2 + f_i(i,j)b_i\lambda_k + f_i(i,k)b_i\lambda_j + f_i(j,k),$$
$$i \neq j, i \neq k,$$
$$f_{i+1}(i,j) = \lambda_i\lambda_j s_{i+1}^{(2)} + s_{i+1}\lambda_i f_i(j) + f_i(i)\lambda_i\lambda_j[2b_i s_{i+1} + b_i^{(2)}] + f_i(i,j)b_i\lambda_i +$$
$$+ \lambda_i\lambda_j b_i^2 f_i(i,i),$$
$$i \neq j,$$
$$f_{i+1}(i,i) = \lambda_i^2 s_{i+1}^{(2)} + f_i(i)\lambda_i^2[2b_i s_{i+1} + b_i^{(2)}] + \lambda_i^2 b_i^2 f_i(i,i),$$
$$i,j,k = \overline{1,N}. \tag{2}$$

As for example, consider an $M/M/1$ type polling system with $N = 5$ queues and exponentially distributed service times and switchover times. This model has $3N+1$ input parameters and N output ones. The neural network model has two hidden layers consisting of 10 neurons (see Fig. 1). We have used about 500 input data to teach the neural network.

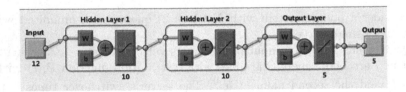

Fig. 1. Neural network for $M/M/1$-type polling system with 5 queues.

Figure 2 presents the dependence of the mean weighted sojourn time

$$V = \sum_{i=1}^{N} \rho_i V_i$$

calculated by the PGF (Probability Generating Function) method [10] and by simulations on the system load ρ. Note that the results of these methods for calculating the system characteristics differ by no more than 0.04%. The training time does not exceed 3 min, and the calculation of characteristics for one data set takes about a second.

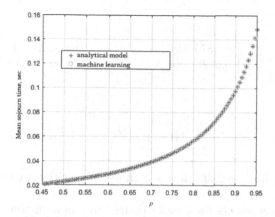

Fig. 2. The results of machine learning for $M/M/1$ type system

4   Machine Learning for a $MAP/M/1$-Type Cyclic Polling System with Gated Service

In this Section we consider a polling system with correlated arrivals described with Markovian Arrival Process (MAP). The input flow to queue i is described by matrices D_0 and D_1 specifying the intensities of transitions of the Markov chain not accompanied and accompanied by an arrival, respectively [11]. Recently, a polling system with correlated arrivals for an arbitrary number of queues has no closed-form solution, therefore, we use the simulation results as input data for machine learning. The model with N queues and W_i states of the MAP for queue i has

$$\sum_{i=1}^{N} W_i(2W_i - 1) + 2N + 1$$

input parameters.

As an example, we consider the system with 4 queues and use approximately 400 input data. Also, denote the MAP arriving to queue i as MAP_i, $i = \overline{1, N}$.

To train the neural network, we use the following set of input data. MAP_i for queue i, $i = \overline{1, 4}$ is defined as follows.

MAP_1:

$$D_0 = t \times \begin{bmatrix} -6 & 2 \\ 0 & -1 \end{bmatrix}, D_1 = t \times \begin{bmatrix} 0 & 4 \\ 0 & 1 \end{bmatrix}$$

and correlation coefficient $c_1 = 0$ where t varies from 0.5 to 50 with step 0.5.

MAP_2:

$$D_0 = t_2 \times \begin{bmatrix} -3 & 0 \\ 0 & -0.6 \end{bmatrix}, D_1 = t_2 \times \begin{bmatrix} 1 & 1 \\ 0.2 & 0.4 \end{bmatrix},$$

$c_2 = 0.07843$.

MAP_3:

$$D_0 = t \times \begin{bmatrix} -1.875 & 0.0625 \\ 0.0625 & -0.25 \end{bmatrix}, D_1 = t \times \begin{bmatrix} 1.8125 & 0 \\ 0 & 0.1875 \end{bmatrix},$$

$c_3 = 0.2704$.

MAP_4:

$$D_0 = t_4 \times \begin{bmatrix} -6 & 2 \\ 0 & -1 \end{bmatrix}, D_1 = t_4 \times \begin{bmatrix} 0 & 4 \\ 0 & 1 \end{bmatrix},$$

$c_4 = 0$.

Service intensities are 0.001 and switchover intensities are 0.001 the same for all queues.

Target data represents the mean sojourn time of a customer in each queue. The mean sojourn times as data for training a neural network were calculated using a software complex [3] for evaluating the performance characteristics of stochastic polling systems.

The network structure is presented as follows:

- The input layer which receives 38 input data has 38 neurons;
- A single hidden layer consisting of 10 neurons;
- An output layer with 4 neurons which determines the output of the neural network.

Hereinafter, the number of neurons in the input and output layers is determined by the number of input and output parameters, respectively; the network also has one hidden layer with 10 neurons (Fig. 3).

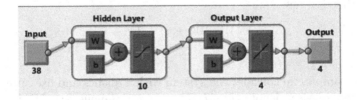

Fig. 3. The structure of the neural network for a $MAP/M/1$ type system

Figure 4 (a) presents the results of calculations of the customer's mean sojourn time obtained using simulation and machine learning. In this case, the relative difference between the results did not exceed 1%.

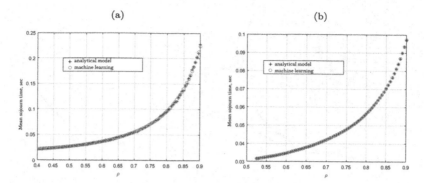

Fig. 4. Results of machine learning for $MAP/M/1$ (a) and $M/M/1$ (b) with adaptive polling and exhaustive service

5   Machine Learning for a $M/M/1$-Type Polling System with Adaptive Cyclic Polling

Now consider a polling system with an adaptive cyclic polling order where the server visits the queues cyclically but skips (does not visit) those that were visited in the previous cycle and were empty at their polling moment. All queues that the server skips in the current cycle will be polled in the next cycle. Such systems are investigated in [12,13] in case of exhaustive and gated service disciplines by the PGF method based on construction of the embedded Markovian process. Using the training data obtained using the formulas [13] for calculating the mean sojourn time system, we construct a neural network and compare the obtained results with the analytical results obtained on a different sample of input data. The data obtained are shown in Fig. 4(a) and 2, the difference between the results is no more than 0.04%. The neural network training time is no more than 3 s.

Then, we consider the case of limited service discipline that does not allow exact analysis due to the specificity of a multidimensional random process describing the behavior of such a system. With limited service, the server can serve no more than a certain (deterministic or random) number of customers per a visit to the queue. For example, consider a system with 4 queues, 1-limited and 2-limited service, that is, the server serves, respectively, no more than one or two customers per a visit to a queue. Figure 4 (b) illustrates the results obtained. The mean sojourn times as data for training the neural network were calculated using the simulation module of the software complex [3]. As follows from Fig. 4 (b) the difference between the simulation results and results obtained from the machine learning is no more than 2 %. The training time does not exceed 3 min, and for each set of input data, the calculation of the output data takes no more than 1 s.

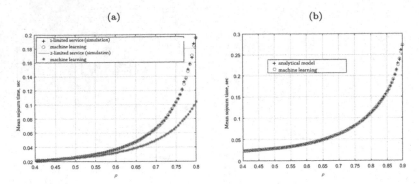

Fig. 5. Results of machine learning for $M/M/1$ (a) with adaptive polling and limited service and $MAP/M/1$ (b) with adaptive polling and exhausting service

6   Machine Learning for $MAP/M/1$-Type System with Adaptive Polling

Let the system have 3 queues, adaptive cyclic order and correlated MAP arrivals. The queue service discipline is exhaustive. MAP_i for queue is given as follows.

MAP_1 is characterized by matrices

$$D_0 = t \times \begin{bmatrix} -6 & 2 \\ 0 & -1 \end{bmatrix}, D_1 = t \times \begin{bmatrix} 0 & 4 \\ 0 & 1 \end{bmatrix}$$

and the correlation coefficient $c_1 = 0$ where t varies from 0.5 to 50 with step 0.5.

MAP_2:

$$D_0 = t \times \begin{bmatrix} -3 & 0 \\ 0 & -0.6 \end{bmatrix}, D_1 = t \times \begin{bmatrix} 1 & 1 \\ 0.2 & 0.4 \end{bmatrix},$$

$c_2 = 0.07843.$

MAP_3:

$$D_0 = t \times \begin{bmatrix} -1.875 & 0.0625 \\ 0.0625 & -0.25 \end{bmatrix}, D_1 = t \times \begin{bmatrix} 1.8125 & 0 \\ 0 & 0.1875 \end{bmatrix},$$

$c_3 = 0.2704.$

For this model, we have 29 input data. Recently, such a polling model does not allow the closed form solution to calculate the performance characteristics, so we use the simulation module of [3] to get the data for machine learning. Comparative analysis is shown on Fig. 5(b).

7   Conclusion

The paper presents the results of machine learning for polling systems with the Poisson and correlated arrivals (MAP). For training a neural network, the analytical results of calculating the characteristics of the systems were used, and we used the simulation results in the case of correlated arrivals. We also provide the comparison between the performance characteristics obtained analytically (or by simulation) and based on machine learning results. It is shown that the results obtained coincide with a high accuracy while the computer model can significantly reduce the time for calculating the characteristics of polling systems comparing to simulation results.

References

1. Vishnevsky, V., Semenova, O.: Polling systems and their application to telecommunication networks. Mathematics 9(2), 117 (2021)
2. Boon, M.A.A., van der Mei, R.D., Winands, E.M.M.: Applications of polling systems. Surv. Oper. Res. Manage. Sci. 16(2), 67–82 (2011)
3. Vishnevsky, V.M., Semenova, O.V., Bui, D.T.: Software complex for evaluating the characteristics of stochastic polling systems: Certificate of state registration of a computer program No. 2019614554 of the Russian Federation; Registered 04 April 2019
4. Cybenko, J.: Approximations by superpositions of a sigmoidal function. Math. Control Signals Syst. 2(4), 303–314 (1989)
5. Sivakami Sundari, M., Palaniammal, S.: Simulation of $M/M/1$ queuing system using ANN. Malaya J. Matematik 1, 279–294 (2015)
6. Sivakami Sundari, M., Palaniammal, S.: An ANN simulation of single server with infinite capacity queuing system. Int. J. Innovative Technol. Exploring Eng. 8(12), 4067–4071 (2019)
7. Csáji, B.C.: Approximation with artificial neural networks. Eötvös Loránd University, Hungary, Faculty of Sciences (2001)
8. Thomas, A., Petridis, M., Walters, S., Malekshahi, S., Morgan, R.: Two hidden layers are usually better than one. In: International Conference on Engineering Applications of Neural Networks (2017). https://doi.org/10.1007/978-3-319-65172-9_24
9. Heaton, J.: Introduction to Neural Networks with Java, Heaton Research, Inc., Chesterfield (2008)
10. Yechiali, U.: Analysis and control of polling systems. In: Donatiello, L., Nelson, R. (eds.) Performance/SIGMETRICS -1993. LNCS, vol. 729, pp. 630–650. Springer, Heidelberg (1993). https://doi.org/10.1007/BFb0013871
11. Dudin, A.N., Klimenok, V.I., Vishnevsky, V.M.: The Theory of Queuing Systems with Correlated Flows. Springer, Cham (2020). https://doi.org/10.1007/978-3-030-32072-0
12. Vishnevsky, V.M., Dudin, A.N., et. al.: Approximate method to study $M/G/1$-type polling system with adaptive polling mechanism. Qual. Technol. Quant. Manage. 9(2), 211–228 (2012)
13. Vishnevsky, V.M., Semenova, O.V., Bui, D.T., Sokolov, A.: Adaptive cyclic polling systems: analysis and application to the broadband wireless networks. In: Vishnevskiy, V.M., Samouylov, K.E., Kozyrev, D.V. (eds.) DCCN 2019. LNCS, vol. 11965, pp. 30–42. Springer, Cham (2019). https://doi.org/10.1007/978-3-030-36614-8_3

Statistical Analysis of psychological Results Tests

M. Y. Voronina🆔 and Y. N. Orlov$^{(\boxtimes)}$🆔

Keldysh Institute of Applied Mathematics,
Miusskaya Square, 4, Moscow 125047, Russia
voronina.miu@phystech.edu

Abstract. In this work, a statistical analysis of the results of psychological testing is carried out in order to establish correlations between the results of surveys of respondents that were not grouped beforehand by some criterion. The goal is also to formalize the construction of patterns (i.e., combining respondents into groups according to their skill level in the following qualities) of typical distributions of quality indicators, which is important for interpreting test results by identifying subgroups and checking for them the Spearman's law of diminishing returns. The effect is that the correlation between responses in the worst group is higher than in the best group.

Keywords: Psychometrical analysis · Accuracy patterns · Correlation · Spearman effect

1 Introduction

The work is based on the testing data of $N = 11335$ respondents. Each respondent was asked 30 questions under each of 10 different methodologies ("question types" P_i). The assessment of each answer had 4 components: productivity, accuracy, speed, efficiency (all of these are "qualities" q_i). The answer to a separate question in each of the four qualities was assessed according to the system "true - 1", "false - 0". The final score for each methodology was obtained as the arithmetic mean of individual answers and took a value from 0 to 1.

The following special qualities of intelligence were tested [1,2]: P_1 – the ability to understand the relationship between concepts or "analogies", P_2 – the ability of numerical thinking or "number series", P_3 – visual memory for abstract contour images or "memory for figures", P_4 – the ability to construct a whole according to available parts or "patterns", P_5 – an assessment of computational skills or "arithmetic counting", P_6 – an assessment of memory for words or "verbal memory", P_7 – finding a lexical equivalent for a sign sequence or "establishing a regularity", P_8 – the ability to reason according to a causal relationship-hereditary connections or "syllogisms", P_9 – the ability to generalize in the analysis of verbal information or "exclusion of the word", P_{10} – the ability to rotate an object in three dimensions or "cubes". Thus, each respondent was assessed with a 10-dimensional vector of accuracy.

© Springer Nature Switzerland AG 2021
V. M. Vishnevskiy et al. (Eds.): DCCN 2021, LNCS 13144, pp. 346–357, 2021.
https://doi.org/10.1007/978-3-030-92507-9_28

2   Distribution Densities of test Scores

Let us compare the assessments of productivity, speed, accuracy, and effectiveness of the full population of respondents without distributing them into groups. The corresponding probability distribution densities in a uniform estimation scale consisting of 10 class intervals are shown in Fig. 1.

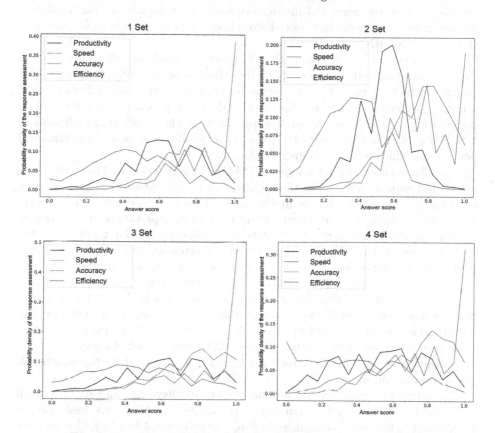

Fig. 1. Densities of distribution of estimates for various qualities

It follows from Fig. 1 that for all blocks of questions there is a certain connection between the distributions of various qualities. Thus, the distribution of efficiency (of some cumulative quality) has a mode located to the left of the maxima of the other distributions. It sits in the region of 0.4. This is the most likely value for this quality. To the right of it is the productivity mode, the corresponding value is 0.6. The accuracy mode, as mentioned earlier, is in the 0.8 range. Speed is most likely in the last class interval. It is difficult to say whether such an equidistant arrangement of modes is related to estimation techniques or is it due to the characteristics of this particular sample. But note that without taking into account the "velocity" parameter, the other three distributions are quite close in shape and can be combined by shifting along the evaluation axis.

3   Spearman Effect

Recently, close attention has been paid to the interpretation of the correlations of responses in relation to psychological tests. Thus, work [1] points out the insufficient methodological foundation of methods for assessing the so-called Spearmen's Law of Diminishing Returns (SLODR, [3]), according to which the correlations between the results of intelligence testing are higher for less intelligent respondents. The papers [4,5] also discuss issues related to metrics for assessing the effect of SLODR.

It is important to emphasize that in practice situations often arise with a disturbance of the indicated tendency, since there is no generally accepted robust metric in which the effect has a mathematically correct explanation. For example, a linear correlation coefficient close to zero does not guarantee that there is no relationship between phenomena. The proximity of the association coefficient to 1 [6,7] also does not guarantee the direct functional connection, if the standard deviation of this coefficient is greater than itself.

Usually, deviation from the expected SLODR effect is declared as an artefacts associated with: differences in the sequence of tasks; self-study during testing; external influences; differences in the intellectual composition of the groups. Thus, the problem is transferred to the field of interpretation of test results within the framework of certain models of intelligence. Without discussing the various theories of intelligence, we will turn our attention to some aspects of the psychometric assessment of the respondent based on the test results.

Let us analyze the pairwise dependence of the accuracy of the answers on the methods used for a plurality of respondents as a whole. Let us first consider the pairwise association between the methods in relation to the first and last quality assessments - "D-students" and "A-students". We divide the entire rating scale into 4 grades evenly with a step of $0,25$: "bad", "satisfactory", "good", "excellent". Accordingly, the first class ("bad") is identified by the number 1, the second 2, and so on.

Let $a(i; j, k)$ – number of respondents, which have assessment i by both methods j and k; $b(i; j, k)$ – number of respondents, which have assessment i by method j, but not having it according to the method k; $c(i; j, k)$ – number of respondents who do not have an assessment i by method j, but having it according to the method k; and finally, $d(i; j, k)$ – number of respondents without an assessment i by both methods. For "D-students" $i = 1$, and for "A-students" $i = 4$. In these terms, the association coefficient is determined by the formula

$$L_{jk}(i) = \frac{a(i; j, k)\, d(i; j, k) - b(i; j, k)\, c(i; j, k)}{a(i; j, k)\, d(i; j, k) + b(i; j, k)\, c(i; j, k)} \tag{1}$$

So, for example, when comparing methods 1 and 2 (analogies and numerical series), it turned out that for "D-students" $L_{12}(1) = 0,97$, and for "A-students" $L_{12}(4) = 0,55$, which is consistent with SLODR views. The same picture is observed for all pairs of combinations of techniques.: $L_{ij}(1) > L_{ij}(4)$.

The question arises: does this effect take place for any scale intervals? Is it true, that if $r < s$, for $L_{ij}(r) > L_{ij}(s)$?

Comparison between "B-students" and "A-students" leads to different results: $L_{12}(3) = 0,32 < L_{12}(4)$. Consequently, the SLODR effect in this metric works partially and is not universal (Fig. 2).

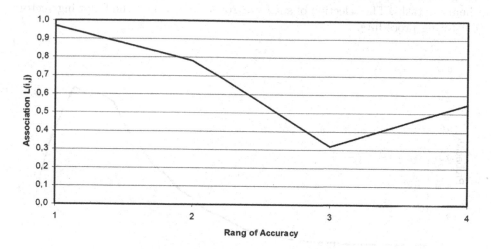

Fig. 2. SLODR violation for association coefficients

The new effect is that the SLODR depends on the scale of response accuracy rang. In the ideal case the dependence of the association coefficient from the rang of respondents has a monotonically decreasing curve. In our example we see, that the lowest group of respondents (first rang of accuracy) has the highest level of association coefficients in accordance with SLODR, but the excellent students has a higher association coefficient than those following them in rang. This effect holds for all pairs of techniques P_i and P_j. So the Spearmen Law for total group of respondents depends on partial content and on scale of accuracy estimation.

It is useful to consider another approach to estimate of SLODR effect for comparison of the lowest and highest scale intervals. Let $N_1^{(1)}(i)$ is a number of respondents, which correspond to the first accuracy interval for P_i, and $N_1^{(2)}(i,j)$ is a number of respondents, corresponding to this interval for both techniques P_i and P_j. Let us designate by $\nu_1^2(i,j)$ the proportion of the intersection set compared to the union set:

$$\nu_1^2(i,j) = \frac{N_1^{(2)}(i,j)}{N_1^{(1)}(i) + N_1^{(1)}(j) - N_1^{(2)}(i,j)}$$

Analogously we can define the proportion $\nu_4^2(i,j)$ for the last accuracy interval. The SLODR effect should be that $\nu_1^2(i,j) > \nu_4^2(i,j)$ for all pairs of techniques. In more general estimation for s types of techniques we expect that $\nu_1^s(i_1, ..., i_s) > \nu_4^s(i_1, ..., i_s)$.

4   Building Patterns of accuracy

Let us now consider the entire set of respondents as a union of non-overlapping subgroups in accordance with the performance of their average responses to 10 thematic tasks. The selection of such subgroups is based on the following vector clustering procedure.

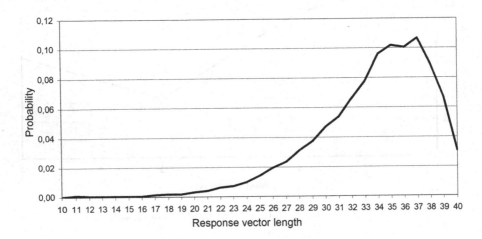

Fig. 3. Distribution of respondents according to the total accuracy of answers

Each respondent is characterized by a 10-dimensional vector of the accuracy of the answers $\mathbf{a}_n = (a_{1,n}, ..., a_{10,n})$. With the chosen scale for assessing the answers, a_{in} takes values from the set of natural numbers $\{1, 2, 3, 4\}$. The length of the response vector is the sum of its components:

$$A_n = \sum_{i=1}^{10} a_{in} \tag{2}$$

Thus, the possible lengths of vectors take integer values from 10 to 40. The distribution of respondents by the length of these vectors is shown in Fig. 3. The right quantile 0,9 is equal to 28, so that almost 90% of the answers are categorized as "good" and "excellent", i.e. have total amount of 30 and above.

The results of the answers of different respondents are also compared in the L_1 norm:

$$d_{kn} = \sum_{i=1}^{10} |a_{in} - a_{ik}| \tag{3}$$

Ideally, respondents belong to the same group if the vectors of the accuracy of their answers match component-wise. However, this approach is ineffective, since then the number of classes in order of magnitude often turns out to be equal to the number of subjects. Therefore, to distinguish a class, we use, firstly, the existing scale of the quality of assessments and, secondly, we the level of closeness of the components of the accuracy vectors of different respondents when answering questions of the same topic. Suppose that the components of vectors from one class cannot differ by more than 1, and the total difference of vectors (norm (3)) does not over top a given level. Clustering will be considered correct if the distribution of distances between any two halves of the selected group has the property of stationarity [8].

The average value of the vector over the group will be called the group pattern.

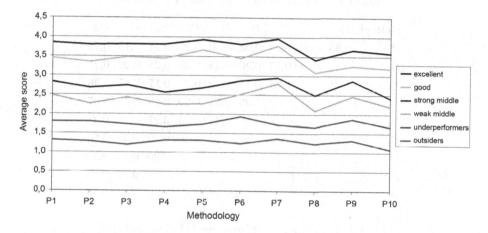

Fig. 4. Homogeneous patterns of groups of respondents

It turned out that about half of the respondents form three homogeneous patterns - "successful", "average" and "bad". The first have grades for all methods from 3 to 4, the second - from 2 to 3 and the third - from 1 to 2. At the same time, the first group included 5319 people, the second 215 and the third only 40. Each of these groups can be distributed into two subgroups-upper and lower-relative to the middle of the length of the evaluation vector A_n. The "excellent" pattern is formed by vectors of "successful" ones with length $A_n \geq 36$. There were 3955 or 35% of such respondents. The "good" pattern is made up of 1364 vectors with conditions $36 > A_n \geq 30$, $a_{in} \geq 3$. "Strong middle" is formed by 129 vectors with conditions $30 > A_n \geq 26$, $3 \geq a_{in} \geq 2$ and "weak middle" is formed by 86 vectors with conditions $26 > A_n > 20$, $3 \geq a_{in} \geq 2$. There are also 15 "underperformers" whose pattern correspond the conditions $A_n \geq 16$, $a_{in} \leq 2$ and 25 "outsiders" with the pattern $A_n < 16$, $a_{in} \leq 2$. These patterns are shown in Fig. 4.

From Fig. 4, it can be seen that the presented patterns have a certain similarity. For example, in all groups, problem P_2 is solved worse than P_1, problem P_8 is worse than P_7, problem P_9 is better than P_8, and problem P_{10} is worse than P_9. Note also that in this example, the "A-students" solve problems P_6 worse than problems P_5, while the "D-students" and "Middle-students", on the contrary, are better.

Let's compare the correlation matrices for the respondents.

Correlation coefficients between the average responses according to different methods are given for some groups in Tables 1, 2, 3.

Table 1. Correlation coefficients for "D-students".

Method	P_2	P_3	P_4	P_5	P_6	P_7	P_8	P_9	P_{10}
P_1	0,35	0,20	0,00	0,05	0,30	0,20	0,31	0,15	0,22
P_2		0,14	0,15	0,20	0,45	0,35	0,25	0,20	0,25
P_3			0,08	0,14	0,20	0,20	0,17	0,47	0,36
P_4				0,25	0,10	0,20	0,29	0,16	0,07
P_5					0,45	0,05	0,25	0,40	0,25
P_6						0,20	0,20	0,35	0,65
P_7							0,20	0,15	0,22
P_8								0,27	0,13
P_9									0,40

Table 2. Correlation coefficients for "Middle-students".

Method	P_2	P_3	P_4	P_5	P_6	P_7	P_8	P_9	P_{10}
P_1	0,28	-0,12	0,13	0,09	0,15	0,26	0,08	0,11	0,00
P_2		0,00	0,13	0,30	0,03	0,19	0,26	0,05	-0,03
P_3			-0,03	-0,04	0,18	-0,03	-0,04	0,04	0,06
P_4				0,24	-0,08	-0,01	-0,01	-0,03	-0,03
P_5					0,12	0,13	0,21	-0,03	-0,06
P_6						0,19	0,14	0,20	-0,06
P_7							0,14	0,15	0,14
P_8								0,14	0,13
P_9									0,07

Table 3. Correlation coefficients for "A-students".

Method	P_2	P_3	P_4	P_5	P_6	P_7	P_8	P_9	P_{10}
P_1	0,16,	0,06	0,13	0,09	0,07	0,12	0,16	0,20	0,10
P_2		0,13	0,12	0,19	0,08	0,10	0,15	0,13	0,14
P_3			0,08	0,07	0,26	0,05	0,04	0,05	0,09
P_4				0,12	0,05	0,10	0,11	0,08	0,18
P_5					0,08	0,11	0,08	0,10	0,08
P_6						0,06	0,08	0,09	0,08
P_7							0,09	0,12	0,12
P_8								0,18	0,16
P_9									0,12

It was found that the correlations of the "D-students" are generally higher than the correlations of the "A-students", which corresponds to the SLODR effect, but 7 pairs (about 15%) don't show this effect. As for the comparison of "middle-students" and "A-students", there are much more disturbance of the SLODR effect: we have 27 pairs that don't show this effect out of 45 (60%). Consequently, the grouping of respondents according to the accuracy of their answers based on their proximity to the patterns of uniform success (Fig. 4) does not give solid foundation to identify the effect of diminishing returns.

Consider now the other half of the respondents who have specific preferences for the types of questions. There are C_{10}^k different ways to choose k from 10 topics, and their total number is equal $2^{10} - 1$. Obviously, considering more than 1000 patterns is not effective. Therefore, we will select those that contain a sufficiently large number of respondents. For this, we will combine some of the topics of the questions. So, the questions of topics 1, 7 and 8 refer to general logical thinking and can be combined into class I "logic". Questions in topics 2 and 5 refer to "mathematics" (grade II), questions in topics 3, 4 and 10 refer to spatial thinking (grade III), and questions in topics 6 and 9 refer to verbal abilities (grade IV). As a result of the analysis, together with the homogeneous patterns selected above, 20 main patterns were formed, to which approximately 90% of the respondents belong. The types of these patterns and their proportion in the total set are shown in Table 4.

The first type includes the homogeneous patterns discussed above. The second type is formed by patterns for respondents who don't know any one topic out of four. Note that no one knows only one topic out of four. The third type includes respondents who know only one topic worse than the other three. The fourth type includes the patterns of those who know only one topic better than the other three. The fifth type is formed by the patterns of respondents who know better half of the topics. The sixth type is one pattern of highly heterogeneous abilities.

Table 4. List of main patterns of respondent groups

N pattern	Pattern proportion	Topic class	I	II	III	IV
1	0,004	Accuracy class	1–2	1–2	1–2	1–2
2	0,019		2–3	2–3	2–3	2–3
3	0,469		3–4	3–4	3–4	3–4
4	0,002		3–4	1–2	3–4	3–4
5	0,005		3–4	3–4	1–2	3–4
6	0,020		2–3	3–4	3–4	3–4
7	0,050		3–4	2–3	3–4	3–4
8	0,082		3–4	3–4	2–3	3–4
9	0,076		3–4	3–4	3–4	2–3
10	0,018		3–4	2–3	2–3	2–3
11	0,017		2–3	3–4	2–3	2–3
12	0,007		2–3	2–3	3–4	2–3
13	0,015		2–3	2–3	2–3	3–4
14	0,038		3–4	3–4	2–3	2–3
15	0,016		3–4	2–3	3–4	2–3
16	0,028		3–4	2–3	2–3	3–4
17	0,010		2–3	3–4	3–4	2–3
18	0,019		2–3	3–4	2–3	3–4
19	0,009		2–3	2–3	3–4	3–4
20	0,004		3–4	3–4	1–2	2–3

Patterns that are more common than others should be highlighted from the heterogeneous patterns. The first three are formed by the patterns of those who know the topics of the three classes out of four "good" and "excellent", and the topic of the remaining class knows "satisfactory". These remaining classes are in descending order: III, IV, II. They are followed by two patterns of those who know the topics of the two classes "good" and "excellent", and the topics of the other two classes - "satisfactory". The pairs of themes that respondents who know on "excellent" are the following: I, II and I, IV.

The structure of the set of respondents is determined by the proportion of the main patterns. The test results of different groups are comparable with each other in a statistical sense, if they have a similar structure according to the main patterns. It is possible that the discrepancies in the estimates of the conditional probabilities of the relationship between the quality of answers to questions of the established topics in different experiments are due to the different composition of the groups of respondents.

The patterns shown in Fig. 5, taking into account 0,7% of the total number of respondents. This is due to the fact that there are very few "D-students" in the original data set. The highest score in these patterns refers to the method P_7. Next are the methods P_6, P_5, P_4. Of those questions on which the subjects showed high results (above 3 on a 4-point scale), problems of the type P_8 ("syllogisms") are solved the worst.

The patterns of "smartest" and "less smart" in Fig. 6-7 together cover about 30% of respondents. The patterns of typical "middle" cover about 10%.

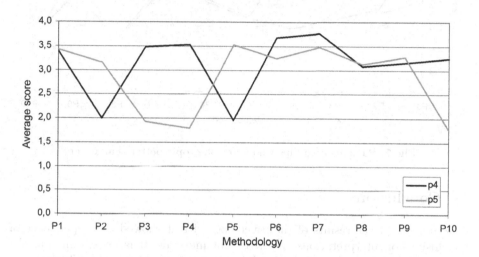

Fig. 5. Patterns of groups who do not know the topic of the same class

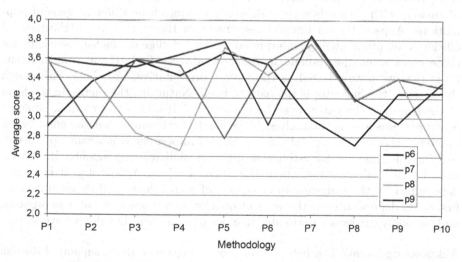

Fig. 6. Patterns of groups that know one topic worse than others

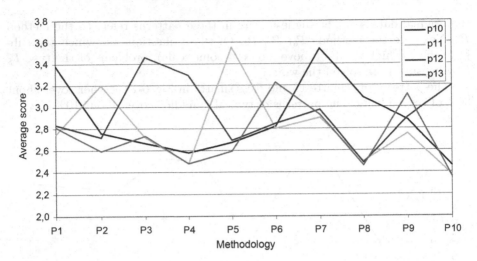

Fig. 7. Patterns of groups who know one topic better than others

5   Conclusion

The analysis of the results of the psychological test showed the dependence of conclusions or interpretations on assessment methods. It is important that the composition of the respondents in the statistical experiment is different each time, therefore, it is not correct to compare the statistics of the distributions of answers with each other, since these are samples from different general populations. Apparently, an integral assessment of the test result in the form of efficiency or other quality, interpreted as an intelligence factor, is useful for those cases when the respondent's answers do not fit into the standard patterns that we have selected as the most representative. For those who match the pattern, the indicator is the pattern itself. Naturally, the metric in the form of the length of the pattern can be applied, but it allows you to rank homogeneous patterns among themselves. If the length is the same, but the answers are different according to separate methods, then, apparently, it seems not correct to bring one number for estimation different skills of respondents. In conclusion, let's say a few words about the SLODR effect. Apparently, it is largely determined by the composition of the set of respondents and is not universal: firstly, its dependence on the level of quality of responses is not monotonous, and, secondly, there is a noticeable number of examples that don't have it.

Acknowledgement. This publication has been prepared with the support of Russian Ministry of Science and Higher Education, agreement №075-15-2020-808

References

1. Korneev, A.A., Krichevets, A.N., Ushakov, D.V.: Spearman's Law of diminishing returns: types of distribution asymmetry and their role in generating artefacts. Siberian Psychol. J. **2019**(71), 24–43 (2019) (in Russian)
2. Sugonyaev, K.V., Radchenko, Y.I.: Spearman's law of diminishing returns: a study on large-scale Russian samples. Psychology **11**(1), 5–21 (2018). (in Russian)
3. Spearmen, C.: The Abilities of Man. MacMillan, New York, 484 p. (1927)
4. Hartmann, P., Reuter, M.: Spearmen's law of diminishing returns tested with two methods. Intelligence **34**(1), 47–62 (2006)
5. Murray A.L., Dixon H., Johnson, W.: Spearmen's law of diminishing returns: a statistical artifact? Intelligence **41**(5), 439–451 (2013)
6. Kobzar, A.I.: Applied Mathematical Statistics. 816 p. Fizmatlit (2006) (in Russian)
7. Kendall, M., Stewart, A.: Statistical Inferences and Connections. vol. 2. Nauka, Moscow, 899 p. (1973) (in Russian)
8. Orlov, Y.N.: Optimal partitioning of the histogram for estimating the sample density of the distribution of a nonstationary time series. Preprints of the Keldysh IPM. no. 14. 26 p. (2013) (in Russian)

Reliability Model of a Homogeneous Hot-Standby k-Out-of-n: G System

H. G. K. Houankpo[1] and Dmitry Kozyrev[1,2]

[1] Peoples' Friendship University of Russia (RUDN University),
6 Miklukho-Maklaya Street, Moscow 117198, Russia
`kozyrev-dv@rudn.ru`
[2] V. A. Trapeznikov Institute of Control Sciences of RAS, 65 Profsoyuznaya Street,
Moscow 117997, Russia

Abstract. We consider the mathematical model of a closed homogenous redundant hot-standby system. This system consists of an arbitrary number of data links with an exponential distribution function (DF) of uptime and a general independent (GI) distribution function of the repair time of its elements with a single repair unit. We also consider the simulation model for the cases where it's not always possible to carry out the mathematical model (to obtain explicit analytic expressions). The system with n components works until k components fail. With introduction of additional variable, explicit analytic expressions are obtained for the steady-state probabilities (SSP) of the system and the steady-state probability of failure-free system operation (PFFSO) using the constant variation method to solving the Kolmogorov differential equations systems. The SSP of the system are obtained with general independent distribution function of uptime using a simulation approach. We built dependence plots of the probability of system uptime against the fast relative speed of recovery; also plots of the reliability function relative to the reliability assessment. Five different distributions were selected for numerical and graphical analysis as Exponential (M), Weibull-Gnedenko (WG), Pareto (PAR), Gamma (G) and Lognormal (LN) distributions. The simulation algorithms were performed in the R programming language.

Keywords: Mathematical model · Steady-state probabilities · Simulation · Hybrid data transmission systems · System reliability · Sensitivity

Notations

A –		Random variable, time to failure of the main component,
B –		Random variable, recovery time of a failed component,
$A(x)$ –		DF of the random variable A,

© Springer Nature Switzerland AG 2021
V. M. Vishnevskiy et al. (Eds.): DCCN 2021, LNCS 13144, pp. 358–368, 2021.
https://doi.org/10.1007/978-3-030-92507-9_29

$$B(x) - \quad \text{DF of the random variable B,}$$
$$b(x) - \quad \text{Probability density function (PDF) of the random variable B,}$$
$$\tilde{b}(\lambda_i) = \int_0^\infty e^{-\lambda_i x} \cdot b(x) dx - \quad \text{Laplace transform of the PDF } b(x) \; ; \; i = \overline{0, k-1},$$
$$EA - \quad \text{Mean uptime of a working component,}$$
$$EB = b - \quad \text{Mean repair time of a failed component,}$$
$$DB - \quad \text{Recovery time variance,}$$
$$c = \frac{\sqrt{DB}}{EB} - \quad \text{Coefficients of variation,}$$
$$\rho = \frac{EA}{EB} - \quad \text{Relative recovery rate,}$$
$$\beta(x) = \frac{b(x)}{1 - B(x)} - \quad \text{Conditional PDF of the residual repair duration of the element being repaired at time } t \text{ (recovery rate) [15],}$$
$$\lambda_i = (n - i) \cdot \alpha - \quad \text{Parameter of the exponential distribution of the uptime of components; } i = \overline{0, k-1}.$$

1 Introduction

Computer and communications networks are constantly evolving due to the results of theoretical and practical problems aimed at improving the availability and reliability of networks and data transmission systems. Researchers are often faced with the development of complex systems among which (including) the system k-out-of-n.

Important works on different k-out-of-n systems has been developed. Cao Wang [1] proposed the performance of civil infrastructure using k-out-of-n systems with identical component deterioration. Xinchen Zhuang, Tianxiang Yu, Zhongchao Sun and Kunling Song [2] focused on Reliability and capacity evaluation of multi-performance multi-state weighted k-out-of-n systems. A recursive approach is developed to evaluate the system reliability more efficiently, and comparison with the existing approach is carried out in various aspects. Tetsushi Yuge [3] investigated the reliability of systems with simultaneous and consecutive failures, where the reliability of a k-out-of-n system and a consecutive k-out-of- n system subjected to shock and considering the simultaneous failures are discussed. Eunkyung Chae, Chan-woo Park, Jeon-gwon Kang [4] studied the reliability analysis of M-out-of-N system with common cause failures for railway, where the reliability of a hot-standby sparing system considering common cause failure has been analyzed and used to improve the accuracy of the system reliability evaluation and Huyang Xu, Yuanchen Fan, Nasser Fard [5] developed the reliability assessment of repairable Load-sharing K-out-of-N: system with flowgraph model.

In our previous works [6–10], we managed to apply a simulation approach to reliability assessment of a redundant system with arbitrary input distributions, and focused on reliability analysis of various homogeneous and heterogeneous systems with different types of redundancy – warm, cold and hot. In these papers, we utilized the supplementary variable method to investigate the sensitivity of steady-state reliability characteristics of some special cases of systems. Further, in [11] we applied the k-out-of-n system model to the reliability study of a hexacopter-based flight module of a tethered unmanned high-altitude platform.

In the current paper, we study the mathematical model of a closed homogeneous hot-standby k-out-of-n system. The purpose of this work is to carry out mathematical modeling of the system, obtain the analytical expressions for its SSP and carry out numerical analysis for a specific example.

2   Mathematical Model of the System

2.1   Description, Assumptions and Statement of the Problem

Let's consider a closed homogenous redundant hot standby system. The system consists of an arbitrary numbers of data sources with an exponential (DF) of uptime and a general independent DF of the repair time of its components with a single repair unit, which according to a modified Kendall's notation [12], we denote as $\langle M_{k<n}/GI/1\rangle$.

We study the dependence of the steady-state probability of failure-free system operation (PFFSO) on the relative recovery rate (RRR) with different coefficient of variation. We introduce the following assumptions regarding the system's operation:

Assumption 1: initially, backup elements participate in the functioning of the system on a par with the main element.

Assumption 2: elements are received for repair one at a time.

Assumption 3: the system with n components fails if k components fail.

The aim is to find the explicit analytical expressions for the steady-state probabilities distribution of the system and for the steady-state PFFSO, both in the general case and for some special cases of distributions.

2.2   Explicit Analytic Expressions of the System

We consider a stochastic process $\{v(t), t \geq 0\}$, where $v(t)$ is the number of failed elements at time t, defined on the phase space $E = \{0, 1, \cdots, k\}$.

In order to solve the stated problem, we use an approach based on the Markovization technique [13]. To describe the behavior of the system using a Markov process, we introduce an additional variable $x(t) \in \mathbb{R}_+^2$—the total time spent on the repair of the failed element, by the time t. We obtain a two dimensional [14] process $(v(t), x(t))$, with an extended set of states $\varepsilon = \{(0), (1, x), \cdots, (k, x)\}$.

Denote by $p_0(t)$ the probability that at time t the system is in state $i = 0$, and by $p_i(t, x)$ the probability density function (in continuous component) that at time t the system is in state $i(i = 1, 2, \cdots, k)$, and the time spent on repairing the failed element is in the range $(x, x + dx)$ (Fig. 1).

$$p_0(t) = P\{v(t) = 0\}, \tag{1}$$

$$p_i(t, x)dx = P\{v(t) = i, x < x(t) < x + dx\}, i = \overline{1, k}. \tag{2}$$

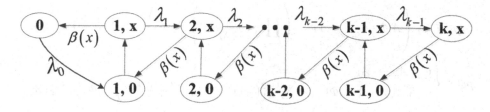

Fig. 1. State transition graph.

Theorem 1. *The SSP of the considered repairable redundant system are:*

$$p_0 = C_1 \cdot \frac{\tilde{b}(\lambda_1)}{\lambda_0}; p_1 = C_1 \cdot \Phi_1;$$

$$p_i = C_1 \left(A_i \Phi_i + \sum_{j=1}^{i-1} (-1)^{i-j} \left(\prod_{k=j}^{i-1} \frac{\lambda_k}{\lambda_j - \lambda_{k+1}} \right) A_j \Phi_j \right); i = \overline{2, k-1}; k \geqslant 3;$$

$$p_k = \begin{cases} C_1 \left(A_k b - A_{k-1} \Phi_{k-1} \right); k = 2; \\ C_1 \left(A_k b - A_{k-1} \Phi_{k-1} + \sum_{j=1}^{k-2} (-1)^{k-j} \left(\prod_{i=j}^{k-2} \frac{\lambda_{i+1}}{\lambda_j - \lambda_{i+1}} \right) A_j \Phi_j \right); k \geqslant 3; \end{cases}$$

Proof. From (1) and (2), with the help of the total probability rule we obtain the Kolmogorov's forward system of differential equations, which allows to evaluate the SSP of the considered system [16]. Using the total probability formula, and passing to the limit as $\Delta \to 0$, we derive the following Kolmogorov differential equations systems:

$$\begin{cases} \lambda_0 \cdot p_0 = \int_0^\infty p_1(x) \cdot \beta(x) dx \\ \frac{dp_1(x)}{dx} = -(\lambda_1 + \beta(x)) \cdot p_1(x) \\ \frac{dp_i(x)}{dx} = -(\lambda_i + \beta(x)) \cdot p_i(x) + \lambda_{i-1} \cdot p_{i-1}(x); i = \overline{2, k-1} \\ \frac{dp_k(x)}{dx} = -p_k(x) \cdot \beta(x) + \lambda_{k-1} \cdot p_{k-1}(x) \end{cases} \tag{3}$$

with boundary condition,

$$p_1(0) dx = \lambda_0 \cdot p_0 + \int_0^\infty p_2(x) \cdot \beta(x) dx, \tag{4}$$

$$p_i(0) dx = \int_0^\infty p_{i-1}(x) \cdot \beta(x) dx; i = \overline{2, k-1}. \tag{5}$$

We suppose that for the described process, there exists a stationary probability distribution as $t \to \infty$.

We proceed to solving the resulting system of balance equations using the constant variation method [17]. From here we find the stationary probabilities for macro-states. In sum, we get the following analytical expressions for the SSP of the repairable system in the following form:

$$p_0 = C_1 \cdot \frac{\tilde{b}(\lambda_1)}{\lambda_0}; p_1 = C_1 \cdot \Phi_1;$$

$$p_i = C_1 \left(A_i \Phi_i + \sum_{j=1}^{i-1} (-1)^{i-j} \left(\prod_{k=j}^{i-1} \frac{\lambda_k}{\lambda_j - \lambda_{k+1}} \right) A_j \Phi_j \right) ; i = \overline{2, k-1}; k \geqslant 3;$$

$$p_k = \begin{cases} C_1 \left(A_k b - A_{k-1} \Phi_{k-1} \right); k = 2; \\ C_1 \left(A_k b - A_{k-1} \Phi_{k-1} + \sum_{j=1}^{k-2} (-1)^{k-j} \left(\prod_{i=j}^{k-2} \frac{\lambda_{i+1}}{\lambda_j - \lambda_{i+1}} \right) A_j \Phi_j \right); k \geqslant 3; \end{cases}$$

Where

$$C_1 = \left(\frac{\tilde{b}(\lambda_1)}{\lambda_0} + \Phi_1 + A_k b - A_{k-1} \Phi_{k-1} \right)^{-1} ; k = 2;$$

$$C_1^{-1} = \frac{\tilde{b}(\lambda_1)}{\lambda_0} + \Phi_1 + \sum_{i=2}^{k-1} \left(A_i \Phi_i + \sum_{j=1}^{i-1} (-1)^{i-j} \left(\prod_{k=j}^{i-1} \frac{\lambda_k}{\lambda_j - \lambda_{k+1}} \right) A_j \Phi_j \right)$$

$$+ \left(A_k b - A_{k-1} \Phi_{k-1} + \sum_{j=1}^{k-2} (-1)^{k-j} \left(\prod_{i=j}^{k-2} \frac{\lambda_{i+1}}{\lambda_j - \lambda_{i+1}} \right) A_j \Phi_j \right) ; k \geqslant 3;$$

$$\Phi_i = \frac{1 - \tilde{b}(\lambda_i)}{\lambda_i}$$

$$A_2 = A_1; k = 2$$

$$A_2 = \left(1 - \left(1 - \frac{\lambda_1}{\lambda_1 - \lambda_2} \right) \lambda_1 \right) \frac{1}{\tilde{b}(\lambda_2)}; k \geqslant 3$$

$$A_{i+1} = (A_i + \sum_{j=1}^{i-1} (-1)^{i-j} \left(\prod_{k=j}^{i-1} \frac{\lambda_k}{\lambda_j - \lambda_{k+1}} \right) A_j$$

$$- \sum_{j=1}^{i} (-1)^{i+1-j} \left(\prod_{k=j}^{i} \frac{\lambda_k}{\lambda_j - \lambda_{k+1}} \right) A_j \tilde{b}(\lambda_j)) \frac{1}{\tilde{b}(\lambda_{i+1})}; i = \overline{2, k-2}; k \geqslant 4$$

$$A_k = A_{k-1}(1 + \tilde{b}(\lambda_{k-1})) + \sum_{j=1}^{k-2} (-1)^{i-1-j} \left(\prod_{i=j}^{k-2} \frac{\lambda_i}{\lambda_j - \lambda_{i+1}} \right) A_j$$

$$+ \sum_{j=1}^{k-2} (-1)^{k-j} \left(\prod_{i=j}^{k-2} \frac{\lambda_{i+1}}{\lambda_j - \lambda_{i+1}} \right) A_j \tilde{b}(\lambda_j); k \geqslant 3 \quad \square$$

From these expressions, there is a dependence of the SSP of the system states on the type of the repair time distribution. However, this dependence becomes vanishingly small with the "rapid" repair of the failed elements, i.e. with the increase in relative recovery rate [6–11].

2.3   Example. Numerical Analysis of Mathematical Modeling

To analyze the numerical results of the model with different known distributions of repair time, where $k = 3$ and $n = 5$, we consider special cases when $EB = b = 1$ with different coefficients of variation $c = \{0.5; 1; 2\}$; and $EA = 25$.

Table 1 shows the PFFSO $1 - p_k$.

Table 1. Values of probability $1 - p_3$ of the system's failure-free operation.

GI	c		
	0.5	1	2
Weibull	0,9985355	0,9968928	0,9911761
Pareto	0,9986520	0,9968809	0,9947930
Gamma	0,9988354	0,9968928	0,9861888
LogNormal	0,9988074	0,9965752	0,9869367

Obviously, for all the distributions under consideration, the greater is the coefficient of variation, the less is the PFFSO.

Figure 2 shows the graphs of the probabilities of failure-free system operation. For graphical results, we consider models with $\rho = 25$.

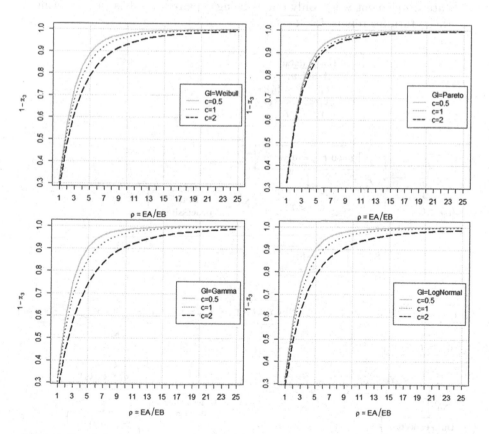

Fig. 2. Graphs of the probability of failure-free system operation $1 - p_3$ versus ρ.

The graphical results also confirm the conclusion about the greater is the coefficient of variation, the less is the PFFSO, and the dependence of the SSP of the system states on the type of the repair time distribution becomes vanishingly small with the "rapid" repair of the failed elements.

3   Simulation Model of the System

In this section, we develop a simulation model which allow us to carry out reliability modeling for the system with a general independent DF of uptime and of repair time of its elements.

3.1   Calculation of the Steady-State Probabilities of the System

Let's define the following states of the simulated system:

· State 0: All elements work;
· State 1: One element failed and is being repaired, the rest is working;
· State i: i element work, only one is being repaired, $i-1$ is (are) waiting its (their) turn for repair; $i = \overline{2-k-1}$

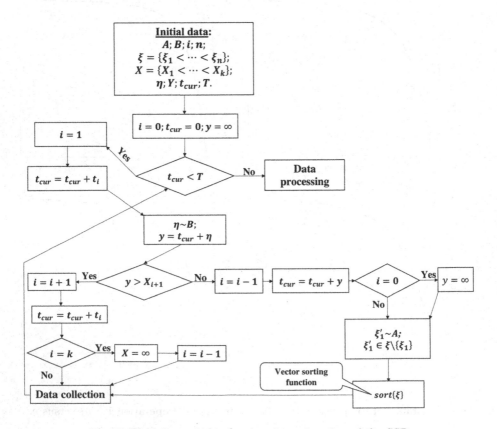

Fig. 3. Flowchart of the algoritm for estimation of the SSP.

· State k: The system has failed, one is being repaired, the rest are waiting their turn for repair.

We consider the simulation clock (double t) that changes when the system's elements has failed or has been repaired. The system state variables (int i, j) define the transition from state i to state j. The service variable (double $t_{nextfail}$) and repair variable (double $t_{nextrepair}$) store respectively the time until the next element failure and the time until the next repair of the failed element. We introduce the main loop iteration counter (int z).

Figure 3 shows the flowchart of the algorithm for estimation of the SSP.

3.2   Calculation of the Assessment of the System Reliability

Figure 4 shows the flowchart of the algoritm for the assessment of the system reliability.

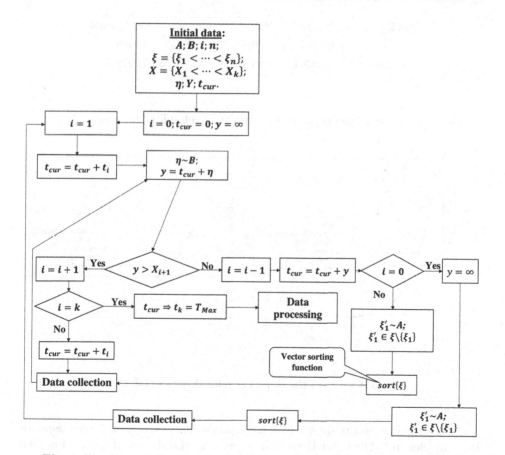

Fig. 4. Flowchart of the algoritm for the assessment of the system reliability.

3.3   Example. Numerical Analysis

In this subsection, we are going to use the coefficient of variation that has great PFFSO for the numerical analysis, then when $c = 0.5$. To analyze the numerical results of the simulation modeling in this subsection with the same distributions of repair time, we consider the following input data: $a1 = 25; b1 = 1; k = 3; n = 5; T = 10000; NG = 1000$.

Table 2 shows the values of probability $1 - p_k$ of failure-free operation of the considered system obtained by the analytical and simulation approaches.

Evidently, the simulation results are in significant agreement with the results obtained by the mathematical modeling; with a very low absolute error value 0.001.

Table 2. Values of probability $1 - p_3$ of the system's failure-free operation obtained by simulation and mathematical modeling.

Model	Weibull	Pareto	Gamma	LogNormal
Mathematical	0,9985355	0.9986520	0.9988354	0.9988074
Simulation	0.9976441	0.9978903	0.9978418	0.9978456

Figure 5 presents the graphs of the empirical reliability function $R^*(t)$.

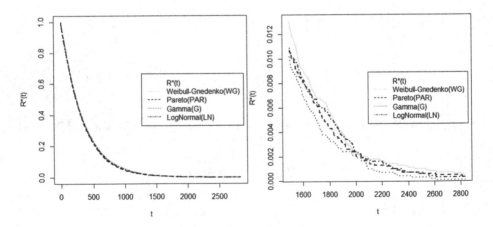

Fig. 5. Graphs of the empirical reliability function $R^*(t)$.

The graphical results show a high asymptotic insensitivity of the corresponding empirical reliability function of the system to the shapes of the uptime and repair time distributions of the system's elements [6].

4   Conclusion

In this work, we obtained closed-form analytical expressions for the steady-state probabilities distribution of system states and for the stationary probability of system uptime in the general case. The resulting formulas show the presence of an explicit dependence of reliability measures of the system on the type of the DF of the repair time of its elements. Numerical analysis showed that the greater is the coefficient of variation, the less is the PFFSO. The graphical results confirm the conclusion about the greater is the coefficient of variation, the less is the PFFSO, and the dependence of the steady-state probabilities of the system states on the type of the repair time distribution becomes vanishingly small with the "rapid" repair of the failed elements.

The simulation model results are in significant agreement with the results obtained by the mathematical modeling; then we can use the simulation in the cases it is not possible to obtain explicit analytical expressions. The elaborated simulation approach allows to expand the area of analytical research in the case of non-exponential distributions of elements' uptime and repair time of failed elements. Also, the graphs of the empirical function of the system reliability were built.

Acknowledgments. The publication has been supported by the RUDN University Strategic Academic Leadership Program. The reported study was funded by RFBR, project No. 20-37-90137 (recipient Dmitry Kozyrev, formal analysis, validation, and recipient H.G.K. Houankpo, methodology and numerical analysis)

References

1. Wang, C.: Time-dependent reliability of (weighted) k-out-of-n systems with identical component deterioration. J. Infrastruct. Preserv. Resil. **2**(1), 1–10 (2021). https://doi.org/10.1186/s43065-021-00018-1
2. Zhuang, X., Yu, T., Sun, Z., Song, K.: Reliability and capacity evaluation of multi-performance multi-state weighted k-out-of-n systems. Commun. Stat. - Simul. Comput. https://doi.org/10.1080/03610918.2020.1788590
3. Tetsushi, Y.: Reliability of systems with simultaneous and consecutive failures. In: 2019 4th International Conference on System Reliability and Safety (ICSRS). https://doi.org/10.1109/ICSRS48664.2019.8987614
4. Chae, E., Park, C.-w., Kang, J.-g.: Reliability analysis of M-out-of-N system with common cause failures for railway. https://doi.org/10.7782/JKSR.2018.21.10.969
5. Xu, H., Fan, Y., Fard, N.: Reliability assessment of repairable load-sharing K-out-of-N: system with flowgraph model. https://doi.org/10.1109/RAM.2018.8463109
6. Houankpo, H.G.K., Kozyrev, D.V., Nibasumba, E., Mouale, M.N.B., Sergeeva, I.A.: A simulation approach to reliability assessment of a redundant system with arbitrary input distributions. In: Vishnevskiy, V.M., Samouylov, K.E., Kozyrev, D.V. (eds.) DCCN 2020. LNCS, vol. 12563, pp. 380–392. Springer, Cham (2020). https://doi.org/10.1007/978-3-030-66471-8_29

7. Houankpo, H.G.K., Kozyrev, D.V., Nibasumba, E., Mouale, M.N.B.: Reliability analysis of a homogeneous hot standby data transmission system. In: Proceedings of the 30th European Safety and Reliability Conference and 15th Probabilistic Safety Assessment and Management Conference (ESREL2020 PSAM15), pp. 1–8 (2020). https://doi.org/10.3850/978-981-14-8593-0_5755-cd
8. Houankpo, H.G.K., Kozyrev, D.V., Nibasumba, E., Mouale, M.N.B.: Mathematical model for reliability analysis of a heterogeneous redundant data transmission system. In: 12th International Congress on Ultra Modern Telecommunications and Control Systems and Workshops (ICUMT), pp. 189–194 (2020). https://doi.org/10.1109/ICUMT51630.2020.9222431
9. Houankpo, H.G.K., Kozyrev, D.: Reliability model of a homogeneous warm-standby data transmission system with general repair time distribution. In: Vishnevskiy, V.M., Samouylov, K.E., Kozyrev, D.V. (eds.) DCCN 2019. LNCS, vol. 11965, pp. 443–454. Springer, Cham (2019). https://doi.org/10.1007/978-3-030-36614-8_34
10. Houankpo, H.G.K., Kozyrev, D.V.: Sensitivity analysis of steady state reliability characteristics of a repairable cold standby data transmission system to the shapes of lifetime and repair time distributions of its elements. In: Samouilov, K.E., Sevastianov, L.A., Kulyabov, D.S. (eds.) Selected Papers of the VII Conference "Information and Telecommunication Technologies and Mathematical Modeling of High-Tech Systems", Moscow, Russia, 24 April 2017, CEUR Workshop Modelings 1995, pp. 107–113 (2017). http://ceur-ws.org/Vol-1995/paper-15-970.pdf
11. Kozyrev, D.V., Phuong, N.D., Houankpo, H.G.K., Sokolov, A.: Reliability evaluation of a hexacopter-based flight module of a tethered unmanned high-altitude platform. In: Vishnevskiy, V.M., Samouylov, K.E., Kozyrev, D.V. (eds.) DCCN 2019. CCIS, vol. 1141, pp. 646–656. Springer, Cham (2019). https://doi.org/10.1007/978-3-030-36625-4_52
12. Kendall, D.G.: Stochastic processes occurring in the theory of queues and their analysis by the method of embedded Markov chains. Ann. Math. Stat. **24**, 338–354 (1953)
13. Parshutina, S., Bogatyrev, V.: Models to support design of highly reliable distributed computer systems with redundant processes of data transmission and handling. In: International Conference "Quality Management, Transport and Information Security, Information Technologies" (IT&QM&IS) (2017). https://doi.org/10.1109/ITMQIS.2017.8085772
14. Teh, T., Lai, C.M., Cheng, Y.H.: Impact of the real-time thermal loading on the bulk electric system reliability. IEEE Trans. Reliab. **66**(4), 11101119 (2017). https://doi.org/10.1109/TR.2017.2740158
15. Lisnianski, A., Laredo, D., Haim, H.B.: Multi-state Markov model for reliability analysis of a combined cycle gas turbine power plant. In: Second International Symposium on Stochastic Models in Reliability Engineering, Life Science and Operations Management (SMRLO) (2016). https://doi.org/10.1109/SMRLO.2016.31
16. Sevastyanov, B.A.: An Ergodic theorem for Markov processes and its application to telephone systems with refusals. Theor. Probab. Appl. **2**, 104–112 (1957)
17. Petrovsky, I.G.: Lectures on the theory of ordinary differential equations (Lektsii po teorii obyknovennykh differentsialnykh uravneniy), Moscow, GITTL (1952). 232 p. (in Russian)

Author Index

Printed in the United States
by Baker & Taylor Publisher Services